T0180836

Convergent Leadership—Divergent Exposures

Franco Oboni · Cesar H. Oboni

Convergent Leadership—Divergent Exposures

Climate Change, Resilience, Vulnerabilities, and Ethics

 Springer

Franco Oboni
Oboni Riskope Associates Inc.
Vancouver, BC, Canada

Cesar H. Oboni
Oboni Riskope Associates Inc.
Vancouver, BC, Canada

With Contribution by
Janis A. Shandro
Vancouver, Canada

ISBN 978-3-030-74932-3   ISBN 978-3-030-74930-9   (eBook)
https://doi.org/10.1007/978-3-030-74930-9

This Springer imprint is published by the registered company Springer Nature Switzerland AG
The registered company address is: Gewerbestrasse 11, 6330 Cham, Switzerland

Preface

Recent climate change-related disasters such as large-scale fires, flooding, hail and dust storms, locusts and epidemics linked to various hazards have demonstrated the need to enhance tactical and strategic planning to foster healthy awareness and reduce reactions based on fear and panic. Climatic divergence from long-term averages and "usual extremes," in terms of both frequencies and intensities, reinforces this need, as a new wave of Green Deals (EU), the Paris Agreement and other efforts from key players around the world will require careful evaluations and risk-informed decisions.

The goal of this book is to allow the healthy and ethical operational, tactical and strategic planning of systems based on sensible estimates of business-as-usual and divergent risks of a various origin and significance. "Systems" are defined as interrelated elements, procedures, organizations, etc., geared toward accomplishing a set of goals or objectives in the field of administration, industry, environment and society.

Consider this book as a medicine to help contain fear and knee-jerk reactions, reduce blunders and enhance value building and ethics, while planning for climate change and other "runaway" hazards. To attain these goals, it is paramount to foster predictability and foreseeability of divergent hazards. Divergent hazards are present today and will always lurk in our future, perhaps in the form of climate change events, potential meteorites/asteroid collisions, sun flares (magnetic pulse), pandemics, super-volcanoes, etc.

With this in mind, we believe that it is in the best interest of any individual, corporation or government to shed some light on these hazard exposures, allowing for sustainable planning and mitigation through ethical decisions. These can be achieved through the application of cutting-edge risk assessment methodologies and risk management practices provided the methodologies are convergent, drillable and updatable and of course allow for adaptation of the responses to diverging hazards.

Vancouver, USA

Franco Oboni
Cesar H. Oboni

Acknowledgements

This book would not have been possible without the cooperation of our network of world-class experts in many fields.

In particular, we wish to thank:

- Dr. Janis A. Shandro, Health Impact Assessment and Risk Management Consultant, for her contributions to Chaps. 5 and 6;
- The fantastic trio of technology consultants: Kevin Prasad, Keith Bedford and Tiana Van Dyk for their support and advice in the field of data management, analytics and machine learning tools discussed in Sect. 7.1.1;
- MDA Geospatial Services Inc. and MAXAR Technologies for the use of satellite imagery included within this book;
- Two friendly lawyers and an insurance executive who wish to remain anonymous for answering our questions and offering advice regarding legal (Sect. 6.1.2) and general insurance matters;
- Our clients, including Fortune 500 companies in the mining, transportation and logistics, forestry, insurance, automotive and food industries as well as lawyers and judges in civil and criminal courts of law, international organizations, civilian administrations and armed forces, and finally large-scale environmental rehabilitation and engineering/construction companies who keep "pushing the envelope" of their requests and are the drivers of our R&D;
- Last but not least, our friend and colleague Claudio Angelino, M.Sc., P. Eng., for his help and constructive criticism on the manuscript.

The authors wish to express their gratitude to Dr. Annett Büttner, Senior Publishing Editor, Springer Earth Sciences, Geography and Environment, and to Kim Williams, Kim Williams Books, Torino, for their timely and constructive criticism.

Finally, the authors wish to express their love and gratitude to their beloved families for their patience and support. We of course include the tribe's next generation (by date of birth): Latika, Niya, Carter, Leon and, the latest little ones, Jayleen and Matilde. We hope this book will contribute to deliver to you and to your generation a better and healthier planet.

About This Book

The primary purpose of this book is to identify and describe those risk assessment approaches and risk management practices that must be implemented at the systems level to develop a path forward to reach societally acceptable risks. In support of this necessary effort, we use positive and negative events examples, real-life case histories and studies, while introducing existing and proven risk management methodological approaches. Throughout the book, anecdotes and technical notes are introduced in text boxes to deliver historic details, or real-life stories or technical details without overcrowding the texts.

This book discussion is divided into five parts.

Part I, State of Affairs, looks at several recent events around the world where tactical and strategic planning were apparently poorly developed and lead to unpleasant outcomes. It then sets the context of divergence with examples drawn from mythological tales and recent events, including a detailed discussion of climate change exposures. The need for clear definitions and compliance with a very precise glossary is stressed, to avoid confusion and misunderstandings under pressure to report risks.

Part II, Divergent Exposures, the Public and Ethics, delves into the differences between business-as-usual and divergent events, credible events and black swans and their relationship with standard levels of industrial mitigation. Attention is then shifted toward metaphoric descriptions of risks that are useful when considering public reactions and a taxonomic description of uncertainties. Notions such as "return period" and "force majeure" are discussed to show how misleading, or plain erroneous, these notions can be in the face of divergent exposures and public expectations. This leads to a review of what the contemporary public seems to want, based on recent conferences and public hearings and to a closure consisting of communication and transparency requirements as well as ethics considerations.

Part III, Convergent Assessment of Exposures, is structured around the necessary steps in a modern convergent risk assessment capable of handling business-as-usual as well as divergent exposures. It therefore looks at system definition, hazard identification, how to define probabilities of events and events' consequences.

Part IV, Tactical and Strategic Planning for Convergent/Divergent Reality, discusses a quintessential element of modern risk-informed decision-making: the risk tolerance threshold in both historic and modern terms. It then reviews the "how to" procedure leading to the ability to prepare rational mitigative roadmaps, distinguishing tactical from strategic families of risks. The discussion also bears on the need to have resilient solutions (if not anti-fragile) and, finally, the need to select financially viable solutions using risk-adjusted cost estimates for projects.

Part V, Convergent Assessment for Divergent Exposures: Case Studies, features discussion of risk-informed decision-making deployments through examples of railroads, wharves, chemical plants in isolation and as a system. It concludes with a look at what we expect might happen in the years and decades to come.

These five principal parts are followed by two **Appendices**. Appendix A helps making sense of probabilities and frequencies when little data are available and rough estimates are required. Appendix B forms the counterpoint to Part IV and many points delivered in Part III, as it concentrates on the DON'TS related to risk assessments. Readers should refer to it every time they feel like shaking their heads while reading Parts III and IV.

In particular, the book shows:

- How critical risks, and their necessary management controls, can be identified for both existing and divergent situations (Parts II and III);
- How cost-effective risk reduction strategies can be identified and included in the design and management of existing infrastructure and new projects (Parts III and IV);
- How risk assessment approaches can provide a solid basis for the justification and prioritization of continuous improvement programs (Part IV);
- How an extant asset portfolio can be risk-prioritized in order to design mitigative measures while appropriately communicating risk to the public (Part V).

Discussion of Key Terms and Notation

The authors invite you to use and freely download a copy of the *Riskope's technical glossary* (https://www.riskope.com/knowledge-centre/tool-box/glossary/). For your ease, we discuss here the definitions of key terms in this book. In the references of each chapter, a text box with these key terms and active links to the technical glossary is provided for facilitated reference. At the end of this discussion, interested readers will find a short note on the scientific notation of numbers used throughout the book.

Act of God. An event with a probability of occurrence below credibility. Probabilities can be quantified down to certain frequency levels. In our day-to-day practice, we consider any event having a frequency rate larger than $1/10^5$ event/years as credible, between $1/10^5$ and $1/10^6$ as poorly credible, and below $1/10^6$ as non-credible, hence an act of God. Below such values, no meaningful uncertainty ranges can be given for single events. Indeed, predictions of most events which require human error frequencies of the order of $1/10^6$ or less are clearly non-credible because the historical data set required to establish such human performance is generally nonexistent. Let us use an extreme example to put things in perspective:

The Big Bang that created our universe occurred some 10^{10} years ago, meaning the history of our universe is about fourteen billion years old (thus, approximately 10^{10} years old). That means a "frequentist" person could think that the "occurrence rate" is $1/10^{10}$ event/years. Thus, any yearly frequency smaller than $1/10^{10}$ really means that the event "occurrence rate" is less likely than our universe! This obviously is our "cosmological" extreme lower bound limit of credibility in risk assessments. As risk assessors, we have to be humble; thus, nuclear probabilistic safety assessment researchers think probabilities can be quantified with reasonable "certainty" down to frequency of $1/10^6$ to $1/10^7$ event/years (CSNI 1990, p. 243). Below such values, no meaningful uncertainty ranges can be given for single events. Therefore, those sources recommend a cutoff frequency of $1/10^7$ for single events. Now we are getting near to what we use. Finally, the definition of a credible accident by the process industry literature, and other industries where major accidents/events are a concern, is "the accident which is within the realm of possibility (i.e., a probability higher than 10^{-6} events/year) and has a propensity to cause significant damage (at least one fatality)" (Khan and Abbasi 2002). Seismic, geological and other geosciences use the threshold value of 10^{-5} to define "maximum credible events." Now we are bang-on!

Black Swan. A metaphor describing an event that comes as a surprise has a major effect and is often inappropriately rationalized after the fact with the benefit of hindsight. The term is based on an ancient saying that presumed black swans did not exist—a saying that became reinterpreted to teach a different lesson after the first European encounter with them. The idea is to explain the role of events that have high profile, maybe hard to predict and escape from normal expectations. NB: We will see in this book that many events that have been given black swan status actually are not. The theory also tends to explain that small probabilities escape "mathematical" analysis (see definition of act of God above) and that psychological biases tend to blind people and have significant consequences (this book is full of references to this phenomenon). We prefer to use the predictability (of likelihood) and foreseeability (of consequences) to characterize events and eventually end up generating divergent hazard exposures. An unpredictable hazard (difficult to predict its magnitude and probability) which is also unforeseeable (difficult to predict its consequences) can be called a black swan if the consequences are very high, trending to catastrophic. However, a "business-as-usual" scenario is considered divergent if its frequency and magnitude differ from long-term "averages." Consequences may or may not vary.

Business-as-Usual. Business-as-usual is defined in this book as an unchanging state of affairs despite the occurrence of non-divergent hazards of any kind (man-made and natural). As defined below, divergent hazards part from long-term averages and "usual extremes," in terms of both frequencies and intensities. Thus, their predictability and foreseeability of their consequences are low. An explosion, even if of significant proportion in a refinery, can be considered as business-as-usual as we know these events do occur with a certain frequency and their damages are rather foreseeable. However, if that refinery explosion would overcome all mitigation and become unpredictable and perhaps also unforeseeable, then the hazard generating that event would be considered a divergent event. If that divergent event would hit several operations simultaneously, then we would enter in the realm of act of God and black swan.

Convergent. Convergent risk assessments integrate areas that are significant to an organization, such as operational risk generated by various hazards or compliance, within a single framework. The goal is to provide a holistic view of risk for the organization eliminating informational siloes and fostering a 360° view of the risk landscape surrounding the organization. This increased visibility fosters the ability to manage sometimes competing goals and interests. Convergent assessment means that all hazards potentially present (technical, man-made, natural, etc.) are considered in the aggregate risk evaluations together with their multi-dimensional consequences.

Divergent. Hazards or exposures become divergent when they part from long-term averages and "usual extremes," in terms of both frequencies and magnitude. For example, a 100-year rain event that occurs three times within a short period can be considered a divergent event.

Drillable. The adjective comes from "drill down" to view results at a more detailed level. Drillable hazard and risk register mean that database information can be retrieved using various queries, such as asking which are the highest (tolerable and intolerable) risks within the company generated by potential earthquake, employee dishonesty, etc.

Foreseeability. The facility to perceive, know in advance, or reasonably anticipate that damage or injury will probably ensue from acts or omissions. A foreseeable event or situation is one that can be known or guessed at before it happens. We apply it to consequences: Can we foresee the damage generated by a (predicted) hazard hit based on present or future mitigation and policies/actions?

"Usual" hazards, including their "usual extremes," are predictable and foreseeable within a relative narrow margin of uncertainty, provided interdependencies and systemic amplification are properly considered.

Divergent hazards are unpredictable insofar as their frequency (at a given magnitude level) "explodes." Their foreseeability may also reduce due to various internal or external reasons (see the refinery example under business-as-usual above) and because "wounded systems" are less robust and consequences may get amplified: Massive fires could lead to erosion, slope failures, etc., in case of rain.

Predictability. The state of knowing what something will be like, when something will happen, etc. We apply it to hazards: Can we predict the magnitude and the frequency of a hazard?

Scalability. Scalability means that the risk register is built in such a way that new projects, alterations, in-depth studies of some structures or elements of a system can be added and developed within the same risk register.

Societal Risk Acceptability. A societal response to risks that can be set by policy for determining what are acceptable risks in a given jurisdiction. Developers and other decision-makers should remember that "risk acceptability" is a threshold set by society and is therefore related to the individuals in a given societal and political jurisdiction rather than determined project by project.

Survivability. In general, the ability to remain alive or continue to exist after a (divergent) hazard hit. The "survival threshold" for business is a financial measure generally higher than the corporate tolerance. It is the cumulative loss that will kill the corporation or the project.

Sustainability. The ability of a system to maintain its functions over long term.

System. A group of interrelated elements, procedures, organizations geared toward achieving a goal or objective. For example, a railroad network is a system with fixed and moving assets that is geared toward carrying passengers and/or freight from multiple points of departure to multiple destinations. Systems can be variously characterized:

- **Fragile system**. A system is fragile if in the event of a failure of some or all of the mitigations upon a hazard hitting the system it collapses and the consequences cannot be contained within a reasonable amount of resources.
- **Ductile system**. A system that gradually damages and absorbs the hazard hits without sudden collapse.

- **Resilient system**. The ability to recover after an accident, the *ability to withstand disruption and rebound quickly* (https://www.riskope.com/2016/11/30/resilience-sustainability-insurance/) after a disturbance or interruption. Resilient design is the result of an intentional effort geared toward enhancing the resilience of systems, buildings, landscapes, communities and regions to natural and man-made hazards and disturbances. These may include sudden or long-term changes induced by climate changes, such as but not limited to:

  - Sea level rise;
  - Heat waves;
  - Regional drought;
  - Massive forest fires.

- **Robust system**. A system that does not break easily, for example, a structure that has more resistance than normally allotted to withstand it loadings (engineers use the factor of safety indicator to define how robust a system is. So, a structure with a factor of safety larger than normally considered is robust). NB: A robust system can be fragile or ductile.
- **Wounded system**. A system that has been damaged and not or poorly repaired. A wounded system is still operational but less robust than in its initial conditions.

Tolerance (To Risks). A threshold defining how much risk can be tolerated (by an individual, an organization, a project). It is generally a perceived threshold as many factors can enter into the decision about what is tolerable or not. Corporate tolerance has to be aligned with societal risk acceptability (i.e., those which can be agreed upon by society).

Uncertainties and Risk. Uncertainties exist in both the probability and the consequences of events. Probability and consequences are the parameters of risk. Uncertainties are included using ranges based on the state of knowledge at deployment time. These will be altered during the life of the system (see "Updatable" below). Uncertainties in risk assessments derive from uncertainties in the data, model and system. In addition, these mix with soft issues like language and cultural barriers. There are, for example, languages where the term "risk" simply does not exist: The closest word the Japanese have in their language is 危険 (*kiken*), which means danger, risk, hazard, peril, jeopardy and pitfall all in one. Thus, it fails to grasp the specific concept of risk and generates confusion. This is the reason Japanese had to borrow the term リスク (*risku*) from other languages. NB: Risk originally comes from Ancient Greek *rhizikon, rhiza* a metaphor for difficult to avoid peril of the sea. From the sixteenth century, the term was also used positively as in middle-high-German *Rysigo* indicating a business and hope for economic success.

Explicitly considering and evaluating data uncertainties helps explain how risk estimates are obtained and appease risk communication difficulties. Enhancing the transparency of uncertainties in risk assessments is of course paramount. One cannot reach this goal if, for example, one selects arbitrary limits for boiler plate risk matrices. Oftentimes, the coloring scheme in such matrices does not bear any resemblance to public (or corporate) risk tolerance. Thus, risk assessment reduces to a binning exercise of "magic numbers": for example, likelihood = 3 * impact = 5, thus risk = 15! Such incompetent approaches are unfortunately the cause of claims that traditional risk management is not sufficient (*Covid-19 Highlights Need For New Approaches To Risk Management* (forbes.com)). In addition, critical overexposures sometimes arise from their use (*Council Post: How To Effectively Manage Risk In The Time Of Covid-19* (forbes.com)). However, countries and corporations relying on serious risk management approaches had the situation covered, just as they had in cases of SARS and other calamities.

Yet, many decision-makers love these oversimplified and misleading approaches as they look intuitive and seem not to require any learning. They also often forget that the multitude of uncertainties, hazards and their consequences on present-day systems is complicated. Thus, it cannot be grasped by oversimplified approaches.

Avoiding *misleading statements* (https://www.riskope.com/2018/05/23/austra lian-law-firm-plans-to-sue-worlds-no-1-miner-bhp/), openly exposing uncertainties, abolishing arbitrary selections are all actions leading to demystifying the risk assessment process. They are there for every party's benefit even if they seem complicated. The public (and management) will indeed be in a better position if they understand why risk estimates are fraught with inevitable uncertainties even if they are based on the best available data and *unbiased interpretation* (https://www.riskope. com/2016/05/26/comments-on-bc-tailings-dams-risk-reduction-audit/). That type of understanding does not come from nice colored matrices. Acknowledging uncertainties in risk assessment is quite different than asserting "unpredictability" or declaring ignorance. Unpredictability is often used as an alibi to avoid difficult discussions, especially when dealing with long-term projects. At the other end of the spectrum, ignoring uncertainty and posturing more certainty than data or experience may justify is a sure way to lose credibility. The spectrum is not binary, not "black or white," nil or one, but is instead a continuum from very high confidence to "guess work" where neither extreme statement is certainly not true. At the end of the day, no matter what efforts are made, some people within the *public will remain dissatisfied* (https://www.nature.com/articles/ncomms 10996). Indeed, the demand for certainty has characterized humans since the time of the Oracles, especially in times of crisis. When asked with this type of *predictions by clients* (https://www.riskope.com/2010/03/05/crisis-forecast-one-year-and-3-months-later-reality-check/) requesting support in their *tactical and strategic planning* (https://www.riskope.com/2019/02/06/perform-operational-tactical-and-strate gic-planning-avoiding-limbic-brain-traps/, we give probabilistic answers, which include:

- Lessons learned from the past;
- Explicit uncertainty considerations;
- Varying degrees of mathematical modeling, compatible with the available data, budget and time.

Updatable. Values of probabilities and consequences as well as their respective uncertainties that can be made current at any time on the basis of data as it is obtained, delivering a new risk landscape of the company.

Scientific Notation of Numbers

This book uses the scientific notation of numbers as follows:

- Positive powers of ten, e.g., 10^2 to indicate 100, 10^3 for 1000 ... 10^6 for 1,000,000. Furthermore, $1.5*10^2 = 150$.
- Negative powers of ten, e.g., 10^{-2} to indicate 1/100, 10^{-3} for 1/1000 ... 10^{-6} for 1/1,000,000.

Thus, for example, $1/10^6 = 10^{-6} = 1/1,000,000$ and $2.3*10^{-6} = 2.3/1,000,000 = 0.0000023$.

The scientific format makes it possible to avoid writing long and confusing series of zeroes in very small numbers or many digits in large numbers.

In general, we tend to use the 1/x to indicate frequencies and the negative power notation for probabilities.

References

CSNI (1990) *Proceedings of the CSNI Workshop on PSA Applications and Limitations. Santa-Fe*, New Mexico, September 4–6, 1990. https://www.oecd-nea.org/jcms/pl_18678

Khan, F, Abbasi S (2002) A Criterion for Developing Credible Accident Scenarios for Risk Assessment, Journal of Loss Prevention in the Process Industries 15(6):467–475, DOI: 10.1016/S0950-4230(02)00050-5

Third-Party Links in this Section

Divergent hazard break traditional risk assessments	https://www.forbes.com/sites/steveculp/2020/06/29/covid-19-highlights-need-for-new-approaches-to-risk-management/?sh=6fe32ed566b8
Traditional risk management is not sufficient	https://www.forbes.com/sites/forbesbusinessdevelopmentcouncil/2020/10/08/how-to-effectively-manage-risk-in-the-time-of-covid-19/?sh=4e37d9e1dce4
Public dissatisfaction	https://www.nature.com/articles/ncomms10996

Contents

Authors and Contributor

About the Authors

Dr. Franco Oboni, Ph.D. is Civil Engineer with over 40 years of experience and has specialized in quantitative risk assessment (QRA) for over 30 years. He leads Riskope, a Vancouver-based practice active internationally providing advice on QRA, enterprise risk management (ERM) and risk-informed decision-making support. His clients include Global 1000 companies, large insurance companies, natural resource entities (mining companies, forestry, etc.), railroads, wharves, governments and suppliers. He consults on risk and crisis mitigation projects, risk and security audits and geo-environmental hazard mitigation studies on four continents. He has authored over fifty papers and has co-authored several books: *Improving Sustainability through Reasonable Risk and Crisis Management* (https://www.riskope.com/knowledge-centre/tool-box/books-by-franco-and-cesar-oboni/) (2007), *The Long Shadow of Human-generated Geohazards: Risks and Crises* (https://www.intechopen.com/books/geohazards-caused-by-human-activity/the-long-shadow-of-human-generated-geohazards-risks-and-crises) (2016) and *Tailings Dam Management for the Twenty-First Century* (2019, Springer). He delivers customized seminars (in English, French, Italian and Spanish) to industrial audiences worldwide. He was co-recipient of the Italian Canadian Chamber of Commerce (Canada West) 2010 Innovation Award.

Cesar H. Oboni has been involved in risk analyses and reviews for numerous facilities and organizations, communities, mining companies (coal mining, gold, copper and zinc mining), military bodies and transportation entities, including "at perpetuity" projects. In addition to his activities in risk assessment and optimum risk estimates, he has been very active in the analysis of special and emerging risks, co-authoring a report on cyber-defense at national scale for an European country. He analyzes and prioritizes residual risk of mining infrastructures, tailings and water treatment plants. His clients include Fortune 500 companies, large mining corporations, UN/ UNDP, transportation and military bodies. He is Author of more than 35 papers published in the proceedings of international conferences and symposiums. He is Co-author of several books: *Improving Sustainability through Reasonable Risk and Crisis Management* (https://www.riskope.com/knowledge-centre/tool-box/books-by-franco-and-cesar-oboni/) (2007), *The Long Shadow of Human-generated Geohazards: Risks and Crises* (https://www.intechopen.com/books/geohazards-caused-by-human-activity/the-long-shadow-of-human-generated-geohazards-risks-and-crises) (2016) and *Tailings Dam Management for the Twenty-First Century* (2019, Springer). He was co-recipient of the 2010 ICCC, the Italian Canadian Chamber of Commerce (Canada West) Innovation Award.

Contributor

Dr. Janis A. Shandro, Ph.D. work and research are focused on a range of topics associated with community and occupational health, safety development and security. She has broad experience across diverse sectors in the areas of effective risk management, strategic health and development investments, and monitoring and evaluation approaches. She has successfully supported the adoption of improved health, safety and social performance in challenging and unique settings as well as developed and implemented a range of programs and capacity building efforts for diverse target groups. Technically, she specializes in identifying and managing social, health, safety and security risks associated with large-scale development projects, incidents and emergency scenarios. She has a proven track record of working collaboratively with communities, governments (indigenous and non), international organizations, multilateral development banks, academic institutions and the private sector in developing and developed nations. Her projects include health/social impact assessment; health and safety; indigenous health; the health and social performance of the global development sector; climate change and health; international performance standards, requirements and safeguard policies; and resettlement/livelihood restoration.

Acronyms

ALARA	As Low As Reasonably Achievable
ALARP	As Low As Reasonably Practicable
BACT	Best Available Control Technology
BI	Business Interruption
CAPEX	Capital Expenditures
CC	Clausius–Clapeyron
CCF	Common Cause Failure
CDC	Center for Disease Control
CDP	Carbon Disclosure Project
CSR	Corporate Social Responsibility
DT	Digital Transformation
EPA	Environmental Protection Agency
EPC	Engineering, Procurement and Construction
ERM	Enterprise Risk Management
ESG	Environmental, Social and Governance
FERC	Federal Energy Regulatory Commission
FMEA	Failure Mode and Effects Analysis
FTA	Failure Tree Analysis
HAZOP	Hazard and Operability
ICMM	International Council on Mining and Metals
IoT	Internet of Things
IPCC	Intergovernmental Panel on Climate Change
ISO 31000	International Standard Organization Standard 31000
ISO 31010	International Standard Organization Document 31010
LCU	Life-Changing Unit
LIA	Little Ice Age
MERS	Middle Eastern Respiratory Syndrome
MFL	Maximum Foreseeable Loss
MPL	Maximum Possible Loss
NI43-101	Canadian Standard of Disclosure for Mineral Projects
NPS	Customer Experience Net Promoter Score
NPV	Net Present Value

ORE	Optimum Risk Estimates (© Oboni Riskope Associates)
PD	Property Damage (insurance)
PIG	Probability Impact Graph
PML	Probable Maximum Loss
R.O.I.	Return on Investment
RIDM	Risk-Informed Decision-Making
RM	Risk Management
SARS	Severe Acute Respiratory Syndrome
SLO	Social License to Operate
SOP	Standard Operating Procedure
TSF	Tailings Storage Facility
UBC	University of British Columbia
UNEP	United Nations Environment Programme
UNISDR	United Nations International Strategy for Disaster Reduction
WHO	World Health Organization

Chapter 1
Introduction

We start this book by looking back a few decades and discussing what has changed in the arena of risk assessment and management. We look at the culprits of many misunderstandings and misjudgments, including misallocation of resources and show the importance of using a very well-defined glossary. For that reason we invite the readers to use the Discussion of Key terms section in the book front matters and links found in the references at the end this chapter.

1.1 A Bit of Historic Perspective and a Few Important Terms

We recently found in our archives the syllabus of the course we gave at University of British Columbia (continuous education, UBC) in 1999 (Fig. 1.1) and decided to compare it with our 2020 MBA and corporate courses, which focus entirely on tactical and strategic planning. The ultimate goal is to create corporate value and societal well-being while facing uncertainties and make ethical and sustainable decisions in front of divergent risk exposures.

At first sight the contents of the syllabus seem quite similar: we were already talking about the connection between risk and crisis management, risk acceptability, public perception, complex consequences, holistic scenarios etc. Many of the subjects we treated were included ten years later (2009) in ISO 31000.

Thus, it is natural to ask what has changed and what has not in these twenty years?

What has not changed is the way risk is calculated. According to Eq. 1.1, risk R can be expressed, in its simplest form as the product of the probability of occurrence of an undesirable event (probability of failure, or p_f) and the damages it potentially causes D (Kaplan and Garrick 1981; Lowrance 1976; Haimes 2009):

$$R = p_f * D \qquad (1.1)$$

© The Author(s), under exclusive license to Springer Nature Switzerland AG 2021
F. Oboni and C. H. Oboni, *Convergent Leadership—Divergent Exposures*,
https://doi.org/10.1007/978-3-030-74930-9_1

Fig. 1.1 University of
British Columbia, risk and
crisis management courses
table of content

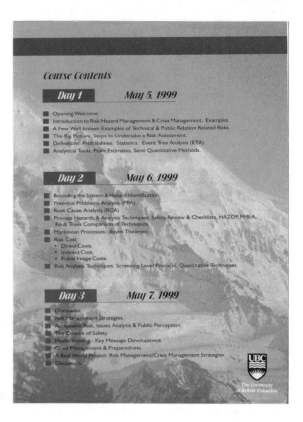

However, there is no universal consensus on the definition of risk, and in recent
years the focus has broadened to cover the whole spectrum of probabilities, consequences and uncertainties (Aven 2012). In this book, instead of D we will generally
use C, which represents the multidimensional consequences of an event as explained
later in the book (Chap. 9).

The changes that have resulted from the broadening of focus cited above may be
difficult to spot, but they are very significant. Here is a "partial" list:

1. Due to the stronger influence of emerging and divergent exposures risks and
    climate changes we have introduced in our courses a strong lesson on force
    majeure, as this clause, present in all commercial contracts, actually generates
    a significant potential risk to all involved parties (see Sect. 4.4).
2. We have become increasingly involved in clearly and rationally defining all the
    terms we use, for lack of clarity and confusion have revealed themselves to be the
    source of horrendous corporate and governmental overspending and blunders.
    Terms such as "strategic", "manageable/unmanageable", "credible", etc., are
    now clearly defined and correspond to concrete and reproducible situations.

3. Over the years we have understood how important it is to focus on the architecture of the hazard/risk registers, to avoid double counting of hazards and consequences, aimed at providing detailed understanding of the risk landscape of any given system from social and corporate/project points of view (see Chaps. 7, 8, 9, and 11).

4. In the last decade or so it has become obvious that common practice risk assessments systematically underestimate the consequences of potential mishaps. This fosters the feeling of unpredictability and un-foreseeability, thus the potential for situations of panic. In this book (see Chap. 9) we explain how holistic consequences can be evaluated and included in a risk assessment avoiding the "paralysis by analysis" syndrome while reducing the potential for panic.

5. Corporate risk tolerance and societal risk acceptability are said to have become "sciences", but while ISO 31000 (2009) and many authors talk about tolerance, they do not discuss how to develop it in real life (see Chap. 10). Rational corporate risk tolerance models have been created, proven and calibrated thanks to hundreds of real-life case studies.

6. Risk tolerance models aim to enable focused mitigation roadmaps and acquire a distinct competitive edge while fostering survivability and sustainability (see Chaps. 11, 12, 13, 14, and 15).

These changes lead us naturally to think about two very successful terms which have become international buzzwords, and both owing to Nassim Nicholas Taleb: "antifragile" (Taleb 2012) and "black swan" (Taleb 2007).

"Fragile" is the opposite of "robust". Adding the prefix "anti-" to fragile, Taleb created a term to describe a system that thrives—its performance actually improves—under stress. Taleb's 2012 book mostly considers the notions of fragility and antifragility in biological, medical, economic and political systems. Under that definition is it difficult to imagine an engineering system that would be antifragile? Engineers of all fields strive to design robust systems that have better and longer life spans despite complex hazards and climate change. The results have been measured in several cases (Oboni and Oboni 2013) displaying, along the way, some surprising results. However, we do not think this is the same as the idea of designing a system that will actually work better when experiencing unexpected or random conditions such as those from divergent hazards. We do not know how to make any industrial system thrive under stress; we do not know, for example, how to make infrastructural projects self-adjusting to hazards (and that would be sci-fi by today's standards). Thus, it seems that the "antifragile concept" is not really applicable to engineering as we know it today, because all we can do is make systems more robust, reliable and resilient (see Sect. 11.1.1), thus avoiding fragility.

The other term is black swan, a term that over the years we have become more and more allergic to because of the use the public has made of it. Reportedly Taleb himself is irritated as well by the misuse of the term black swan (Avishai 2020) referred to the last pandemic.

The problem is that the term has caught the imagination of many, perhaps explaining its world-wide success, generating a slanted image that many have jumped to use, as it allows "iffy" constructs and offers an excuse for a classic human behavior: doing nothing and procrastinating.

Let's explain the point: black swans are defined by Taleb as "an unpredictable or unforeseen event, typically ones with extreme consequences". However, we have read *for example* (https://www.dailymaverick.co.za/article/2015-01-06-the-black-swans-that-could-come-our-way-in-2015/) that "geopolitical black swan events, such as the Arab Spring and the Japanese tsunami (Fig. 1.2), have further complicated the market dynamics". That is a blatant misuse of the term. Indeed, similar events that have occurred over fifteen times *in the last two centuries* (https://en.wikipedia.org/wiki/List_of_economic_crises), such as the 2008 economic meltdown, are far from being black swans. We humans have a very short, selective, memory, making the financial crisis of 2008 appear to many of us as an unpredictable and unforeseen event despite such a high recurrence of similar events. That's WRONG!

We note that black swans are not the only members of the modern risk mythology. Indeed, in 2009 they were joined by dragon-kings (Sornette 2009), events of extreme impact events (kings) with a unique origin (dragon) relative to other events from the same system due to positive feed-back, tipping points and possible bifurcation. Sachs et al. (2012) modeled natural hazard dragon-kings such as earthquakes, volcanic eruptions, landslides, floods, and self-organized criticality phenomena (Bak et al. 1987) such as wildfires.

They were followed by gray rhinos, a concept introduced by Michele Wucker (2016), commentator and policy analyst specializing in the world economy and crisis

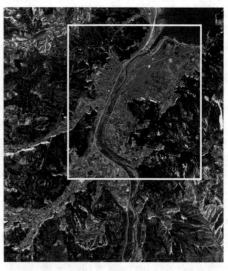

Fig. 1.2 Japan Tsunami, South of Miyako, Iwate Prefecture. Left: before/Right: after. Notice the reach of the tsunami water along the valleys and the resulting devastation. Left: Satellite image © 2021 Maxar Technologies. Right: RADARSAT-2 Data and Products © MDA Geospatial Services Inc. (2021) - All Rights Reserved. RADARSAT is an official mark of the Canadian Space Agency

anticipation at the World Economic Forum Annual Meeting in Davos in 2013. The concept gained strong following as gray rhinos were defined as highly probable, high impact yet neglected threats that may occur after a series of warnings and clear evidence. Gray rhinos were soon to be followed by yet another mythological beast, *the black elephant* (https://www.straitstimes.com/opinion/the-black-elephant-challe nge-for-governments), created in 2017 by Peter Ho. We were thrilled to learn that a black swan and a gray rhino gave birth to the black elephant. This creature is a problem that is actually visible to everyone, but no one wants to deal with. As a result, it gets ignored and when it bursts everyone acts surprised and calls it black swan.

We do not think that "neglected threats that may occur after a series of warnings" and "problems that are actually visible to everyone, but no one wants to deal with" ought to enter in the Olympus menagerie. We think that if we want to move forward and create value while reducing exposures, we need to identify and quantify probabilities of occurrence and consequences of mishaps in our complex, interdependent systems, especially if events start diverging from business-as-usual.

As we will show in this book, there are relatively simple means to obtain sophisticated results, provided strong attention is given to the glossary and communication, and appropriate techniques are deployed.

After reviewing economic recession and mythological animals let us discuss the very real Fukushima tsunami (Fig. 1.2). We note that the sea defenses were designed for the *"maximum agreed event* (C:\Users\fobon\Desktop\Divergence and conver gence\120620e0102.pdf (tepco.co.jp))" of 5.6 m. Various studies recommended considering a 15 m high tsunami wave, but these were dismissed as unrealistic or "uncertain" (Clarke and Eddy 2017) albeit the region had seen three major earthquakes (magnitudes greater than 8: Sanriku, 1869,1896 and 1933) with waves higher than 20 m. The U.S. *Nuclear Regulatory Commission* (https://en.wikipedia.org/ wiki/Nuclear_Regulatory_Commission) had studied emergency power loss in their *NUREG-1150* (https://web.archive.org/web/20131105020856/http:/www.nrc.gov/ reading-rm/doc-collections/nuregs/staff/sr1150/v1/sr1150v1part-2.pdf) 1991 report clearly showing that the probability of losing emergency power during an earthquake was well in the realm of credibility. Finally the power plant design trusted only one line of defense (hence the project was not robust and reliable but fragile) leaving the electrical commands liable to flood in the underground of the plant (yet, engineers generally know that trusting one line of defense, the properties of one material, device, etc., does not make good sense) despite a series of patches started in 2008.

Let's put on our risk assessment hat. Chemical and other industries often consider 1/1000,000 the limit of credibility, so anything that occurs at a higher rate or likelihood is not a black swan, especially if the designer claims to have designed for it. Let's note that some researchers (for instance, Molina 1991) think probabilities can be quantified with reasonable certainty down to frequency levels of $1/10^6$ to $1/10^7$ event/year. Below such values no meaningful estimates can be given for single events. Therefore, those researchers recommend a cut-off frequency of $1/10^7$ event/year for single events. Indeed, predictions of most events which require human error frequencies of the order of $1/10^7$ are clearly non-credible because the historical data set required to establish such human performance is generally non-existent. Based on the summary above, considering Fukushima a black swan was dead wrong especially since tsunami stone markers existed in Japan alerting passers-by to the possible *floods' levels* (https://www.nytimes.com/2011/04/21/world/asia/21stones.html).

So, we are left with a blatant gap between business-as-usual and black swans, leading us to define an intermediate class of exposures we call "divergent". We immediately put forward a word of caution: there is no linearity in these definitions, and they always have to be discussed very carefully. What we know, however, is that if the gap is not filled, then we will keep seeing poorly-made risk assessments which censor exposures both in terms of probabilities and consequences in order to narrow the realm of "credible" exposures and abandon to fate the so-called black swan.

1.2   The Culprits of All Our Evils

From the previous discussion of what has changed in the last twenty years and the misuse of Taleb's terms it can be concluded that risk perception, communication, corporate prestige, psychological factors, excuses and denials are important ingredients if not the main culprits of all out evils. They must not be forgotten in the discussion related to divergent hazards and convergent platforms aiming at a better understanding of the holistic risk landscape surrounding a corporation, a society, or a project.

Those same ingredients are those that render people and decision-makers blind (generating the desire to create mythological beasts) and make it difficult to properly plan for divergent exposures at the tactical and strategic levels. Fear and possibly panic will ensue when the public, workers, and others feel their destiny is unforeseeable and unpredictable.

In this book we endeavor to show that the bar of foreseeability and predictability can be raised quite significantly by applying reasonable convergent approaches, thus greatly enhancing the sustainability of our societies, corporations, and projects facing divergent hazard.

If risk perception, corporate prestige, and psychological factors are not properly managed they are important sources of irrational decisions. It all sounds so simple, so easy, yet, when it comes to actually doing things, few companies, boards of directors, governments, or individuals actually give themselves the means to embrace rational and scientific approaches to risk and crisis management using clearly defined tolerance criteria in order to develop sensible roadmaps to tactical and strategic divergent exposures mitigation and sustainability.

Indeed, the actions of the board of directors and other levels of corporate and governmental spheres can often seem like the result of cognitive biases, and other psychological effects, as illustrated by Anecdote 1.1.

Anecdote 1.1: Railroad Executives' Seminar

Roughly ten years ago we were invited by a very famous and prestigious European railway company to give a seminar to top management. NB: the company name is covered by confidentiality and we have also slightly altered the story to further protect the identities of the implied parties. Here is the story.

European railway companies are state-owned enterprises. Thus, their boards generally have c-suite executives who either are administrative veterans or politicians, men and women of great experience but traditionally poorly inclined to follow the rapid changes imposed by our fast-evolving world.

The subject of the seminar was, of course, risk management. We were surprised that our client would be ready to depart from the traditional all-hazard management, the obsolete approach inherited from the royal railroads' era. In the good old days, rail transportation was indeed a royal privilege, thanks to royal trains, and therefore rail network also reflected the prestige and power of kingdoms, above and beyond functionality.

We soon learned that the "old boys" of that particular administration had either retired or were promoted to higher functions, and the new CEO was very eager to promote "new ways", including risk management, in the company. We were very honored to have him sit in the seminar.

We started sharing our numerous experiences with US and Canadian railroads risk management, wharves and ship loaders risk assessments, large-scale logistic risks, and country-wide approaches. Our intention was to then expand on other horizons, ending with a panorama of strategic and tactical/operational risks before entering into rational prioritization and risk-based decision making.

The CEO, however, quickly erupted with an aggressive statement. It sounded roughly like "Well, I appreciate that you come from the technical side (a disdainful grin appeared on his face…) and like to talk about these "technical/natural risks" (disdain again…), but we are here to hear about business risks, strategic risks, high-level understanding (a happy grin now…)…". Believe me, if he did not use the black swan stereotype, it was just because no one had invented it yet! Without losing emphasis he ended up saying that at the corporate level they would like to use modern approaches (he even mentioned an off-the shelf methodology based on cards…), but they were certainly not interested in knowing about "technical details" such as probabilities and consequence functions, tolerance, etc.

Our reaction was simple. We said: "Sir, as far as we understand your company's business is running freight and passenger trains. We understand our mission is to support your business' objectives by bringing in good risk management and decision-making tools. Only a rational and unbiased approach will tell which ones are the major risks your company is facing. Intuition is not enough, and it actually proves generally wrong. Risks are defined by their probabilities of occurrence and their consequences. They should be compared to your company's tolerance for rational prioritization. That's way too difficult for intuition: you need to perform a study to prioritize your risks. We can say that the so-called high-level risks you mention depend on political decisions. Shall we call them political hazards? Political decisions are slow (some might even never happen) by definition, especially when they would impact a state-owned corporation. Instead, natural hazards are blind. They do not need to be re-elected, and climate is evolving faster than we might think. That's why we believe it is not rational to exclude natural phenomena from the analyses".

After that exchange, the CEO left, and we finished the seminar. We knew we would never work for that company/board/CEO again. We assumed they went shopping for a deck of cards and started discussing high-level strategic risks. Hopefully they were not Tarot cards.

Not more than six months later a landslide (a natural culprit of the "natural/technical kind") took out one of the major double track lines in the country. The company was shamed by the media as they had to shuttle people on buses around the natural phenomenon. It took roughly another year for a similar phenomenon to occur elsewhere on the network, triggered, this time, by "anomalous rainfalls". Remember, black swans were not invented yet!

No need to say, no one ever called us back. A few weeks later, we learned by a request for proposal from that country that:

(a) the Ministry of Transportation had "taken away" from the railroad company their risk management function (a loss of freedom, like being put under monitoring by a higher authority).

(b) The Ministry of Transportation had put together a risk management approach that goes down to the single switch, single signal, single rockfall detail at country-wide scale (so much for looking into strategic, high-level risks only…NOT!). This is pretty much what we had done for many other clients but was thrown away by the CEO.

(c) The Ministry of Transportation had developed a pilot study and was now in the request for proposal stage for a custom-tailored computer application at the country scale.

So, you might ask, what are the conclusions of this anecdote? Here they are, under the form of simple questions based on the concept of loss. So, why lose:

- Ten years of precious time?
- Your freedom as you sink deep in crises?
- Loads of money?
- Your vision, blinded by irrational approaches?

Closing point. Cognitive biases of any kind may be very costly, and they are paramount to explain many failures of human endeavors and projects. In this book we will explore how to foster sustainable thinking by reducing biases and better evaluating risks.

1.3   Clear Definitions Are Key

A major cause of trouble in the risk field is the lack of compliance to a precise glossary and semantic confusion. Insisting on the need for a clear glossary sounds pedantic, but we started in Sect. 1.1, Eq. (1.1) to define the term risk. Indeed, every day we see examples of key figures in the corporate or administrative world confusing risk, probability, consequences, etc. Thus, we strongly invite the readers to read in detail the glossary and reference to it in order to ensure we are all "talking in the same language". A box with links to the glossary and key terms is present at the end of each Chapter of this book.

Many professionals, including medical doctors and engineers of all flavors, are trained to solve a problem in the most efficient way. They generally know about secondary effects, but these are not their first concern. Ask an engineer to design a robust dam and he will. The problem is that "robust" means a different thing to an engineer than to the public (see Sect. 5.2, 5.3). Indeed, he or she may forget that defenses built to save the dam may create other nasty problems downstream. Clear definitions are indeed key, and Anecdote 1.2 is a sad example of why.

Anecdote 1.2: Unclear failure criteria
A relatively recent event in the Solomon Islands illustrates the problem of unclear definitions (Albert et al. 2016). At the Solomon Islands' Gold Ridge Mine, arsenic, cyanide and other heavy metals, perhaps selenium and mercury, are present in the tailings. Excess water accumulated behind the dam, which overflowed uncontrollably after heavy rain. In 2014 and 2015 the catastrophic flooding brought by *Tropical Cyclone ITA* (http://www.unocha.org/top-stories/all-stories/solomon-islands-worst-flooding-history#_blank) and a *close call* (http://www.radioaustralia.net.au/international/radio/program/pacific-beat/potential-gold-ridge-mine-disaster-raised-in-solomon-islands-ads/1400037#_blank) raised alarm at various levels. 8000 people live downstream. Tens of millions of litres of contaminated water reportedly escaped from the dam via its weir. A weir is a structure built across a water course or on top of a dam allowing water to flow steadily over its top. Weirs built on top (at the crest) of a dam are there to allow excess water accumulated behind the dam to run away. That is, of course, without damaging the dam by uncontrolled over-topping. That event altered the life of the 8000 downstream residents. Indeed, health authorities in the Solomon Islands released a statement advising downstream communities not to use rivers for drinking, cooking or bathing.

The environmentalist point of view was obviously that the dam did not protect the environment from the chemicals contained in the tailings' reservoir, therefore the tailings dam has "failed" because of the weir. For the dam engineer the weir and its associated spillway relieved the pressure and reduced the likelihood of a dam collapse. The system worked perfectly! For dam engineers a failure is a dam collapse such as those seen in the reports of recent major dam failures at *Samarco* (https://www.riskope.com/2016/06/29/sam arco-fundao-dam-failure-mode-and-effects-analysis-fmea/) and *Mount Polley* (https://www.riskope.com/2015/02/04/mount-polley-dam-breach-discussing-tailings-dam-failure-frequency-and-portfolio-risk/). The weir functioning was a success from the dam engineer's point of view.

In the case of the Solomon Islands' Gold Ridge dam it is obvious that a performance/success criteria should have been established for the level of contaminants under normal storage conditions such that any discharged water through the weir would meet "acceptable" levels. However, could a failure criterion be the exceedance of water quality objectives at a specified point, i.e., just before the nearest drinking water source downstream?

Closing point. What are the criteria for defining a tailings dam structural failure? For example, has the dam "failed" if:

1. It continues to contain all material despite being damaged?
2. It contains spills, but all solid and water impacts are limited to

   a.   within property limits?
   b.   within an acceptable inundation or exceedance areas?

3. It impacts areas beyond the acceptable inundation line?

In our courses and book, *Tailings Dam Management for the Twenty-First Century* (Oboni and Oboni 2020) we insist on the necessity of clearly defining:

- the system that is the object of the risk assessment (see Chap. 6);
- the success and failure criteria for that system;
- the consequences leading to evaluating the risks.

These points should be clearly defined before attempting any risk characterization. Thus we need to understand what the client means by failure. Does it mean that there will not be a catastrophic failure like Samarco, Mount Polley or *Brumadinho* (https://www.riskope.com/2019/04/10/brumadinho-tailings-dam-failures-and-boeing-737-max-8-tragedy-offer-striking-similarities/) (Hatje et al. 2017; Byrne et al. 2015; Thompson et al. 2020)? Or does it mean there will not be release of contaminants in the environment through seepage or minor failures? Or does it mean something else again?

We know many feasibility studies which adopt only a very narrow definition, thus falling in the trap of an information silo. This oversight leaves room for conflicts with local residents (Albert et al. 2016), and this may lead to crises that will inevitably damage CSR (see Chap. 5, Sect. 6.1.2) and SLO.

Nowadays the following can be noted and will lead to the need for increasingly clearer and transparent definitions.

After each failure, the mining (Roche et al. 2017) and other industries see codes evolve and imposes tougher criteria and specifications (Directive 1982). Meanwhile

FMEA (see Appendix B) remains the common practice risk assessment methodology. FMEA lacks the finesse needed to predict the progress "toward zero failures" goal stated by experts in the aftermath of recent tragedies (Oboni and Oboni 2016). The FMEA we see are often confusing documents where consequences, hazards and risks are neither clearly defined nor understood.

In many fields, the effects of today's risk mitigation programs will only slowly become visible. That is because the world portfolio will contain mitigated and unmitigated (legacy) projects of all types. During that time, the public will perceive at best a status quo unless clear definitions are put forward.

It will be exceedingly difficult to evaluate future progress, as factors such as climate change, divergent exposures, seismicity, increases in population, changes of land use will further complicate the situation. Thus, public outcry and hostility toward industries and new infrastructures, fueled by the diffusion of information and communication technology will likely increase.

Furthermore, the same factors will complicate scenarios and their consequences. We believe that will occur in any portfolio, likely in any country around the world. When it comes to discussing consequences, it is paramount to remember that there is no simple rule such as "small hazard means small consequences". An example will show this. On 27 May 2011 the Hoeganaes Gallatin facility in Tennessee (USA) experienced a hydrogen gas explosion, followed by metal dust flash fires, leading to three fatalities and two injuries. A leaking gas was not tested and mistaken for nitrogen, and in a bid to repair the leak, the removal of a gas-line trench cover using a forklift equipped with a chain on its forks created sparks which ignited the hydrogen and metal dust (CSB 2012). Thus a very destructive accident in the US was generated by the "small hazard" of a forklift hitting a gas line.

As regards the larger structures, today the industry better investigates, designs, builds and manages/monitors modern ones better than any older ones. Based on our experience, we see their probability of failure decrease, especially for the critical ones. Of course, larger structures may lead to way larger consequences than the failure of smaller structures. Nevertheless, we cannot forget that the dam in Val di Stava, Italy, which collapsed in 1985 and was one of the deadliest tailings catastrophes, reportedly had only a small volume of 300,000 m^3 (Chandler and Tosatti 1995) for mining standards. All that to say that the interplay of probability and consequences does not necessarily mean the risks are larger (or smaller). Indeed, we clearly showed in a 2015 paper (Caldwell et al. 2015) that, in the realm of waste containment structures, if we look at volume released versus lives lost it would be preposterous to see any correlations.

In summary:

- Industry should avoid simplistic statements equating bigger structures to bigger risks.
- As structures become larger, their probability of failure should reduce. That reduction necessitates acting on all aspects of a system's life, including:

  – investigation;
  – design;

- construction;
- service life/production;
- monitoring;
- inspections and management.

as all of these contribute to the chance of failure.

- Population increases, land use shifts, environmental constraints increase consequences of failures and therefore increase risks. This is extremely important to note as some industrial systems, such as waste storage facilities, have a long-life span during service and post-service.

1.4   The Book Plan

This book discussion is split in five parts.

PART I, State of Affairs, looks at several recent events around the world where tactical and strategic planning were apparently poorly developed and lead to unpleasant outcomes (Chap. 3.3). Chapter 3 then sets the context of divergence using mythological tales and recent events, including a detailed discussion on climate change exposures (Sect. 3.2). The need for clear definitions and compliance to a very precise glossary is stressed, to avoid confusion and misunderstandings under pressure to report risks (Sect. 3.3).

PART II, Divergent Exposures, the Public and Ethics, delves into the differences between business-as-usual and divergent events (Chap. 4), credible events and black swans and their relationship with standard levels of industrial mitigation (Sect. 4.1). Attention is then shifted toward metaphoric description of risks (Sect. 4.2) that are useful when considering public reactions and a taxonomic description of uncertainties. Notions such as return period and force majeure are discussed (Sects. 4.3 and 4.4) to show how misleading, or plain erroneous, these notions can be in the face of divergent exposures and public expectations. Chapter 5 discusses health, well-being and resiliency of business and for people (Sect. 5.2), followed by a review of what the contemporary public seems to want (Sect. 5.3), based on recent conferences and public hearings, and to a closure consisting of requirements of communication and transparency as well as considerations of ethics (Sect. 5.4).

PART III, Convergent Assessment of Exposures, is structured around the necessary steps to perform a modern convergent risk assessment capable of handling business-as-usual as well as divergent exposures. Thus, it looks at system definition (Chap. 6), hazard identification (Chap. 7), how to define probabilities of events (Chap. 8) and the consequences of events (Chap. 9).

PART IV, Tactical and Strategic Planning for Convergent/Divergent Reality, discusses a quintessential element of modern risk-informed decision making: the risk tolerance threshold both in historic and modern terms (Chap. 10). Chapter 11 then reviews the "how to" procedure and spells out simple rules to develop resilient solutions (Sect. 11.1.1). Using risk tolerance gives risk assessments the ability to prepare rational mitigative roadmaps, distinguishing tactical from strategic families of risks (Chap. 12). Finally we touch on risk adjusted project cost-estimates (Sect. 12.2).

PART V, Convergent Assessment for Divergent Exposures: Case Studies, features risk-informed decision-making deployments examples on Railroads, wharves, chemical plants in isolation and as a system and concludes with a look into what we expect should happen in terms of leadership re-orientation before delving in the book conclusions.

These five principal parts are followed by two **Appendices**. Appendix A helps making sense of probabilities and frequencies when little data are available and rough estimates are required. Appendix B forms the counterpoint to Part IV and many points delivered in Part III, as it concentrates on the DON'TS related to risk assessments. Readers should refer to it every time they feel like shaking their heads while reading the book.

In particular we show:

- How critical risks, and their necessary management controls, can be identified for both existing and divergent situations (Part II and III);
- How cost-effective risk reduction strategies can be identified and included in the design and management of existing infrastructure and new projects (Part III and IV);
- How risk assessment approaches can provide a solid basis for the justification and prioritization of continuous improvement, risk mitigation and resilience fostering programs, namely Risk Informed Decision Making (RIDM) (Part IV);
- How an extant asset portfolio can be risk-prioritized in order to design mitigative measures while appropriately communicating risk to the public (Part IV).

Appendix

Links to more information about the Key terms from the Authors	
A,B	*Act of God* (https://www.riskope.com/2020/12/09/act-of-god-in-probabilistic-risk-assessment/) *Black swan* (https://www.riskope.com/2011/06/14/black-swan-mania-using-buzzwords-can-be-a-dangerous-habit/) *Business-as-usual* (https://www.riskope.com/2021/01/13/business-as-usual-definition-in-risk-assessment/)

(continued)

(continued)

C,D	*Convergent* (https://www.riskope.com/2021/01/20/convergent-risk-assessments/) *Divergent* (https://www.riskope.com/2020/11/18/tactical-and-strategic-planning-to-mitigate-divergent-events/) *Drillable* (https://www.riskope.com/2020/01/15/probability-impact-graphs-do-not-fly/)
F	*Foreseeability/foreseeable* (https://www.riskope.com/2021/01/06/foreseeability-and-predictability-in-risk-assessments/) *Fragile/fragility* (https://www.riskope.com/2020/04/01/antifragile-resilient-solutions-for-tactical-and-strategic-planning/)
P,R	*Predictability/predictable* (https://www.riskope.com/2021/01/06/foreseeability-and-predictability-in-risk-assessments/) *Resilient* (https://www.riskope.com/2016/11/23/resilience-cannot-based-instinctual-decision-making/) *Resilience* (https://www.riskope.com/2016/11/23/resilience-cannot-based-instinctual-decision-making/)
S	*Scalable* (https://www.riskope.com/2015/04/16/how-system-definition-and-interdependencies-allow-transparent-and-scalable-risk-assessments/) *Societal risk acceptability* (https://www.riskope.com/2014/01/09/aspects-of-risk-tolerance-manageable-vs-unmanageable-risks-in-relation-to-critical-decisions-perpetuity-projects-public-opposition/) *Sustainability/sustainable* (https://www.riskope.com/2019/01/16/improving-sustainability-through-reasonable-risk-and-crisis-management/) *Survivability* (https://www.riskope.com/2011/03/17/ale-fmea-fmeca-qualitative-methods-is-it-really-what-we-need/) *System* (https://www.riskope.com/2017/07/26/three-ways-to-enhancing-your-risk-registers/)
T,U	*Tolerance* (https://www.riskope.com/2020/04/29/risk-tolerance-thresholds/) *Uncertainty/uncertainties* (https://www.riskope.com/2015/12/10/3-decision-making-truths-derived-from-uncertainty-taxonomy-scheme-of-classification-and-a-road-sign/) *Updatable* (https://www.riskope.com/2020/01/07/climate-adaptation-and-risk-assessment/)

(continued)

(continued)

Other linked information (https://www.riskope.com/blog-news/) search Riskope blog and use the search box

Third Parties links in this section:	
Geopolitical black swan events	https://www.dailymaverick.co.za/article/2015-01-06-the-black-swans-that-could-come-our-way-in-2015/
List of economic crises	https://en.wikipedia.org/wiki/List_of_economic_crises
Black elephant	https://www.straitstimes.com/opinion/the-black-elephant-challenge-for-governments
Maximum agreed event	*120620e0102.pdf (tepco.co.jp)* (https://www.tepco.co.jp/en/press/corp-com/release/betu12_e/images/120620e0102.pdf)
NUREG-1150	https://web.archive.org/web/20131105020856/http://www.nrc.gov/reading-rm/doc-collections/nuregs/staff/sr1150/v1/sr1150v1part-2.pdf
Flood levels	https://www.nytimes.com/2011/04/21/world/asia/21stones.html
Tropical Cyclone ITA	https://www.unocha.org/top-stories/all-stories/solomon-islands-worst-flooding-history#_blank
Close call	https://www.radioaustralia.net.au/international/radio/program/pacific-beat/potential-gold-ridge-mine-disaster-raised-in-solomon-islands-ads/1400037#_blank

References

Albert S, Grinham A, Pikacha P, Boseto D, Kera J (2016) Baseline assessment of water quality and aquatic ecology downstream of Gold Ridge Mine, Solomon Islands, February 2016

Aven T (2012) The risk concept—historical and recent development trends. Reliab Eng Syst Saf 99:33–44

Avishai B (2020) The pandemic isn't a black swan but a portent of a more fragile global system. The New Yorker, April 21, 2020

Bak P, Tang C, Wiesenfeld K (1987) Self-organized criticality: an explanation of 1/f noise. Phys Rev Lett 59(4):381–384

Byrne P, Hudson-Edwards K, Macklin M, Brewer P, Bird G, Williams R (2015) The long-term environmental impacts of the Mount Polley mine tailings spill, British Columbia, Canada. In: European geophysical union annual symposium. Vienna. https://doi.org/10.1130/abs/2016AM-278498

Caldwell JA, Oboni F, Oboni C (2015) Tailings facility failures in 2014 and an update on failure statistics. Proc Tailings Mine Waste 2015:25–28

Chandler RJ, Tosatti G (1995) The Stava tailings dams failure, Italy, July 1985. Proc Inst Civ Eng Geotech Eng 113(2):67–79

Clarke RA, Eddy RP (2017) Warnings: finding Cassandras to stop catastrophe. Harper Collins, p 84

Directive (1982) Council Directive 82/501/EEC on the major-accident hazards of certain industrial activities. https://eur-lex.europa.eu/legal-content/EN/TXT/?uri=CELEX%3A31982L0501

CSB (2012) Investigation report. Metal dust flash fires and hydrogen explosion, Report No. 2011-4-I-TN. Hoeganaes Corporation, Gallatin, TN

Haimes YY (2009) On the complex definition of risk: a systems-based approach. Risk Anal 29:1647–1654

Hatje V, Pedreira RM, de Rezende CE, Schettini CAF, de Souza GC, Marin DC, Hackspacher PC (2017) The environmental impacts of one of the largest tailing dam failures worldwide. Sci Rep 7(1):1–13

[ISO 31000:2009]. International Organization for Standardization (2009) ISO 31000, Risk management—principles and guidelines. International Standards Organization. https://www.iso.org/iso-31000-risk-management.html

Kaplan S, Garrick BJ (1981) On the quantitative definition of risk. Risk Anal 1(1):11–27

Lowrance W (1976) Of acceptable risk—science and the determination of safety, Technical Report. William Kaufmann Inc., Los Altos, CA

Molina T (1991) CSNI workshop on PSA applications and limitations (No. NUREG/CP–0115). Nuclear Regulatory Commission

Oboni C, Oboni F (2013) Factual and foreseeable reliability of tailings dams and nuclear reactors—a societal acceptability perspective. In: Tailings and mine waste 2013

Oboni F, Oboni C (2016) A systemic look at tailings dams failure process. In Tailings and mine waste

Oboni F, Oboni C (2020) *Tailings Dam Management for the Twenty-First Century* (https://link.spr inger.com/book/10.1007/978-3-030-19447-5#about). Springer, Cham. https://doi.org/10.1007/978-3-030-19447-5_11. ISBN: 978-3-030-19446-8

Roche C, Thygesen K, Baker E (2017) Mine tailings storage: safety is no accident. A UNEP Rapid Response Assessment. United Nations Environment Programme and GRID-Arendal, Nairobi and Arendal

Sachs MK, Yoder MR, Turcotte DL, Rundle JB, Malamud BD (2012) Black swans, power laws, and dragon-kings: earthquakes, volcanic eruptions, landslides, wildfires, floods, and SOC models. Eur Phys J Spec Top 205:167–182. https://doi.org/10.1140/epjst/e2012-01569-3

Sornette D (2009) Dragon-Kings, Black swans and the prediction of crises. Int J Terraspace Sci Eng 1(3):1–17

Taleb NN (2007) The Black Swan: the impact of the highly improbable, 2nd edn. Random House

Taleb NN (2012) Antifragile: things that gain from disorder, 3rd edn. Random House

Thompson F, de Oliveira BC, Cordeiro MC, Masi BP, Rangel TP, Paz P, Freitas T, Lopes G, Silva BS, Cabral AS, Soares M (2020) Severe impacts of the Brumadinho dam failure (Minas Gerais, Brazil) on the water quality of the Paraopeba River. Sci Total Environ 705, Article 135914

Wucker M (2016) The Gray Rhino: how to recognize and act on the obvious dangers we ignore. St. Martin's Press

Part I
State of Affairs

In Part I, we will reply to a number of very important questions:

What constitutes a priority (Sect. 2.1) when tackling the complex decisions geared toward fostering sustainability, values and corporate as well as societal well-being?

Why have so many projects and endeavors had fatal flaws (Sect. 2.2), from inception, that lead them to crash when exposed to divergent hazards?

Has Mankind always had to cope with divergence (Sect. 3.1)?

Is there increasing (Sect. 3.2) public pressure to become more transparent (Sect. 3.3) on divergent exposures, including some "exotic" exposure that make most people smile in disbelief?

And, finally and most importantly:

What are the goals of convergent leadership in a divergent risk world (Sect. 3.4)?

Chapter 2
Mankind, Risks and Planning

Some authors (Kabadayi and Osvath 2017) state that the key difference between human intelligence and animal intelligence is that human perform "*conscious and not infrequent planning for the future*" (https://www.stat.berkeley.edu/~aldous/Real-World/phil_uncertainty.html). Indeed, in various manners and to various magnitudes humans have altered the earth's environment for millennia, in what many call the Anthropocene epoch, following conscious planning. Of course, there is no specific starting date for the Anthropocene, and some place its beginning as late as the industrial era. We believe Anthropocene began way earlier. Also, some voices have risen stating that only "now" we are able to evaluate the effects of what we're doing. The era of this "new understanding" is called by some authors the Sapiezoic era, in which mankind not only consciously plans its actions but is also supposed to evaluate the effects of those actions. As a result of this evolution, today we should think in terms of tactical and strategic planning while facing the great uncertainties, complex systems and divergent risk exposures that characterize our planet. Systems are defined in this book as sets of elements working together and/or interconnected, geared toward accomplishing a goal, an objective in the field of administration, industry, environment and society. A box with links to key terms is included in the references at the end of this chapter to facilitate the read.

Planning in the face of uncertainty requires some notion of what is likely or unlikely to happen and what the consequences of the occurrence may be, i.e., an evaluation of risk. From here, it only seems natural that we must all have some intuitive notion of "likelihood", often related to a frequentist approach. That enhances predictability, which is a positive result. When we think about something being "likely" or "unlikely", we are consciously recognizing its opposite, i.e., unpredictability or uncertainty. In many instances we think only qualitatively to likelihood of failure, maybe adding some rough scale to the words (little, quite, very, extremely, etc.).

Historically, and in its simplest form, in many industries and areas of knowledge (Laplace 1902; Hopkin 2018; Stamatelatos et al. 2011; Kapurch 2010; Papoulis and Pillai 2002) risk R has been expressed (Eq. 2.1) as probability of failure times

© The Author(s), under exclusive license to Springer Nature Switzerland AG 2021
F. Oboni and C. H. Oboni, *Convergent Leadership—Divergent Exposures*,
https://doi.org/10.1007/978-3-030-74930-9_2

consequences in general, or more restrictively cost of consequences:

$$R = p_f * C \tag{2.1}$$

However, some authors (Slovic 1987; Loosemore et al. 2012) have attempted to introduce more complex structures. Any efficient portfolio risk management (mitigation) effort should be developed following an Enterprise Risk Management (ERM) logic (Jaafari 2001), and not only a p_f reduction logic. Indeed, looking at only one part of the risk—for example, only at the probability or only at the consequence—leads to wasting or misallocating limited resources (Chowdhury and Flentje 2003).

Of course, risks generated by any human endeavor should also match public acceptance criteria (Fischhoff et al. 1978), perhaps once some mitigation are implemented. As we will see later, the public should have a say in how risk assessments are built and communicated (Oboni and Oboni 2020) (see Chap. 5).

Risks lurk in any endeavor, including under normal operating conditions. Business-as-usual includes the variability of any parameter as considered and specified in the design as "normal operating conditions" and does not represent a hazard. For example, a $\pm 10\%$ price variation of the oil in a project could be considered as "business-as-usual" if so specified, whereas an event pushing a $+30\%$ hike would be a hazard still within the realm of business-as-usual because such an increase is within usual extremes. The probability of a catastrophic accident within world-wide benchmarking for a certain type of accident is also business-as-usual, whereas an accident "above benchmarks" is a hazard which could become divergent. The hazard and its consequences are always subject to uncertainties. Scenarios with low predictability and foreseeability become divergent.

When people are asked about future plans they oftentimes reply in terms like: "I'm going to …." or "I prefer to stay home" or "maybe …." or "probably…" or "seems difficult…". What those replies indicate is that deep inside people attribute a likelihood scale to those replies which then is used to formulate their reaction/action plan. Some could argue that at this level likelihood and uncertainty are almost synonyms and can be used indifferently, but that is far from the truth. To cover this point, the Intergovernmental Panel on Climate Change (IPCC) issued authoritative analyses of scientific understanding of climate change (Mastrandrea et al. 2011). As future predictions involve uncertainty, and the panel wants the authors to be consistent in how they qualify uncertainty, IPCC provides technical Guidance *Notes for Lead Authors* (https://www.ipcc.ch/site/assets/uploads/2017/08/AR5_Uncertainty_Guidance_Note.pdf), from which Table 2.1 was adapted.

Table 2.1 addresses the issue of uncertainty and possible mathematical modelling. It shows that within a complex setting such as future climate change any asserted numerical probability is (at best) an output from some complicated model in which all these different kinds of uncertainty are present. This point is obvious, but quite far away from mathematics or probability textbooks!

After these preliminary considerations, we will now review a few examples of real-life issues where tactical and strategic planning was blatantly missed perhaps due to poor evaluation or consideration for uncertainties, too simplistic reasoning

Table 2.1 Knowledge level versus sources of uncertainty and typical approaches or considerations for parameters evaluations

Knowledge level	Indicative examples of uncertainty sources	Typical approaches or considerations for parameters evaluations
Unpredictability	Projections of human behavior not easily amenable to prediction (e.g. evolution of political systems). Chaotic components of complex systems	Use of scenarios spanning a plausible range, clearly stating assumptions, limits considered, and subjective judgments. Ranges derived from ensembles of model runs
Structural uncertainty	Inadequate models, incomplete or competing conceptual frameworks, lack of agreement on model structure, ambiguous system boundaries or definitions, significant processes or relationships wrongly specified or not considered	Specify assumptions and system definitions clearly, compare models with observations for a range of conditions, assess maturity of the underlying science and degree to which understanding is based on fundamental concepts tested in other areas
Value uncertainty	Missing, inaccurate or non-representative data, inappropriate spatial or temporal resolution, poorly known or changing model parameters	Analysis of statistical properties of sets of values (observations, model ensemble results, etc.); bootstrap and hierarchical statistical tests; comparison of models with observations

and lack of effort to enhance predictability and foreseeability when facing potential divergent exposures.

The goal of this book is to allow operational, tactical and strategic planning based on sensible estimates of business-as-usual and divergent risks of various origin and significance. The issue of stranded assets which may result, for example, in companies not being able to burn all their coal due to carbon emissions restrictions, is only briefly introduced in Sect. 5.1.

2.1 What Constitutes a Priority?

In a world where socio-political affairs and climate changes evolve at a dynamic, fast pace, which risk comes first and for whom? This is a difficult question to answer because there are widespread biases, censoring and confusion in risk approaches. These are ubiquitous, at all levels, and generate the need for the mythological menagerie cited in Sect. 1.1 and advanced research on decision-making processes in business, as discussed later in this section. We always warn our clients against biases and censoring to avoid corporate blunders and make them aware of the costly confusion of common practice. Anecdote 2.1 is a perfect illustration of bias, censoring and confusion.

Anecdote 2.1: An Interesting Conversation with an Agro-Industrial Client
During a kick-off discussion with agro-industrial clients interested in risk management, they stated: "As to the risk management issue, we will focus on the three major risks which challenge our agro-industrial business, i.e., pricing, foreign currency and credit control". We are accustomed to this type of statement, common when a client approaches for the first time risk management. In this case we see:
- arbitrary censoring of reality to three "risks";
- biasing by considering them as major without any proof of it;
- confusing the notions of risk and hazard.

Calling risks the hazards lurking on their businesses is a very common mistake clients make. It may seem we are splitting hairs, but experience shows this is a source of:

- major blunders in risk management;
- misallocation of funds;
- waste of capital and management time.

All hazards have a certain likelihood of going out of the business-as-usual range, and possibly diverging. Rain, for example, is a hazard, but only once it goes out of the "business-as-usual" range. Once hazards get to a magnitude beyond "business-as-usual", they have a certain chance of generating (unpleasant) multidimensional consequences. By "multidimensional" we mean, for example, that a hazard could bring simultaneously a physical, reputational and financial hit (see Chap. 9). Risks are evaluated once both the hazard and its consequences are simultaneously taken into account.

A proper risk assessment should have a horizontal overview of all the hazards and all their consequences. It is the deed of a risk specialist, not of hazard specialists.

Back to our clients. As a proof of the above, they seemed to forget at least one hazard their business is exposed to, as they do not own, or 100% control, their production sites. That's another classic mistake managers make, i.e., censoring and biasing their approach by refusing to start with a proper holistic approach and the definition of the system.

Obviously, in order to serve these clients we need to seek pricing/currency/credit hazard specialists. They will deliver the data we need to present the client an unbiased holistic view of the risks. The assessment will include an evaluation of the uncertainties of the hazards and the potential consequences. Finally, the prioritization of the risks will include those uncertainties. But more importantly, biases or censoring will not taint the risk assessment. Additionally, it will most certainly include some hazards the client is not thinking about, for example, land ownership, climate change, etc.

Closing point. We could certainly help these clients well beyond what he can imagine, using the support of hazard specialists we will liaise with. We know by experience that clients need specific risk management support in order to avoid money squandering and corporate blunders.

We welcome the fact that research on bad decision-making habits won the Nobel Prize in economics in 2002 for Daniel Kahnemann and in 2017 for Richard Thaler. We will discuss why this research is useful in explaining the difficulties some organizations have in supporting risk-management programs and tell why we are so enthusiastic about the research by Prof. Thaler of the University of Chicago (Thaler and Ganser 2015).

The field of behavioral economics (Kahneman et al. 1991; Tversky and Kahneman 1991) studies how human judgment is often clouded by prejudices and misconceptions. One example is the availability heuristic (availability bias), a mental shortcut we all tend to take when we rely on immediate examples coming to our mind when evaluating a decision. We fall into the trap that if something can be recalled, it must be important, or likely more important than alternative solutions which are not as readily recalled. That's most likely why the 2008 recession was considered (erroneously) a black swan: because most people did not remember or were not even born in 1929! The black swan "fad", as we have discussed earlier (Sect. 1.1), is in fact based on mankind's having a short memory and considering the latest events to be unique, unheard of, or unprecedented.

Sometimes we must use availability heuristics because available data are indeed very scarce and too recent. However reliable statistical evidence will systematically outperform intuition when looking back in time to past events. Looking backwards, however, is not enough. Actually, it is critically limiting and incomplete when we need to manage risks of corporations and projects. A good risk assessment must be forward looking. It should examine both classic and hypothetical scenarios to support management decisions, including scenarios that have not yet occurred, or have not yet occurred with significant magnitudes, i.e., do not display divergent risks.

Over the last five decades or so the risk management community has settled on a well-established common practice representing risk assessments results with probability impact graphs (PIGs), risk matrices and risk heat map (see Appendix B). These have shown themselves to have a number of staggering intrinsic conceptual errors, with potential dramatic consequences on their users. Voices opposing this common practice have trouble being heard. Kahneman and Tversky also explored ways we humans have found to introduce irrelevant criteria in decision-making, including the continued mainstream reliance on inappropriate techniques, complacence with their results, or using intuition to correct perceived fallacies.

As a matter of fact, Kahneman and Tversky have explored in detail how human judgement is distorted when making decisions under uncertainty. Humans tend to be risk-averse when facing the prospect of a gain, and paradoxically risk-prone when facing the prospect of a loss. This is true even if the loss is almost certain to occur! So, using improper methods, which almost surely lead to confusion, losses, and poor planning, sits well with mainstream human nature.

Classic economists used to assume that rational decision-making occurs because people desire to increase their economic well-being following the concepts promoted by Adam Smith, in *An Inquiry Into the Nature and Causes of the Wealth of Nations*. Hence the concept of "perfect equilibrium" between offer and demand dominated the discipline. However, modern economists know that these basic assumptions are not always true and the real world is way more complicated: individuals do not always make rational utility-maximizing decisions. Indeed, human nature can be quite irrational and, thus, human decision-making complex. For instance the research of Richard Thaler (Thaler and Sunstein 2009; Thaler and Ganser 2015) basically shows that *"in order to do good economics, you have to keep in mind that people are human"* (https://www.nytimes.com/2017/10/09/business/nobel-eco nomics-richard-thaler.html). He has found that decision-makers (investors, within his research framework) simplify their environment by conceptually separating accounts, thus focusing on the narrow impact of each single decision, blinding themselves to the overall (negative) effects. In other words, they do not perform convergent risk assessments. This human trait results in limited rationality, social preferences and a lack of self-control with widespread consequences. Thus, Thaler divides the population into "humans" and "econs". The second category is populated by people capable of rational optimum decision-making and constitutes a small minority. Reportedly a number of countries, including the USA, New Zealand, Australia, Italy and the UK have incorporated Thaler's concepts in their tax and public health policies to some extent (Lunn 2013).

The similarities between the behavioral economics research and some difficulties organizations encounter in the application of risk management are staggering.

For instance:

- In order to succeed with risk management programs you have to keep in mind that people are human and cognitive biases very real.
- Decision-makers like to simplify their risk environment by creating and maintaining information siloes, i.e., the nemesis of convergent approaches. Furthermore, the "high-level" syndrome is an epidemic: everyone requires high-level summaries, no one wants to delve into details, whereas experience shows that with risk, the devil lies in details.
- It is a general practice to perform simplifications which conceptually separate accounts, thus focusing on the narrow impact of each single decision, blinding everyone to the overall (negative), interdependent effects.

These human behaviors inevitably result in limited rationality and social preferences. Added to that is a lack of self-control, up to the point of influencing the risk assessment community as a whole and exposing the corporate world to unidentified overexposures.

Adapting Thaler's classification to risk management, we can divide humans and corporations into two categories: "self-blinding" and "risk-aware". Here too the second category is populated by a small minority capable of rational optimum Risk-Informed Decision-Making (RIDM, see Sect. 12.2).

Within a population of humans and corporations that seemingly practice and apply risk management, "self-blinding" and "risk-aware" merrily co-exist. Time, changing conditions, climate changes and other systemic divergent shocks will make the difference visible in the long run.

Long-term risk mitigation plans at a country-wide scale are difficult to evaluate. However, their effects can be measured, even though the elements of humanity's global risk equation have and will change radically in the future (Sharman 2007). The often-perceived worrisome trajectory is generally exacerbated by media and informational pressures. Among these changes we can cite:

- *Global climate change* (**natural hazards**). The intensity/magnitude, probability and annual distribution of many natural hazards may change because of global climate change.
- *Higher air and water contamination* (**man-made hazards**). More people, industries, agro-industrial production areas may lead to increased pressure on ecosystems, contamination of soils, water and air as well as the dissemination of modified substances and organisms.
- *Population increase* (**human targets**). Larger cities, denser population mean higher number of exposed people, or small groups feeling they have to assume risks to make life in the cities more agreeable, thus fueling turmoil and possibly eco-terrorism.
- *Growth of infrastructure value and density* (**physical support targets**). Sophisticated and complex infrastructures and infrastructure networks, vital for the economy and social life, are increasing in number and density all over the world.
- *Increasing pressure on medical and social organizations to apply ever more sophisticated and costly life saving techniques and devices* (**risk acceptability**). Acceptability levels are decreasing. Some consider any death to be the result of someone's mistake. The egocentric view of many industrialized (or soon to be) countries/societies leads them to forget that humans are not invincible. Indeed, it is impossible to prolong every single life. Any attempt in that direction will inevitably fail because simply no society can afford that level of care.

By analyzing historic choices and decisions under the lens of behavioral science one could conclude that adequate risk evaluation had scarcely any relevance in history. Some well-known historic blunders corroborate that impression, but these were decisions where critical hazards were not properly identified. Another behavioral paradox of our society is the general awe we demonstrate towards risk takers (but only when they are lucky).

Alternative choices and decisions can only be compared by pairing them with their risk profile, from cradle to grave. The conditions in which our predecessors operated were very different. They often were desperate, within no social freedom, under duress of regimes etc. As a result, choices and options were in some cases nonexistent or a mere illusion. Paradoxically however, while today the choices available to an individual seem wider, decision makers' options are often way more limited. For example, no one in an elected position can actually afford to propose a tough choice, as that would seal his or her political fate. As examples, we can cite retirement age and policies, public health care, counter terrorism, etc. decisions.

What lies ahead in the very near future are tough choices. For example, when shrinking resources will make it impossible to shelter everyone in a country from natural or man-made disasters, we will require transparency and clear rules in order to avoid upheavals and turmoil (Norman 2009).

Very often in our world, decisions, planning and development of large governmental and/or humanitarian efforts occur without any measuring of their efficiency. That way, we waste vast amounts of public/donation money. Saving a highly visible life at certain enormous cost, rather than possibly quietly saving thousands, often works the best from a political/elective standpoint today. However, future decision makers will not be able to go that way.

As an example, let's take a new environmental protection law. Generally such a law (or international ban) that imposes incredibly costly environmental cleanup programs to save a few lives will have more intuitive and emotional backing than a law accepting a higher level of pollutants (implying a higher number of victims due to residual hazards) but freeing funds to mitigate another less publicly vivid, or media-genic, but nevertheless very real hazard. Humanity does not seem to be ready to accept proper, sensible, sustainable, and transparent risk evaluations. No matter what the real costs of its choices are. Thus, our society very often rejects risk-based approaches applied to society, rather than to corporations, as excessively cynical.

However, the real question is: can we measure the effects of large-scale risk mitigation programs and properly define what constitutes a priority? The answer is affirmative. We can measure long-term risk mitigation plans at the local and country-wide scales, and calibrate risk mitigation investments. It is not an easy task, but it can and should be done.

Risk approaches call for more equitable and risk-informed choices in the allotment of funds and mitigation. Thus, they are more ethical and democratic. Further, some will argue that this thinking does not apply to major strategic decisions. Examples of those are:

- fighting poverty in third-world countries;
- reducing production of opium and other intoxicating substances in poor countries;
- building coastal protection to protect cities and industries from raising sea-levels.

These may well remain the exclusive sphere of politicians and diplomacy and governmental and international agencies. However, RIDM should at least apply to the tactical and operational levels (for example, choosing where and how to fight poverty to maximize effects, how to reduce production of drugs, where and how to enhance coastal protection, etc.). If risk-informed decision-making was systematically applied to support this type of decisions, we would already have gone a step further towards a sustainable future.

2.2   Real-Life Examples of Tactical and Strategic Planning

In this section we will present some real-life examples to illustrate what we mean by tactical and strategic planning. We start with Technical Note 2.1, the first part of a series of notes illustrating how a country wide risk assessment can be developed for a specific hazard.

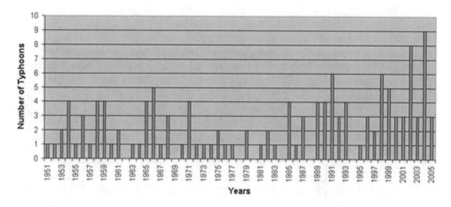

Fig. 2.1   Number of typhoons per year in Japan (1951–2005)

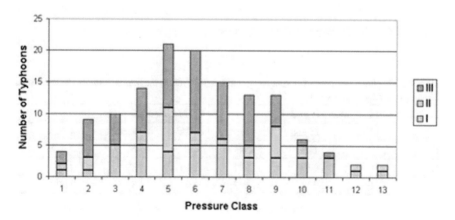

Fig. 2.2   Number of Typhoons per Pressure Class in each Period

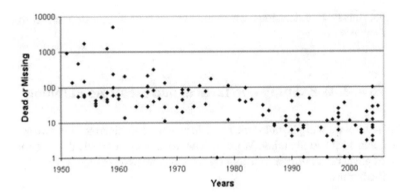

Fig. 2.3   Victims (dead or missing) due to typhoons in Japan versus years

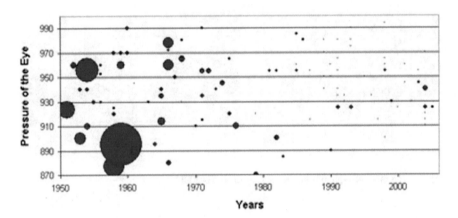

Fig. 2.4   Casualties versus pressure at the eye versus years

Table 2.2   Typhoon casualties per year and per event (minimum, median, maximum) in Japan over the three considered periods

Period	Casualties per year			Casualties per event		
	Min	Median	Max	Min	Median	Max
I	0	135	5492	0	61	5098
II	0	44	217	1	32	169
III	0	23	225	0	8	99

Technical Note 2.1: An a Posteriori Evaluation of Japanese Typhoon Mitigation

This section summarizes an original analysis we developed in our book *Improving Sustainability through Reasonable Risk and Crisis Management* (Oboni and Oboni 2007), which discusses the need to integrate potential risks in a transparent and quantitative way in any decision and human endeavor related to tactical and strategic planning. Integrating risks in decisions also means being able to measure and evaluate the effects of a long-term mitigation program in order to understand if it actually represented a wise social investment for a country, an organization, a community.

The analysis of the Japanese typhoon mitigation program is based on the period going from the program's inception in 1951–2005. Let's note in passing that there have been more typhoons on average in the last ten years than *the average from 1980 to 2010* (https://www.jma.go.jp/jma/jma-eng/jma-center/rsmc-hp-pub-eg/climatology.html), while the population has also, of course, increased. We will discuss the effects of these changes below.

The Japanese program was designed to protect the population from the floods generated by typhoons and other high-intensity meteorological phenomena. Any program for any natural or man-made hazard could be evaluated using the same methodological development. Examples of other hazards include tsunamis, droughts, sea-level rises, earthquakes, explosive remnants of war (including land mines), terrorism, etc.

In our study the evaluation of the efficiency of the mitigation program was carried out by determining the investment necessary to save a life.

The data we used were all derived from publicly available information sources spanning more than half a century, i.e. from *1951 (data set start) to 2005* (http://agora.ex.nii.ac.jp/digital-typhoon/). When data were missing or unavailable, assumptions and hypotheses were made to bridge the informational gap and still allow the discussion to go forward.

The study followed the general steps of a Quantitative Risk Assessment (QRA), as it:

- defined the system to be studied (Japan as one national target of the hazard);
- identified the hazards (typhoon) and characterized it in terms of probability of occurrence and magnitude;
- evaluated consequences (flooding with its human and infrastructure (houses) damages or losses).

In this Technical Note we focus on the preliminary phases and hazard identification, leaving the rest of the study to be summarized later on in Technical Notes 11.1 and 12.1.

Frequency and Probability of Occurrence

Within the considered period from 1951 to 2005, 133 typhoons were recorded in Japan (Fig. 2.1).

The data set available in Japan allows the evaluation of frequencies and probabilities of events of various magnitudes. At this stage, however, the generic frequency and probability will be looked at.

The calculation of the frequency is quite simple: the number of typhoons divided by the period under consideration, thus $133/55 = 2.42$ events/yr. Frequency measures the number of occurrences of a given phenomenon per unit of time, whereas probability measures the chances a given phenomenon will occur during a certain period (see Appendix A).

For the sake of this study the initial 55-year period was split into three periods of roughly equal duration as follows: Period I, 1951–1969 (19 years); Period II, 1970–1987 (18 years); Period III, 1988–2005 (18 years).

It can be noted that Period I featured a total of 40 typhoons while the two other periods saw the occurrence of respectively 26 and 67 typhoons. By using simple mathematical rules (see Sect. 8.1.3) that make it possible to convert a frequency in a probability, it can be calculated that the probability of seeing no typhoons "next year" at the end of Period III (2006) is 0.02 while it was 0.1 at the end of Period I (1970), respectively 0.23 at the end Period II (1988).

As a result the probability of at least one typhoon "next year" was approximately $1 - 0.1 = 0.9$ in 1970, 0.77 in 1988 and a staggering 0.98 for 2006!

Now let us consider the magnitude of typhoons. Lots of data were needed to compare the strength of these phenomena, including:

- the size (diameter in kilometers): the bigger, the stronger;
- the maximum sustained wind: the faster, the stronger;
- the pressure at the eye: the lower, the stronger.

For historic and monitoring reasons, not all the records present all of the data, thus it was necessary to use as a typhoon categorization parameter the only data available for the full set of typhoons: the atmospheric pressure at the eye.

Pressure data useful for this study could be used to generate typhoon magnitude categories over the three periods (official typhoons categories could not be applied because of lack of data for the older events). Thirteen classes were defined on the basis of the pressure at the eye, class 1 being the highest eye pressure (least dangerous) and class 13 being the lowest eye pressure (most dangerous).

Figure 2.2 depicts stacks of number of occurrences of typhoons in periods I, II and III in each of the thirteen pressure classes. As it can be seen the most recent period III did not have any class 12, 13 events and a very small number of class 10,11, compared to the prior periods I, II. Thus the data set showed that typhoons became more frequent but less powerful over the observation time.

As we saw above, other factors far outweigh this data interpretation, so we cannot claim to be in the position to establish any verifiable strength-time relationship. Thus, the assumption that the destructive strength of a typhoon is not a function of the year of occurrence was made in this study.

Cost of Consequences

The evaluation of loss of human life due to natural hazards in the world varies from a single injured person to multiple fatalities up to the order of 100,000–500,000 in case of large historic earthquakes (Lee and Jones 2004). As an example, the 1923 Great Kantō earthquake was the deadliest disaster in Japan (Imaizumi et al. 2016) with over 100,000 casualties. From the entire period of observation in Japan, we can derive a fair amount of data about victims from each typhoon. However, as for other hazard hits, the terms "death" (or casualty, or fatality) and "injured" both cover a wide spectrum of adverse consequences, adding a layer of uncertainties to the published estimates. Moreover, no international definition standards exist, leading to great uncertainties when statistics and old records are used in a study.

Statistics on deaths can also be considered uncertain as the time of death should be carefully recorded, but generally is not (are we looking strictly at deaths on site, during the event, or deaths that occurred as a delayed consequence, in the aftermath of an event, etc.?). These considerations lead us to accept the available data as only approximate at best.

Figure 2.3 displays the general downwards trend of victims over time. If we want to get a realistic image of the human impact of typhoons over the entire 55-year period, we could calculate the average number of deaths per typhoon, $14,659/133 = 110$, a number that seems too high with respect to the large majority of typhoons causing few or no victims, as the average is raised by a few typhoons that occurred before 1960. Thus, it was suggested that the median will be used in this study (i.e., the value that divides the sample in exactly two halves), 20 victims per event, as the most realistic parameter to quantify the central tendency of human impact of the whole set of data, or of a group of typhoons. Considering median, minimum and maximum casualty count for the three periods, Table 2.2 was derived.

As Table 2.2 shows, the median values clearly decrease with time and are generally shifted towards the minimum because only a few typhoons generate high casualty counts. At this point it becomes interesting to compare Table 2.2 with *recent events* (https://mainichi.jp/english/articles/20191112/p2a/00m/0na/004000c). For example, in 2019 typhoons Hagibis and Bualoi, in rapid sequence, jointly caused 103 fatalities. 72 people died because of floods generated by the typhoons, of which 22 indoor and 50 outdoor. 30 victims in the outdoor category died in their cars while travelling home or toward a shelter, or while trying to inspect damages.

Figure 2.4 links pressure at the eye, year of occurrence and casualty count, with the diameter of the bubbles representing the number of deaths per event.

We can see that low pressure at the eye is not the only predictive indicator of typhoon's strength and resulting casualty count, especially when considering the simultaneous and opposite effects of mitigative measures implemented in Japan and demographic growth. The prior points lead to the conclusion that a study like this one (Japanese national scale) can only look at macroscopic relationships, generalized to the entire country, because of the complexity of the environment in which the hazards hits, complicated by human interference, demographic growth, and the completely stochastic behavior of the hazard (typhoons) in terms of their paths, intensity, etc.

Were the study to be concerned with a detailed area, then the record would become insufficient as 54 years of monitoring cannot offer a statistical basis for each area of Japan. This is where experience and a lot of thinking have to be developed by a risk assessment team, which must complete the often irrelevant and censored statistics with proper probabilities estimates including dynamically evolving systems (climate changes, etc.). The limit where incomplete statistics and probabilities estimates merge is the day-to-day area of work of risk management experts.

Closing point. In addition to all the points developed above, over the last 54 years a significant change in population density and land use has occurred, leading us to believe that, in absence of any mitigation, the same strength typhoon having the same trajectory would result in *a lot more victims now than it did 50 years ago* [C:\Users\fobon\Desktop\Divergence and convergence\PPT - Natural Catastrophe Risk Management Policy in Japan PowerPoint Presentation - ID:4450064 (slideserve.com)]. Based on publicly available data the population of Japan has increased from approximately 83 M in 1950 to 127 M in 2007 (i.e., an increase of 53%), which cannot be ignored when evaluating comparative risks or discussing societal benefits versus mitigative investments. Accordingly, some recent events have generated more casualties than expected, showing that past trends and mitigation programs can be easily outdone. Past does not equal future in this type of analysis: divergence is lurking.

2.2.1 Hurricanes and Related Flooding

In this next section we will talk about "modern ideas" about hazards related to flooding that actually arose in Rome two thousand years ago (Anecdote 2.2).

Anecdote 2.2: Tiber River Management Authority
Let's talk about these "modern ideas" that actually arose in Rome two thousand years ago.

The first *Curator alvei Tiberis* (Tiber River Management Authority) was established by Emperor Augustus during a general restructuring of Roman administrations in charge of roads, water, fire protection and police. Under this new management concept, the watercourse was cleaned of debris that could have become hazardous during a flood, as a first-level mitigative effort.

Despite the efforts, during the reign of Emperor Tiberius, the Tiber river again had, in 15 AD, a catastrophic flood, reported by Tacitus. The Roman central administration was caught by surprise by this tragic event, which destroyed many infrastructures and generated numerous fatalities.

In the aftermath of the event two commissioners were asked to formulate a mitigative plan as the event(s) apparently exceeded the societal tolerance threshold of those times. The Roman Senate immediately debated over the road-map to risk mitigation, and in particular about two opposing strategies: one was a classic "do nothing" or status quo approach; the other one was a "brute force" approach encompassing the construction of dams and even a Tiber River diversion. For a number of political and religious reasons the second approach was rejected by the Roman Senate, and the status quo option was selected. Even at that time political reasons were already making the collective memory very short!

However, Emperor Tiberius, in the second year of his reign, pushed aside the Senate's decision, created a commission with five "friendly" Senators, and forcefully passed go-ahead decrees for the mitigations. As the hydraulic works were already very significant, the commission decided, as an after-thought, to further widen the sections as to enable navigation! The original mandate of the *Curator alvei Tiberis* was extended to include the dykes, banks, wastewater collections and drainages. Roads were installed along the dykes to make it possible to tow freight barges.

Records of those times indicate the frequency of the Tiber River flooding in the two centuries which followed these mitigations was significantly reduced.

Closing point. Twenty centuries have passed. It takes large catastrophes for us humans to make decisions, overcome local interests and cognitive biases. Today we have better ways of supporting decision-making but, unfortunately, not many of us take advantage of this knowledge.

Hurricane Hazel killed a dozen people around Toronto in 1954. In the aftermath, flood plains development was banned and conservation authorities were created. Nowadays, after the 1954 experience in Toronto, Hurricane Katrina (Fig. 2.5) in 2005 and Hurricane Sandy (Anecdote 2.3) in 2012 in the US, flooding in Calgary in 2013, and other examples around the world, it seems a good time to ask ourselves if our society and our governments are doing the right things, enough of them, soon enough? Let's summarize a few of these examples and review their consequences:

- The 2005 flood in Alberta reportedly cost insurers approximately 300 M CAD.

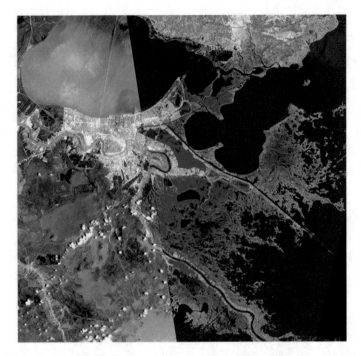

Fig. 2.5 Hurricane Katrina, New Orleans, Lousiana. A comparison of a Landsat 5 (base case June 19th 2005) and a RADARSAT-1 (flooded case Sept. 1st 2005) scene allowed to identify flood water. A representation of Hurricane Katrina flood water was obtained by comparison (blue water layer in the right overlay). RADARSAT © Canadian Space Agency/Agence spatiale Canadienne 2005, distributed by MDA Geospatial Services Inc. All Rights Reserved. RADARSAT is an official mark of the Canadian Space Agency

- The 2013 *flooding in Calgary* (https://www.cbc.ca/news/canada/calgary/alb erta-floods-costliest-natural-disaster-in-canadian-history-1.1864599) reportedly resulted in the biggest *flood coverage* payout (https://www.canadianunderwr iter.ca/insurance/alberta-floods-costliest-insured-natural-disaster-in-canadian-history-ib-1002612421/) in Alberta history. Total damage estimates exceeded 5B CAD and in terms of insurable damages, made the 2013 Alberta floods also the costliest disaster in Canadian history at 1.9B CAD, until the occurrence of the *2016 Fort McMurray wildfire* (https://en.wikipedia.org/wiki/2016_Fort_McM urray_wildfire) with an estimated impact of 3.7B CAD (Statistics Canada).
- The amount of severe weather claims insurers have covered has raised ten times in the past decade, with catastrophic losses over 1B CAD in 2011 and 2012.
- New York City mayor Michael Bloomberg committed 20B USD to protective work in the aftermath of Hurricane Sandy (see below), roughly 50% of the estimated reconstruction cost.

It becomes apparent that extreme climate events and the extent of potential targets, both people and infrastructure, are increasing at mind-boggling speed adding to unforeseeability. Insurers are battling against old risk assessment approaches, often too crude to allow them to develop healthy (sustainable) business options.

Any change at the corporate or governmental level requires political courage, good public communication, and democratic consensus. Changes may well be very unpopular (for example, suppressing insurance coverage to overexposed properties), but if risk is properly evaluated, communicated, tactical and strategic solutions may be developed that are better than status quo.

Politicians, however, generally find it more to their advantage to exploit a "heroic" reaction to the devastation than quiet spending for proactive mitigation, although exceptions emphasizing strategies and planning for risk reduction do exist (Forino et al. 2017). The short-term view of "success" many politicians and CEOs seem to apply goes something like this: "If nothing happens during my term, I will be remembered as a good guy who spent money on [insert the name of the project]? If something nasty happens during my term I can cry "Unheard of!", "Unprecedented!", "First time in history!" (even if this is not true), still be a hero, and people will vote for me again". Meanwhile insurers and their clients have to find reasonable ways to cope with their need for protection against extreme events, do reasonable business, cope with adverse outcomes, and stay alive physically and economically.

We believe that better risk assessment tools, quantitative approaches and detailed examinations of possibly divergent hazards and risks are fundamental elements to increase resilience, sustainability and, in the end, save lives.

Anecdote 2.3: Hurricane Sandy

Let's now look at hurricane Sandy (Blake et al. 2013), cited above, and its infamous consequences for New Jersey trains, focusing on the faulty decision-making processes. We look first at how the story probably developed, then how it could have developed in a well-balanced risk-culture. We will adopt a "fairy tale" style as we take numerous shortcuts to summarize the story and focus on the issue we want to discuss.

Once upon a time a railroad network needed to build a shelter for locomotives and passenger cars. A nice flat location was found. It was 20 ft above the rivers. No other specific siting study was commissioned and no one would blame that, as once upon a time no one was thinking about climate change and hurricanes wandering in those locations (although there were legends about an older era during which they had occurred).

Many happy years passed by and despite no unpleasant occurrences, some "general" studies revealed that the area might sustain damage in case of some exceptional event. No one listened. No one prepared a risk assessment, business continuity plan, or crisis plan.

Then one day a massive hurricane started drifting towards the area. There was a lot of pressure to prepare for and minimize damage, and great concern for the health and safety of workers and passengers.

It was decided to stop the trains and shelter them for later use. Where to shelter them was not even a question: there was a shelter! No one remembered the unpleasant warnings of the "general" studies. Once the trains were in the shelter, they would be safe, no questions asked.

Sandy gave an eloquent demonstration a couple days later:

- All train service and stations were shut down at 12:01 a.m. on 29 October.
- Three weeks later, rail service was still not fully restored *and passengers were running out of patience* (http://newyork.cbslocal.com/2012/11/19/exp erts-riders-wonder-why-nj-transit-stored-trains-in-flood-prone-areas/).

Now, please rewind the tape of the tale to the beginning.... play it to "No one listened".

At that juncture start replacing with the following:

Risks were measured, integrated into strategic planning of the network to create value. Business continuity plan and crisis plans were prepared and trained for, drills were performed, the resilience of the system, including the interdependencies to other critical infrastructures, was enhanced.

Then one day a massive hurricane started drifting towards the area. There was a lot of pressure to prepare for and minimize damage, and great concern for the health and safety of workers and passengers.

All the prepared procedures, plans, and mitigations were deployed. Where to shelter the trains was not even a question: there was a rational plan for minimizing risks. Not only the situation was well documented, but managers had a very clear vision of the multifaceted situation.

Sandy passed by and decided to seek revenge somewhere else.

Closing point. This anecdote shows what consequences may arise from decisions that may seem practical but may not have been supported by detailed risk analyses.

2.2.2 Volcanic Ash Cloud

The Eyjafjöll volcanic eruption in 2010 (Davies 2011) caused great turmoil at international scale as commercial flights over the Atlantic and Europe were suspended for safety reasons. In an article subtitled "Risk-management lessons from the volcanic ash cloud", the *Economist* (2010) reported:

Conventional thinking about risk management holds that risks are mainly local and routine—that it is possible to list all the negative events that could happen, determine their probability based on past experience.

This brief excerpt highlights two important facts:

1. Risks that should be included in a risk assessment are not mainly local and routine and should instead derive from a very wide spectrum hazard identification phase. Not only business-as-usual risks, but also divergent risks should be analyzed. Sadly, it is true that many corporations and international agencies fail to perform proper hazard identification and therefore have a very biased and censored risk landscape image.
2. As we have been teaching for years, "future does not equal past", especially in this world of dynamic climate and societal changes. Insurers typically work in an actuarial way, that is, using statistics only, with the result that some end up bankrupt when divergent exposures occur.

Many calculate what economists call "expected loss", i.e., the probability multiplied by the cost if it happens (Eq. 2.1). If the probability of a very disturbing large-scale event is set very low, then the expected loss may become insignificant. That gives a false sense of security, leading managers to assume that this type of risk is manageable with the same resources as more "normal" ones. In Part IV, V we will show how this difficulty can be bypassed by applying rational methodologies. Even the use of a PML (Probable Maximum Loss) estimate can lead to this type of misunderstanding if the probabilities (see Chap. 8) or even the consequences (see Chap. 9) are poorly evaluated.

We will not delve into the intricacies of technical lingo here. Rather, we point out that there is only one valid approach to understanding if a risk is truly significant: the comparison between each scenario's risk and the tolerance thresholds, both corporately or societally (Chap. 10). Again, managers relying on improper methods and measurements will most certainly fail to manage properly. This sounds familiar, doesn't it? So, at the end of the day, some of the statements by the economists pointing at generalized difficulties to handle some risk scenarios are correct. However, the reason is not because the problem is insoluble, but rather because companies (and their managers) refuse to look at the world with the proper instruments! From what we have read, it seems that, unfortunately, some academic personalities also refuse to see that rational and scientific approaches exist and could be applied.

Furthermore, low probability events are often confused with unpredictable events. The discussion on predictability warrants an example: a tornado in Salt Lake City was unpredictable, by scientific consensus, until one happened. From that moment on, while the phenomenon was not repeated, it became a predictable though extremely low probability event. But hazardous industries defined threshold for "human predictability/credibility" quite a while ago (Rulis 1986; Nehnevajsa 1984). It lies somewhere between one in hundred thousand -10^{-5} and one in a million -10^{-6}. Many years ago, we demonstrated that the probability of failure resulting from code design in many countries, for a variety of civil structures, floats around these values. However numerous economists and financial experts seem to dare say that, for example, a phenomenon such as a financial crisis, which happened with various magnitudes seventeen times in two hundred years, and happened with very significant magnitude in 1929 and again in 2008–2009, is unpredictable; they even go so far as refer inappropriately to such events as black swans (see Sect. 1.1).

2.2.3 Rain, Storms and Flooding

Let's start this section with Technical Note 2.2 discussing the forecast of torrential rains in Houston. It shows how the term black swan has been abused (again) and what a scientific-statistical approach concluded (after the fact). We will also show with examples from the world, how these problems can be solved.

Technical Note 2.2: Houston Torrential Rains Forecasts

Michael Wehner of Lawrence Berkeley National Laboratory in Berkeley, California, a co-author of the Geophysical Research Letters study, stated *"Harvey was an essentially unforeseen, classic black swan event* (https://www.sciencenewsforstudents.org/article/fingerprint-climate-change-shows-some-extreme-weather)". Even in 2017, he noted, the chances of that kind of rain dump on Texas were only 1 in 3000.

However, observations since 1880 over the region show a clear positive trend in the intensity of extreme precipitation (between 12 and 22%), roughly two times the increase of the moisture-holding capacity of the atmosphere expected for 1 °C warming according to the Clausius-Clapeyron (CC) relation (Van Oldenborgh et al. 2017). This would indicate that the moisture flux was increased by both the moisture content and stronger winds or updrafts driven by the heat of condensation of the moisture. They also analyzed extreme rainfall in the Houston area in three ensembles of 25 km resolution models. The first also showed 2 × CC scaling, the second 1 × CC scaling and the third did not have a realistic representation of extreme rainfall on the Gulf Coast.

Extrapolating these results to the 2017 event, they concluded that global warming made the precipitation between 8 and 19% more intense, or equivalently made such an event 1.5–5 times more likely. This analysis makes clear that extreme rainfall events along the Gulf Coast are on the rise.

Closing point. There is a history of severely underestimated probabilities of events oftentimes due to cognitive biases. Invoking black swan seems to be a classic and unfortunate result of this trend.

Vulnerable territories and well-known hazards are the ingredients of events such as Hurricanes Harvey (2017, Class 4 hurricane) and Irma (2017, Class 5 hurricane) (USNHC 2018), and any hurricane heading towards the Gulf Coast, Florida and the southeastern coast of the Atlantic Ocean. With the flat topography and meandering rivers, it is difficult to protect those areas from flooding, create buffers of any significant volume even if the event is not divergent or a black swan. Indeed, one does not need to be a rocket scientist to imagine that a stronger than usual meteorological event can create havoc in those areas as it has already done several times. Let's look at facts: we read Harvey was "unprecedented", but Class 5 hurricanes like Irma occurred 10 times in the last 20 years with an array of different casualty rates and damages. So, definitely, no black swans in this area. Further, high-density industrial operations, hazardous materials, and toxic compounds storage facilities are obvious boosters to the magnitude and duration of resulting consequences.

In summary, vulnerable territories, well known hazards are due to a combination of:

- calamitous climate events;
- unforgiving flat topography (Houston is reportedly locally sinking by 2.2 inches per year due to pumping of oil, water and other factors);
- facilities handling or storing hazardous material;
- non-resilient power grids;
- insufficient land-use control and zoning, allowing development in flood prone areas;
- no built-in resilience in various implemented systems.

Reportedly the design criteria of the city of Houston flood system is a 100-year storm (see Technical Note 2.2 and Sect. 4.1). By definition a 100-year storm has an average frequency of 1/100 year meaning a yearly probability of ~ 1% to hit. However, if we look at the facts, we had 8 events in the last 27 years. The frequency is then almost 30% and the return period is 3.5-years. Thus, as it is possible to apply the Poisson process model (see Sect. 8.1.3), one can conclude that:

- The probability evaluated for 1 event stands at 22% per year.
- Having two events per year has 3% probability of occurrence.
- Finally, the yearly probability of having no comparable storm is only 74%.

Going a bit farther into technical details, the exceedance probability (see Technical Note 8.2) is the probability of an event being greater than or equal to a given value, for example, exceeding the ill-evaluated 100-year storm above. Thus, the probability that the next storm will be the worst ever recorded is $1/(8 + 1) = 0.11$ or 11%. This value is impressive, and it is not an opinion: it derives from solid mathematics. Now, that 100-year storm reportedly considers a total rainfall of 13 inches in 24 h. The problem is that such rain events have apparently occurred as stated above, more than 8 times in the last 27 years, a blatant example of divergence. Thus nobody should invoke a black swan. Let's only talk about a clearly predictable event with very poorly assessed multi-dimensional consequences (high un-foreseeability due to lack of effort).

At this point we have to stress the usual confusion made between frequency, probability, and return period. We will come back later on this subject (see Sects. 4.3 and Appendix A).

It is not even necessary to invoke climate change when dealing with vulnerable territories and well-known hazards! However, experts have been adamant in stressing that climate change generates warmer temperatures in the Gulf of Mexico, thus fueling more powerful hurricanes. Hurricanes create water surges along the coastline and those disrupt river flows, preventing normal drainage. During Hurricane Katrina, the level of the Gulf of Mexico surged by some 28 feet. Coastal surge during Harvey was also no news.

It is interesting to note how uncertain the loss estimate was:

- A disaster modeler with *Enki Research quantified Harvey's cost* (https://www. cbsnews.com/news/harvey-may-be-one-of-the-costliest-storms-in-u-s-history/) as mounting to 30B USD.
- *NOAA estimated 125B USD loss* (https://coast.noaa.gov/states/fast-facts/hurric ane-costs.html), when power grid, transportation and other elements that support the region's energy sector were included.
- An insurance analyst at *Imperial Capital said the final tally might be as high as 100B USD* (https://www.bloomberg.com/news/articles/2017-08-27/harvey-s-cost-reaches-catastrophe-as-modelers-see-many-uninsured), with less than a third of Harvey's losses insured.

30B, 125B, 100B: as usual, there was great uncertainty afflicting those estimates. However, *experience* (https://qz.com/1066985/hurricane-harveys-costs-may-be-much-higher-than-katrina-says-a-government-expert/) has shown that cognitive biases tend to underestimate the tally in the immediate aftermath of a catastrophe. Again, the huge price tab should not have been a surprise. It was foreseeable, within ample margins of uncertainty, and highly predictable.

Today, it is estimated (Peterson 2019; Johnson and Watson 2014) that Hurricane Harvey had total costs of 125B USD, whereas Hurricane Katrina holds the record with an approximate cost of 161B USD.

As per the costs of future protection, we know that in New Orleans flood control improvements after Hurricane Katrina reached 14.2B USD. However, those improvements appear to have achieved dubious results, as flooding occurred again recently when the pumping system failed, overwhelmed by rainfall. Engineers in the US estimate it could cost hundreds of billions of dollars to build a flood protection system for Houston. It would involve land-use restrictions, new flood barriers and measures already successfully implemented elsewhere. Here are a few examples:

- **The Netherlands**. The Dutch system attempts to defend Amsterdam and Rotterdam from a 10,000-year storm event. It is interesting to compare this design criteria with Houston's 100-year criteria.
- **Japan**. Tokyo (Kanno 2007; Kimura et al. 2009) has enormous underground storage for excess water to prevent flooding and contamination. The Metropolitan Area Outer Underground Discharge Channel (首都圏外郭放水路, Shutoken Gaikaku Hōsuiro), an underground water infrastructure project in Kasukabe, Saitama, Japan, is the world's largest *underground flood water diversion facility* (https://asia.nikkei.com/Economy/Natural-disasters/Und erground-temple-saves-Tokyo-from-typhoon-flood), geared towards mitigating overflowing of the city's major waterways and rivers during rain and typhoon seasons. Construction lasted 13 years, at a cost, at the time, of 3B USD. During

typhoon Hagibis, mentioned above, the system ran to capacity. Hence it has been decided to increase the volume from 2.56 M m^3 presently available to 8 M m^3.

- **The US**. Chicago (Scalise and Fitzpatrick 2012) Phase I, operational since 2006, has a capacity of 8.7 M m^3 in tunnels. Phase II, including storage volumes, will significantly increase the volume and should be operational by 2029.
- **Italy**. Venice at end of October 2018 witnessed an extraordinary *acqua alta* (literally, high water) which submerged approximately 70% of the city and was reportedly the *fifth event of such magnitude* (https://www.timesofisrael.com/at-least-11-dead-as-violent-storms-slam-into-italy/) in 924 years. *Aqua alta* is a complex event combining high tide and wind-driven surge in the Venetian Lagoon: southerly winds hinder the tide outflow and "push" open-sea water into the lagoon. *Aqua alta* is not uniform through the city, as vertical building settlements in the lagoon were not uniform, the effect of varying foundation conditions as well as buildings of different mass and size. Indeed, Venice stands on floating foundations made of short wooden piles, sometimes as many as 10 per square meter, driven down to a weak strata of lagoon sediments (Bettiol et al. 2016). Storms, tides, perils of the sea and natural settlement of the foundations have forever been part of Venetian life (Indirli et al. 2014). An ambitious mega-project called MOSE (the acryonym stands for MOdulo Sperimentale Elettromeccanico, but is also a play on the name Mose, Italian for Moses, referring to the parting of the waters) became operational in October 2020 and successfully saved the town from two events during that autumn before having a mishap due to preliminary calibration hiccup (Pirazzoli and Umgiesser 2003, 2006). MOSE's movable gates should protect the city from this type of events from now on. We will soon enough realize if climate change will increase the frequency of similar extreme events and completely justify the enormous initial price estimate, over-budget and delays of MOSE.

2.2.4 Design, Procedures and Monitoring

In this section we discuss how to apply resilience and reliability concepts to Oroville Dam. We will stay away from numbers, as we did not have access to detailed information on the dam but will focus on concepts.

A catchment area of 9340 km^2 brings water to the Feather River upstream of the Oroville Dam, which serves the Feather River Valley mainly for water supply, hydroelectricity generation and flood control. The dam's design and building procedure complied with codes. The embankment was completed on 6 October 1967 after delays generated by a series of mishaps, including a railroad accident and a catastrophic flood in 1964 that resulted in a peak flow of 7100 m^3/s above the Oroville Dam site. The Oroville Dam design also complied with seismic codes. A complex monitoring network similar to those in hydro dams in Europe (water pressure and settlements, deformations) completed the system. Furthermore, two main safety features were built in:

- a service spillway with maneuverable gates followed by a concrete channel designed to lead escaping waters to the original valley bottom,
- an emergency spillway, a reinforced low point threshold, at a higher elevation than the service spillway, but lower than the dam crest. The emergency spillway is a "last ditch" protection against overtopping, a potential cause of dam breaches. The emergency spillway leads escaping waters to the natural slope below the dam toe, where natural soil, vegetation is present, in such a way that escaping waters will find their way back to the river.

An inundation study and evacuation plans took account of the population at the time. They covered the dam breach and other failure scenarios.

Three *environmental groups filed a motion in 2005* (https://www.eenews.net/stories/1060050023) with the Federal Energy Regulatory Commission (FERC), requesting a concrete protection for the dam's earthen emergency spillway as "in the event of extreme rain and flooding... heavy erosion" would occur. Regulators considered the request unnecessary or excessive. In 2006, a senior civil engineer sent a memo to his managers stating that the emergency spillway met FERC's guidelines for an emergency spillway, which specify that during a rare flood event, it is acceptable for the emergency spillway to sustain significant damage.

The question is, of course, what is "rare"? Over time, the spillway denomination changed from "emergency" to "auxiliary" without any corresponding physical upgrade. Perhaps someone felt that the events the spillway had to deal with were not so rare? Various media noted that the now auxiliary spillway began carrying water for the first time in 1968. People seem to believe that once in forty years is "rare", but that's preposterous, especially in view of the fast-changing climate conditions in this world and considering the criticality of the system (exposed population, infrastructure, environment).

The spillway was built in concrete, like many other structures of this type. Now, a famous James Bond theme song says "diamonds are forever", but concrete has a very definite life duration. Concrete can crack because of:

- shrinkage;
- chemical reactions exacerbated by air pollution;
- subtle (or not so subtle) changes in the foundation conditions, such as settlement, deformation, seismic events, etc.;
- changing levels of the water table (as in a severe meteorological cycle after a long drought, for example).

The spillway started cracking in 2013, leading the local newspaper, the *Sacramento Bee*, to interview a senior civil engineer with the Department of Water Resources. He promptly explained the phenomenon is common and *"There were some patches needed and so we made repairs and everything checked out"* (https://sacramento.cbslocal.com/2017/02/10/maintenance-records-show-oroville-dam-spillway-previously-patched/).

After the exceptional 2012–2014 drought in California (Grigg 2014; Diffenbaugh et al. 2015), relentless storms brought very strong, continuous precipitations. On 7

February 2017 the service spillway gates were activated. The goal was a flood control release of about 1400 m³/s.

Was it too late? Answering that gate management question is not within the scope of this discussion. However, we note that many historic floods around the world were allegedly the result of dam operators having "sticky finger" in the gate opening decision. The reason for the sticky finger syndrome is often simple: water is a precious resource (whether used for drinking, energy, agriculture); the decision to allow it to run off, especially in drought-stricken regions, is a tough one, loaded with responsibility and potential consequences for the decision-maker. At the Oroville Dam, when the gates were finally opened and the spillway flow reached a high volume, the concrete started eroding [for any of the reasons cited above or due to concrete cavitation (Lee and Hoopes 1996)]. At that point water got "under and around" the original concrete channel spiraling toward a self-boosting chain-reaction of failure. Thus the hope that using the damaged spillway could drain the lake fast enough to avoid use of the auxiliary spillway proved false. It became necessary to lower the discharge from 1800 to 1600 m³/s due to potential damage to downstream infrastructures. As a result, water in the lake rose and at a certain point in time, the emergency spillway level was reached. Water escaping the emergency spillway started eroding the slopes downstream. That lead to very serious concern, triggered evacuation plans, panic, etc. Media widely reported these events. In theory it seemed that the resilience and reliability concepts supporting the Oroville Dam were sound and bullet-proof as well as code compliant, but the facts tell another story. So, let us ask a few questions:

- Are the terms "unheard of" and "rare" reasonable for events occurring once every 40–60 years? (Do not forget that there was a first catastrophic flood of 1964). The answer is simple: Certainly not.
- Shouldn't the probability of this type of mishap be way lower, considering that such an event can alter the life of nearly 200,000 people (see Technical Note 9.1)? Absolutely yes.
- Is it reasonable to hide behind codes that state that an emergency spillway could sustain heavy erosion and damages, without wondering what that means (multi-dimensional consequences, see Chap. 9) from a technical and social point of view (see Sect. 5.2)? Absolutely no.
- Do you think it is reasonable to have one officer make a statement such as, "There were some patches needed and so we made repairs and everything checked out", without a serious risk analysis of the damaged/repaired system and its implications for the public? Again, our answer is no.
- If the criticality of the structure changes due to the population density increase (as indeed happened), should additional mitigations be planned? Definitely yes.

In the meantime, a common belief is proven wrong, once again, by this example: even if a system is designed to withstand all credible worst-case accidents, it is still NOT "by definition" safe against any credible accident (see Appendix B). We need to be smarter.

Personal human experience, bounded rationality and limiting of credible scenarios are important in the context of hazard and risk identification and certainly played a role in this case and many others. Hazard identification is a fundamental phase of a risk assessment (see Chap. 7). It intervenes after the definition of the system (Chap. 6) and the objectives (see Anecdote 1.2). Often we hear that practitioners consider only credible scenarios in a risk identification exercise. Limiting any approach to credible scenarios is a deception and can lead to disaster. Incidentally, this might be one of the reasons why numerous unrealistic risk assessments describe rosy scenarios. We always object that considering only credible scenarios constitutes blatant censorship. Indeed, risk analysis should prioritize the scenarios rather than censoring them. The prioritization resulting from the risk assessment will take care of eliminating non-credible or meaningless scenarios. Prioritization is a result, not an arbitrary decision!

Thus, it is worth focusing on arrogance and censorship. It takes 10,000 h (or approximately 10 years or more) of practice, to reach expert level in a subject matter, as shown by Malcolm Gladwell in his bestseller *Outliers* (Gladwell 2008). So when an expert says that a failure can't happen, or that a failure mode is non-credible, it means that it is "unheard of" during the expert's 10,000 h of observation/practice. Those 10,000 h may or may not have occurred with real hands-on experience and in the system's real environment, but we all know that detailed environmental conditions can affect the likelihood of failure considerably. As an aside, consider that if one takes ten experts and sits them around a table, the sum of experience will not considerably alter the final result: they may very well all agree, and erroneously, that a failure is non-credible because their observation times were "simultaneous" (see Sect. 7.1.2)! In particular, one should beware of one-size-fits-all lists of hazards, lists of failure modes, list of risks, etc. For instance, developing a correct taxonomy of the failure modes would require detailed and complex forensic analyses that are, unfortunately, not generally performed in infrastructural and industrial application, with to the exception of civil aviation. Thus, the 10,000 h of "expert observation" may unfortunately rely on scant reports, hearsay and uncertain data.

When, in the aftermath of some of some catastrophic failure panels of experts develop detailed and scientific forensic analyses they generally—and we would say correctly—determine a *"set of circumstances" that lead to failure* (https://www.min ing.com/why-samarco-tailings-dam-failed/), and at that moment the tragic truth of censoring emerges. Generally speaking, the deterioration of the system leading to failure is not caused by one single failure mode but rather by a combination of triggers. A combination of "minor" failure modes can create a catastrophic situation or, as many would say, "a perfect storm". Thus it becomes apparent that the consideration of failure modes, and in particular credible failure modes, is inadequate: failure modes do not explain why failures occur, but they explain, and not even completely, how a failure could occur under the influence of an extremely limited set of triggers. The causes of failures are way more complex than failure modes. The question is: why does a model known to be flawed remain in use? Perhaps is it because they are used by groups of experts that easily agree as they have too similar backgrounds and rely on the same knowledge base?

At the other end of the spectrum from the Oroville Dam disaster, we find examples drawn from the cyber world and IT. There is no way to escape reality: even the cyber world of crypto-mining gets hit by real-life natural disaster. We are sure very few thought that *flooding could cripple the synthetic world* (https://cryptoslate.com/china-floods-wipe-out-crypto-mining-farms-bitcoin-hashrate-drops/) of crypto-currencies (see Sect. 3.2). In Anecdote 1.2 we wrote about a CEO who thought that our approach to multi-hazard risks was too technical and not sufficiently business oriented. That CEO was taught a very hard lesson by Mother Nature, a lesson that should be learnt by others, and in other arenas. As you can see, censoring is a dangerous business, in any endeavor. Nowadays, with bit coin and other crypto-currencies, investors worldwide are taking huge amount of capital to the cyber sphere, blinded to real-world natural hazards. People forget that good old real-life natural or man-made disasters (Oboni and Oboni 2016) can have disastrous consequences on any industry, business, or infrastructure, including, of course, hi-tech.

Meteorological events and *climate change have also the potential to alter functionalities* (https://www.npr.org/2018/07/16/627254166/rising-seas-could-cause-problems-for-internet-infrastructure?t=1531810096653), as recently reported by a study related to the US coastal areas. Here are a few interesting cases. In 2012, Hurricane Sandy flood related blackouts *drowned underground data cables* (https://www.cnet.com/news/hurricane-sandy-disrupts-wireless-and-int ernet-services/) provoking a "catastrophic failure". In 2015, a heat wave in Australia overcame the cooling system of a data center, *stopping internet service for hours* (https://www.itnews.com.au/news/iinets-perth-data-centre-melts-in-heatwave-399128). Finally, after Hurricane Irma hit Florida in 2017, the *Miami Herald* reported *weeks long Internet blackout* (https://www.miamiherald.com/news/weather/hurricane/article174827041.html). Recent news also reported flooding in China destroying mining hardware provoking heavy economic losses. These problems are clearly the result of poorly selected design parameters and mitigation.

As a final example, we can recount what happened at the Giant Mine (NWT, Canada) arsenic dioxide storage. From the late 1940s to the 1990s, arsenic trioxide was produced as by-product of gold extraction by Giant Mine in Canada, NWT. Giant Mine dumped that by-product into man-made underground chambers. The permafrost contained the contaminant, and the perception was seepage was therefore under control. In those times, this was perceived as a "risk free" solution. Unfortunately, because of permafrost loss, those *chambers are no longer as good at containing the arsenic trioxide* (https://www.aadnc-aandc.gc.ca/eng/110 0100027413/1100100027417) (Clark and Raven 2004), and the government estimated having to *spend up to 1B CAD in mitigations* (https://magazine.cim.org/en/news/2018/federal-government-awards-giant-mine-remediation-contract/) to lower the seepage risks (Jamieson 2014) (see Sect. 5.3.3).

These are the reasons why, in every risk assessment we perform, we look convergently at natural and man-made hazards. Common siloed approach, keeping risks split by source and then treating them separately, will inevitably lead to costly misunderstanding and garbling the road map to sustainable mitigation.

2.3 Summary of Examples of Tactical and Strategic Planning

Let's recap the examples we just described to draw some lessons we will apply later. Table 2.3 displays the various examples, their deficiencies, and their consequences.

Table 2.3 shows that in these examples the deficiencies were technical, behavioral/cognitive and informational, and related to voluntary or involuntary ignorance of risk exposures. The consequences ranged from severe business interruption to private and public losses, infrastructure losses, human losses (fatalities and evacuations) and long-term environmental effects and were thus clearly multi-dimensional (See Chap. 9).

From the discussions above we can see the need to explore various concepts that are ancillary but fundamental to the robustness of the divergence/convergence discourse:

- **Perpetuity.** Human projects, especially those involving hazardous materials and their storage (Roche et al. 2017), demand continued stewardship, management and monitoring and, finally, public information during production and closure. Long-term decision making cannot be based on censored and biased scenarios, or ignore uncertainties. That would be unethical and hazardous even in the short term. One cannot forget that 100 years is already a long time: a century encompasses three to four generations and depending on the jurisdiction 15–25 presidential elections. The Pyramid of Cheops is 4500 years old, still way shorter than perpetuity! Perpetuity is very long indeed. Proper design and careful selections can help significantly, but they cannot be considered sufficient for longer terms.

Table 2.3 Examples of tactical and strategic planning with their deficiencies and consequences

Example	Sect.	Deficiencies	Consequences
Tiber River (15AD), Hurricane Hazel (1954), Calgary flooding (2013)	2.2.1	Lack of consideration for "new" recurring hazards, late decision-making	Catastrophic losses
Trains and Hurricane Sandy (2011)	2.2.1	Inappropriate siting with respect to hazards	Weeks-long delays in restoring service
Icelandic volcanic cloud (2010)	2.2.2	No risk assessment, no SOPs, no consideration for a divergent risk	Air traffic lockdown for several days and transcontinental scale
Flooding: Hurricane Harvey (Houston, 2017) and Hurricane Irma (Florida, 2017)	2.2.3	Poor design parameters, lack of action in mitigating	Flooding, contamination, widespread losses
Oroville Dam spillway (2014) Giant Mine (1970–)	2.2.4	Normalization of deviance, poor communication, overconfidence	Structural damage, evacuation, damage to the environment

Adaptability and due consideration of divergence in a convergent assessment are paramount.

- **Societal acceptability**. Accidents have the potential to create strong societal outcry and reactions, and therefore societal acceptability is not constant in time. A region that is barren and unoccupied today may become inhabited and/or critical in the future. As social perception changes, regulations, which follow them to a certain degree, will change. We must think "atmospheric and social climate change" if we want to bring value to projects. Projects must include specific Plan Bs to enable future adaptations. Today, even a catastrophic explosion of a refinery may be considered as business-as-usual by a large oil company, because it will not affect their production and sales by more than a certain percentage, but societal reaction is a different story.

- **Goal-based restorations**. Restorations projects should be evaluated so that they will protect future generations, allowing them to understand and interact, possibly to their advantage, without taking unnecessary risks. Restoration projects should be resilient and sustainable, have clear monitoring policies and updating procedures based on risk-informed decision making. Walk-away projects for the long term are an illusion, but adaptive risk-informed decisions will make a long-term project sustainable.

- **Site closure and long-term responsibility**. Long-term goals require a transparent approach to project evaluation and alternative selections. Their achievement requires:
- careful definition of the system;
- use of unequivocal language;
- study of interdependent scenarios without arbitrary censoring and bias;
- evaluation of consequences considering their multi-dimensional aspects;
- use of explicit risk tolerance to prioritize and solve issues.

Proper design and careful selections can help significantly. However, we cannot consider them sufficient for longer terms as they may lack adaptability, resilience and the capacity to cope with divergent exposures.

Hence, in the next chapter we will focus our attention on "setting the scene" of divergence from a risk point of view.

Appendixes

Links to more information about the key terms from the authors	
A, B	*Act of god* (https://www.riskope.com/2020/12/09/act-of-god-in-probabilistic-risk-assessment/) *black swan* (https://www.riskope.com/2011/06/14/black-swan-mania-using-buzzwords-can-be-a-dangerous-habit/) *business-as-usual* (https://www.riskope.com/2011/06/14/black-swan-mania-using-buzzwords-can-be-a-dangerous-habit/)

(continued)

(continued)

Links to more information about the key terms from the authors	
C, D	*Convergent* (https://www.riskope.com/2021/01/20/convergent-risk-assessments/) *divergent* (https://www.riskope.com/2020/11/18/tactical-and-strategic-planning-to-mitigate-divergent-events/) *drillable* (https://www.riskope.com/2020/04/01/antifragile-resilient-solutions-for-tactical-and-strategic-planning/)
F	*Foreseeability/foreseeable* (https://www.riskope.com/2021/01/06/foreseeability-and-predictability-in-risk-assessments/) *Fragile/fragility* (https://www.riskope.com/2020/04/01/antifragile-resilient-solutions-for-tactical-and-strategic-planning/)
P, R	*Predictability/predictable resilient* (https://www.riskope.com/2021/01/06/foreseeability-and-predictability-in-risk-assessments/), *resilience* (https://www.riskope.com/2016/11/23/resilience-cannot-based-instinctual-decision-making/)
S	*Scalable* (https://www.riskope.com/2015/04/16/how-system-definition-and-interdependencies-allow-transparent-and-scalable-risk-assessments/) *societal risk acceptability* (https://www.riskope.com/2014/01/09/aspects-of-risk-tolerance-manageable-vs-unmanageable-risks-in-relation-to-critical-decisions-perpetuity-projects-public-opposition/) *sustainability/sustainable* (https://www.riskope.com/2019/01/16/improving-sustainability-through-reasonable-risk-and-crisis-management/) *survivability* (https://www.riskope.com/2011/03/17/ale-fmea-fmeca-qualitative-methods-is-it-really-what-we-need/) *system* (https://www.riskope.com/2017/07/26/three-ways-to-enhancing-your-risk-registers/)
T, U	*Tolerance* (https://www.riskope.com/2020/04/29/risk-tolerance-thresholds/) *uncertainty/uncertainties* (https://www.riskope.com/2015/12/10/3-decision-making-truths-derived-from-uncertainty-taxonomy-scheme-of-classification-and-a-road-sign/) *updatable* (https://www.riskope.com/2020/01/07/climate-adaptation-and-risk-assessment/)

Other linked information (https://www.riskope.com/blog-news/) search Riskope blog and use the search box

Third parties links in this section	
Planning for the future	https://www.stat.berkeley.edu/~aldous/Real-World/phil_uncertainty.html
Note for lead authors	https://www.ipcc.ch/site/assets/uploads/2017/08/AR5_Uncertainty_Guidance_Note.pdf
Economics and humans	https://www.nytimes.com/2017/10/09/business/nobel-economics-richard-thaler.html
Climate averages Japan	https://www.jma.go.jp/jma/jma-eng/jma-center/rsmc-hp-pub-eg/climatology.html
1951–2005 Japan	http://agora.ex.nii.ac.jp/digital-typhoon/
Recent typhoons Japan	https://mainichi.jp/english/articles/20191112/p2a/00m/0na/004000c
Natural catastrophes risk management policy Japan	*PPT—Natural Catastrophe Risk Management Policy in Japan PowerPoint Presentation—ID: 4450064* (slideserve.com) (https://www.slideserve.com/silvio/natural-catastrophe-risk-management-policy-in-japan)

(continued)

(continued)

Third parties links in this section	
Fort McMurray wildfire	https://en.wikipedia.org/wiki/2016_Fort_McMurray_wildfire
Floods in calgary	https://www.cbc.ca/news/canada/calgary/alberta-floods-costliest-natural-disaster-in-canadian-history-1.1864599
Calgary flood payout	https://www.canadianunderwriter.ca/insurance/alberta-floods-costliest-insured-natural-disaster-in-canadian-history-ib-1002612421/
NJ passengers out of patience	http://newyork.cbslocal.com/2012/11/19/experts-riders-wonder-why-nj-transit-stored-trains-in-flood-prone-areas/
Harveys unforeseen event?	*Fingerprint of climate change shows up in some extreme weather| Science News for Students*
Harveys costs	https://www.cbsnews.com/news/harvey-may-be-one-of-the-costliest-storms-in-u-s-history/ https://coast.noaa.gov/states/fast-facts/hurricane-costs.html https://www.bloomberg.com/news/articles/2017-08-27/harvey-s-cost-reaches-catastrophe-as-modelers-see-many-uninsured
Tokyo flood protection	https://asia.nikkei.com/Economy/Natural-disasters/Underground-temple-saves-Tokyo-from-typhoon-flood
Venice flood	https://www.timesofisrael.com/at-least-11-dead-as-violent-storms-slam-into-italy/
Oroville requests	https://www.eenews.net/stories/1060050023
Oroville patches	https://sacramento.cbslocal.com/2017/02/10/maintenance-records-show-oroville-dam-spillway-previously-patched/
Set of circum-stances led to failure	https://www.mining.com/why-samarco-tailings-dam-failed/
Flooding and crypto currencies	https://cryptoslate.com/china-floods-wipe-out-crypto-mining-farms-bitcoin-hashrate-drops/
Climate change alters functionalities	https://www.npr.org/2018/07/16/627254166/rising-seas-could-cause-problems-for-internet-infrastructure?t=1531810096653
Drowning underground cables	https://www.cnet.com/news/hurricane-sandy-disrupts-wireless-and-internet-services/
Heatwave stops internet	https://www.itnews.com.au/news/iinets-perth-data-centre-melts-in-heatwave-399128
Irma internet blackout	https://www.miamiherald.com/news/weather/hurricane/article174827041.html
Permafrost melt releases arsenic	https://www.aadnc-aandc.gc.ca/eng/1100100027413/1100100027417 https://magazine.cim.org/en/news/2018/federal-government-awards-giant-mine-remediation-contract/

References

Bettiol G, Ceccato F, Pigouni AE, Modena C, Simonini P (2016) Effect on the structure in elevation of wood deterioration on small-pile foundation: numerical analyses. Int J Architectural Heritage 10(1):44–54

Blake E, Kimberlain TB, Berg RJ, Cangialosi J, Beven J (2013) Hurricane sandy, tropical cyclone report. National Hurricane Center

Chowdhury R, Flentje P (2003) Role of slope reliability analysis in landslide risk management. Bull Eng Geol Environ 62(1):41–46

Clark ID, Raven KG (2004) Sources and circulation of water and arsenic in the Giant Mine, Yellowknife, NWT, Canada. Isot Environ Health Stud 40(2): 115-128

Davies M (2011) Filtered dry stacked tailings–the fundamentals. In: Proceedings tailings and mine waste. Vancouver, BC, pp 6–9

Diffenbaugh NS, Swain DL, Touma D (2015) Anthropogenic warming has increased drought risk in California. Proc Natl Acad Sci 112(13):3931–3936

Economist (2010) Business.view: not up in the air, 20 April 2010. https://www.economist.com/business/2010/04/20/not-up-in-the-air

Fischhoff B, Slovic P, Lichtenstein S, Read S, Combs B (1978) How safe is safe enough? A psychometric study of attitudes towards technological risks and benefits. Policy Sci 9(2):127–152

Forino G, von Meding J, Brewer G, van Niekerk D (2017) Climate change adaptation and disaster risk reduction integration: strategies, policies, and plans in three Australian Local Governments. Int J Disaster Risk Reduction 24

Gladwell M (2008). Outliers: the story of success. Little, Brown

Grigg NS (2014) The 2011–2012 drought in the United States: new lessons from a record event. Int J Water Resour Dev 30(2):183–199

Hopkin P (2018) Fundamentals of risk management: understanding, evaluating and implementing effective risk management. Kogan Page Publishers

Imaizumi A, Ito K, Okazaki T (2016) Impact of natural disasters on industrial agglomeration: the case of the Great Kantō Earthquake in 1923. Explor Econ Hist 60:52–68

Indirli M, Knezić S, Borg RP, Kaluarachchi Y, Romagnoli F (2014) Venice and its territory: multi-hazard scenarios, vulnerability assessment, disaster resilience, and mitigation. In: Proceedings enhancing resilience of historic cities to flooding and anthropogenic impacts: success and failures in the Italian experience. Accademia dei Lincei, Rome, 4–5 Nov 2014

Jaafari A (2001) Management of risks, uncertainties and opportunities on projects: time for a fundamental shift. Int J Proj Manag 19(2):89–101

Jamieson HE (2014) The legacy of arsenic contamination from mining and processing refractory gold ore at Giant Mine, Yellowknife, Northwest Territories, Canada. Rev Mineral Geochem 79(1):533–551

Johnson ME, Watson CC (2014) Et tu "Brute Force"? No! A statistically based approach to catastrophe modeling. Topics in statistical simulation. Springer, New York, pp 291–298

Kabadayi C, Osvath M (2017) Ravens parallel great apes in flexible planning for tool-use and bartering. Science 357(6347):202–204

Kahneman D, Knetsch JL, Thaler RH (1991) Anomalies: the endowment effect, loss aversion, and status quo bias. J Econ Perspect 5(1):193–206

Kanno H (2007) Inside one of largest capacity drainage pump stations in the world. World Pumps 2007(484):14–15

Kapurch SJ (ed.) (2010) NASA systems engineering handbook. Diane Publishing

Kimura K, Tanaka H, Shibasaki S, Kitamura H (2009) Introduction of the metropolitan area outer underground discharge channel. J Inst Electr Eng Jpn 129(4):200–203

Lee EM, Jones DKC (2004) Landslide risk assessment. Thomas Telford

Lee W, Hoopes JA (1996) Prediction of cavitation damage for spillways. J Hydraul Eng 122(9):481–488

Loosemore M, Raftery J, Reilly C, Higgon D (2012) Risk management in projects. Routledge

Laplace PS (1902) A philosophical essay on probabilities. Wiley

Lunn P (2013) Behavioral economics and regulatory policy, organisation for economic co-operation and development GOV/RPC(2013) 15. https://www.oecd.org/officialdocuments/publicdisplaydo cumentpdf/?cote=GOV/RPC(2013)15&docLanguage=En

Mastrandrea MD, Mach KJ, Plattner GK, Edenhofer O, Stocker TF, Field CB, Ebi KL, Matschoss PR (2011) The IPCC AR5 guidance note on consistent treatment of uncertainties: a common approach across the working groups. Clim Change 108(4):675–691

Nehnevajsa J (1984) Low-probability/high-consequence risks: issues in credibility and acceptance. In: Waller R (ed) Low-probability high-consequence risk analysis. Springer, Berlin, pp 521–529

Norman B (2009, June) Planning for coastal climate change. An insight into international and national approaches. Government Department of Planning and Community Development and Department of Sustainability and Environment, Melbourne

Oboni F, Oboni C (2007) Improving sustainability through reasonable risk and crisis management. Switzerland. ISBN 978-0-9784462-0-8

Oboni F, Oboni C (2016) The long shadow of human-generated geohazards: risks and crises. In: Farid A (ed) Geohazards caused by human activity. https://www.intechopen.com/books/geohazards-caused-by-human-activity/the-long-shadow-of-human-generated-geohazards-risks-and-crises

Oboni F, Oboni C (2020) Tailings dam management for the twenty-first century. Springer, Cham. https://doi.org/10.1007/978-3-030-19447-5_11. ISBN978-3-030-19446-8

Papoulis A, Pillai SU (2002) Probability, random variables, and stochastic processes. Tata McGraw-Hill Education

Peterson J (2019) A new coast: strategies for responding to devastating storms and rising seas. Island Press

Pirazzoli PA, Umgiesser G (2006) The projected "MOSE" barriers against flooding in Venice (Italy) and the Expected Global Sea-level Rise. J Mar Environ Eng 8(3):247–261

Pirazzoli PA, Umgiesser G (2003) Is the 'Mose' project to save Venice already obsolete? EGS-AGU-EUG Joint Assembly, Abstracts from the meeting held in Nice, France, 6–11 Apr 2003 (abstract id. 3393)

Roche C, Thygesen K, Baker E (2017) Mine tailings storage: safety is no accident. A UNEP Rapid Response Assessment. United Nations Environment Programme and GRID-Arendal, Nairobi and Arendal

Rulis AM (1986) De minimis and the threshold of regulation. "De Minimis and the Threshold of Regulation". In: Felix C (ed) Proceedings of the 1986 conference for food protection, food protection technology. Lewis Publishing, pp 29–37

Scalise C, Fitzpatrick K (2012) Chicago deep tunnel design and construction. In Structures congress, vol 2012, pp 1485–1495. 10.1061/9780784412367.132

Sharman R (2007) In Financial Times, Special Report. Risk Management, 1st May 1st 2007

Slovic P (1987) Perception of risk. Science 236(4799):280–285

Stamatelatos M, Dezfuli H, Apostolakis G, Everline C, Guarro S, Mathias D, Mosleh A, Paulos T, Riha D, Smith C, Vesely W (2011) Probabilistic risk assessment procedures guide for NASA managers and practitioners. Office of Safety and Mission Assurance NASA Headquarters

Thaler RH, Ganser LJ (2015) Misbehaving: the making of behavioral economics. WW Norton, New York

Thaler RH, Sunstein CR (2009) Nudge: improving decisions about health, wealth, and happiness. Penguin

Tversky A, Kahneman D (1991) Loss aversion in riskless choice: a reference-dependent model. Quar J Econ 106(4):1039–1061

[USNHC] United States National Hurricane Center (2018) Costliest U.S. tropical cyclones tables updated. https://www.nhc.noaa.gov/news/UpdatedCostliest.pdf

Van Oldenborgh GJ, Van Der Wiel K, Sebastian A, Singh R, Arrighi J, Otto F, Haustein K, Li S, Vecchi G, Cullen H (2017) Attribution of extreme rainfall from Hurricane Harvey. Environ Res Lett 12(12):124009

Chapter 3
The Context of Divergence

In this section we aim at setting the stage for the Divergence and Convergence discourse by reviewing another set of real-life examples and commenting on them from the risk point of view. A box with links to key terms is included in the references at the end of this chapter to facilitate the read.

We start with a classic horror film and then turn into a happier story but not yet a happy ending, due to the human factors we saw in action in Chap. 1.

This book does not deal specifically with issues of stranded assets, but they are briefly discussed in Sect. 5.1.

3.1 Mythological, Biblical and Recent Catastrophes

We have always been of the opinion that mythology—ancient Greek Gods, the disappearance of Atlantis, the Plagues of Egypt and many more legends—were actually accounts of real divergent events, reinterpreted, distorted and made more vivid by millennia of oral transmission, re-interpretation and the sheer thrill of magnifying horror stories, a characteristic still very present in today's media. The thought comes back to us at each new report of massive oil spills, red tides, masses of plastic debris floating in our oceans, watercourses and lakes, and the like. Tailings dams and other bulk-waste failures also received lots of coverage in the last decade, thanks to some very graphic accidents.

In a very distant future our successors (maybe humans, maybe someone else) will likely find some sort of record of these events, possibly lumped up in a single epic account about water pollution at the transition between the electromechanical and the cyber-informational eras during the Sapiezoic era of the Anthropocene epoch (see Chap. 2. It is likely that dissociated events we see happening in our world today will by then have become amalgamated into a mega-event, a planetary plague.

Mankind can act to avoid catastrophic failures as well as those that will occur in the future, avoiding a long string of mega-events that could feed the legends of the future. To reach that goal we have to set well-balanced, rational objectives. This is not a political campaign; this is the beginning of us humans writing a different book for our successors. Indeed, unless we make robust decisions today, archaeologists in a distant future will likely find books or some records relating events that are occurring before our eyes.

3.1.1 The Ten Plagues of Egypt: The Return

Isn't it interesting that a horror film entitled "Ten Plagues of Egypt: The Return" movie seems to be playing in all theaters near our homes? And yet we do not understand the difference between the trailers for upcoming releases to be and the film being shown today. Meanwhile we feel the need to develop theories about black swans and other esoteric concepts. And, by the way, not many of us spend any time planning for potential mitigations.

Is there indeed a relationship between The Ten Plagues of Egypt and our world? Let's look for hints of the ten plagues here and now.

Plague 1 in Egypt saw the clear waters of rivers turn to blood. Today we also see dramatic changes in waters. One cause of blood-red water is Burgundy blood bacteria, a toxic species of cyanobacteria. Red is not the only color to taint our waters. In 2015 the media displayed the mind-boggling pictures of colored waters oozing out of one (small) closed mine in *Colorado, tainting miles of river courses* (https://time.com/3991302/colorado-waste-water-spill/).

Plague 2 Saw frogs leave their watery habitats and invade cities, houses, even beds. As incredible as it sounds today, instances of frogs and other animals raining down from the heavens have been reported throughout history. It is generally believed that these phenomena are caused by extreme weather conditions such as tornadic winds or deluges. As such extreme weather occurrences are more and more likely due to climate change, invasions by frogs might not be unthinkable!

Plague 3 was an invasion by insects (some say fleas, others gnats). Today killer bees, murder hornets, dengue mosquitoes, and tiger mosquitoes are all on the rise. In Egyptian times the infestation by insects might have been caused by their natural predators, the frogs, having left waterways, allowing insects to multiply and thrive, an interesting case of interdependency (See Sect. 8.2.2).

Plague 4 was havoc created by wild animals. Today animals such as polar bears and other species are migrating to escape from territories were the food chain is disrupted, as well as the dangers of poaching and humans taking over their habitats. It is not uncommon to read of dangerous situations created when animals stray into human territory and vice-versa.

Plague 5 was the widespread death of livestock, and therefore a drastic reduction of the food supply. Today's livestock are threatened by a number of diseases: hoof and mouth disease, mad cow disease, swine flu, avian flu, necessitating massive culls of infected livestock. Some of these are diseases that can cross from animals to humans, as the COVID-19 pandemic likely did. Further, possibly compounded by climate change and ocean acidification as well as overfishing, fish are being depleted too, sometimes to extinction. Batrachia and entire species of fishes are actually already "disappearing" from our ecosystem today.

Plague 6 was boils. Moses was instructed by God to take handfuls of soot from a kiln, and throw it toward the sky within sight of Pharaoh; the fine dust caused painful boils on people and animals. Today exposure to the fine particulates in soot, dust and smog caused by the burning of fossil fuels are known to cause cancer and other diseases in humans.

Plague 7 was a rain of fiery hail. Today showers of rock and fire caused by volcanic eruption are not at all far-fetched. We have seen a number of strong volcano eruptions, maybe not stronger than in the past, but certainly deserving a lot of attention because of their impact on human activities, including, of course, commercial aviation (see Sect. 2.2.2). It would be sufficient to have a few simultaneous eruptions around the world—for example a selection from Iceland, Italy (Etna, or Vesuvius), Japan (Sakurajima), Kamchakta, Southeast Asia (Taal), the Philippines (Merapi), Indonesia, Ulawun, Papua New Guinea, the Andean Volcanos, the Cascade Volcanos, Mexico (Popocatépetl), and Colombia (Galeras)—to lock the world into no-travel mode with immense consequences.

Plague 8 was an invasion of locusts. Recent locusts invasions have been declared "unprecedented" (Gronewold 2020) but, frankly, we disagree. There were similar events in the US in 1875, in Ottoman Syria in 1915, in Madagascar in 2013 and in Argentina in 2016. Here are a few more recent examples of locusts invasions:

- In 2017, there was a locust invasion in Crimea during summer.
- In 2020 South Africa and Namibia also reported such an invasion.
- Also in 2020, Pakistan and Somalia declared national emergency over locust swarms.

Plague 9 Saw Egypt plunged into darkness for three days. Today, this could be the result of a large volcano eruption, a mega sandstorm, air pollution/smoke (China, Southeast Asia, India) or a nuclear accident of the largest magnitude. Similar events have occurred on our planet since recorded history, and that is not such a long time after all. Today darkness could be caused by a simple lack of electricity, which could occur either by a solar storm like the one of 1859, the Carrington event (see Sect. 3.2.2), a massive cyberattack, or, as we experienced in Vancouver in 2016, strong winds capable of toppling trees that bring down electrical cables.

Plague 10 was the death of all firstborn, humans and animals. No one ever wants to see a recurrence of that. Unfortunately, in recent years the CDC and WHO issued warnings related to avian flu, MERS, SARS, and plague in the US. Ebola in West

Africa and Zika have affected countless lives, without the need to talk about the COVID-19 pandemic.

Future reinterpretation could be one event causing the death of scores of people, re-interpreted, lumped-up, for more vivid impression as "first child and first-born animals". Then, of course, there are the new emerging strains of flu, and possibly a return of bubonic plague (an outbreak was reported in 2020 in *Inner Mongolia* (https://edition.cnn.com/2020/08/07/asia/china-mongolia-bubonic-plague-death-intl-hnk-scli-scn/index.html)).

Anecdote 3.1: The 2003 SARS epidemic and a "smoky" conversation

The story of the SARS outbreak in Toronto, which began on 7 March 2003, resulted in extraordinary public health and infection control measures for that time. In a four-week period, nineteen individuals developed SARS, including eleven health care workers. The hospital's response included establishing a leadership command team and a SARS isolation unit, implementing mental health support interventions for patients and staff, overcoming problems with logistics and communication, and overcoming resistance to directives. Patients with SARS reported experiencing fear, loneliness, boredom and anger, and they worried about the effects of quarantine and contagion on family members and friends. They experienced anxiety about fever and the effects of insomnia. Staff was adversely affected by fear of contagion and of infecting family, friends and colleagues. Caring for health care workers as patients and colleagues was emotionally difficult. Uncertainty and stigmatization were prominent themes for both staff and patients (Skowronski et al. 2005).

Earlier that year, SARS had claimed more than 50 victims across Asia. In February 2003, the disease spread to Hong Kong: of the 1755 people who were infected, 299 died. The disease raised questions about tech companies' operations in Hong Kong. As Asian authorities scrambled to contain the outbreak of SARS, citizens in the region turned to the Internet and mobile communications to protest public health policy and spread word of traditional Asian remedies for the deadly virus (Maunder et al. 2003).

During the 2003 SARS outbreak I found myself in Tokyo discussing how we could deliver emergency planning and crisis management in that difficult time to a Japanese organization that had operations all over the world, including China.

The man in front of me was very elegant and had impeccable manners, if only had he been smoking less. His corner office in the headquarters of his corporation, was indeed slowly delivering to its occupants, including me, an unbreathable cocktail of fine dusts and nicotine- and tar-laden particles that probably still stains my lungs today.

The green tea, brought in by a secretary who knelt in front of us as she served in observance of long-standing traditions, was helping me to endure the chemical attack. I was at my third cup, and the man was telling me how critical each one of their expatriate managers and their families in continental China was for operations.

When I asked if they had crisis management/emergency planning in place the man replied swiftly: "No", sipped another little green tea, and lit another cigarette.

I immediately asked him, frankly puzzled: "But, what will you do if the worse happens and your managers and/or their families fall sick there?".

The reply came down like a dagger: "…Then we will send in a second wave of personnel!".

So, what has changed between 2003 and today? As far as we can see, not much. There are still organizations who remain totally unprepared and put off until later the decisions about what to do, when perhaps the bulk of their personnel will be home infected with a new virus.

That's precisely not the way a responsible corporation/manager prepares for a very likely hazard which may bring in serious consequences, hence risks!

An epidemic—or worse, a pandemic—is a crisis like any other, and techniques/rules for survival exist and can be set up in advance.

The first step is to systematically anticipate, prepare for and respond to threats. Companies prepared for crises suffer less and recover more quickly than firms that are vulnerable and unprepared.

To be one of the long-term survivors:

- Recognize the barriers preventing prediction of risk scenarios, including cognitive and psychological biases, information siloes, following prestige, and arrogance.
- Develop formal risk assessments and crisis/emergency plans
- Then, if and when risks materialize, follow your pre-defined plans, contain the crisis by acting decisively and quickly.

Closing point. Preparation is key, it should not be a last-minute thought. Improvising during a crisis often leads to bad decision. This is why in hazardous industries SOP abound.

Now, let us imagine how we could rewrite the screenplay so that the film becomes "Ten Plagues of Egypt: The Survival", and world-events do not go down in history as plagues, but rather as stories of sustainable survival. On one hand the solutions seem within our reach. Pierre Louis Maupertuis, who defined the principle of least action (Maupertuis 1744), gave us the first expression of the concept of optimization, or the creation of systems maximizing efficiency or functionality, so all we need to do is optimize our decisions and mitigate our burgeoning plagues! However, the problem is that those views can only make action stationary. For example, Maupertuis's principle determines the shape of the elliptic trajectory on which an object moves under the influence of gravity as a stationary path. This is an idea that is present also in economics and competition studies such as the Theory of Oligopoly (Cournot 1838), the theory of games and economic behavior (Copeland 1945), and equilibrium point and non-cooperative games (Nash 1950, 1951).

The assumption of stationarity needed for "perfect" information, and the requirement for the rational behavior of all stake holders, can checkmate all of the above, despite the numerous attempts to refine those solutions. Chaos theory and modern applied mathematics, like those promoted by Ekeland (1988), certainly constitute research fields full of promises and capable of providing us with formidable tools to rewrite the future of world history. (Incidentally, Ekeland was a consultant to Michael Crichton and Steven Spielberg on "Jurassic Park", and in particular to Jeff Goldblum, who portrayed a chaos mathematician fighting a Hollywood-concocted divergence.) However, at our level—meaning personal, corporate, perhaps governmental—we can opt to implement well-designed and thought-out multi-hazard risk assessments, where:

- we fight bias and censoring as our worst enemies (see Sect. 2.1);
- analysts consider consequences to be multi-dimensional (see Chap. 9), as they truly are (see Sects. 2.2.3 and 2.3), thus reducing un-foreseeability;
- corporate responsibility and business ethics (see Chap. 5) are carried out according to the highest standards;
- we build scenarios thinking to the unthinkable and thus reduce unpredictability.

In today's world, clogged with codes and rules, we do not advice our clients to design simply to pass compliance tests. We advise them to do what they need to do to sustain their organizations and profits in a durable way, so that they will not be remembered as the creators of those future "Ten Plagues".

It doesn't matter if you sell cars, mortgages, cruises, or oil; nor does it matter if your customer experience net promoter score had stellar values up to yesterday; finally, it does not matter if your industry is a one of a kind, or you are the largest, the best, etc.: if the images projected by the media or the news about your latest mishap are vivid enough to stir a public opinion tsunami, you will be badly hit. For example, what if the value of your shares were to drop by 35% in 2 days? In fact, your whole industry's arena could shrink and be boycotted by the masses. You could discover that, as they say in French, you just sawed off the branch you were sitting on.

We read, in disbelief like most people did, the headlines of the *Volkswagen clever-plan-turned-debacle* (https://www.bbc.com/news/business-343 24772) to make diesel cars appear to comply with EPA standards and at the same time more appealing to thrill-seeking drivers. Matt DeLorenzo, managing editor at Kelley Blue Book's KBB.com, had this to say:

> Not only is this a black eye and a huge problem for Volkswagen. From an industry perspective it may set back diesel technology as a means for auto makers to reach the requirements for high fuel economy. We may have reached a tipping point where now *diesels will become more expensive to make than hybrids* (https://www.marketwatch.com/story/volkswagen-loses-14-billion-in-value-as-scandal-related-to-emissions-tests-deepens-2015-09-21).

Overnight Volkswagen's badge of excellence in recognition of the fuel efficiency of their cars and how green their new diesel engines turned into a badge of shame for dishonesty and manipulation.

In fact, a big crisis will damage your organization and your projects, even if you are simply in the same business arena of the culprit, by interdependence. This is why you should have an action plan ready, just in case: so you can turn a potential crisis into a positive outcome for your business.

3.1.2 Super-Volcanos

When thinking about the eruptions of super-volcanos, we tend to think such events are very remote: a black swan. But the fact is that such an event can happen tomorrow, and we have to be prepared. Technical note 3.1 gives examples and consequences.

Technical Note 3.1: Italian Quakes and Volcanic Eruptions

Italy is both a seismic and a volcanic territory. The last eruption at the Phlegraean Fields caldera near Naples was in 1538; the last at Vesuvius was in 1944 (the historic one that destroyed Pompeii was in 79 AD); Mt. Etna is highly active with very significant eruptions in 1669, 1910, 1923, and 1928 all the way to more recent significant events in 1981, 1983, 1991, 1993, 2001 and 2002, and 2020.

The Reggio and Messina quake of 1783 resulted in a swarm of 5 strong quakes ($M > 5.9$). The event may be considered as a black swan for the two cities worst hit, but not on either the regional nor global level.

The most important recent Italian catastrophic earthquakes were:

- 1968, Valle del Belice (Sicily);
- 1976, Friuli;
- 1980, Irpinia (Campania-Basilicata);
- 1984, San Donato Val di Comino (Lazio);
- 1990, Carlentini (Sicily);
- 1997, Umbria e Marche;
- 2002, Molise and Apulia;
- 2009, L'Aquila (Abruzzo);
- 2012, Emilia;
- 2016–2017, Central Italy.

In fact, there are good estimates of deterministic and probabilistic scenarios of direct damages, for large potential events in Italy. However, there are no estimates for:

- short-term and long-term economic damages;
- how to limit the damage to (say) 1% GDP;
- what financial/legislative instruments are needed to ensure a rapid recovery;
- the costs/benefits of relocating 500,000 people or more.

Closing point It is difficult to claim a black swan in these conditions as some, but not all communities in the area of Naples have evacuation plans prepared for a possible explosion of the *Phlegraean Fields caldera* (http://www.regione.campania.it/regione/it/tematiche/magazine-protezione-civile/rischio-vulcanico-campi-flegrei?page=1).

There are around twenty known super-volcanoes on this planet. One of these is Tambora, a stratovolcano in Indonesia that is still active today. Its eruption in 1815 is the most powerful volcanic eruption in recorded history and summed up to be one of the greatest world catastrophes. It killed tens of thousands of people and plunged South East Asia into darkness, due to an umbrella of volcanic ash, for a week. That year summer did not really happen in the region. Tambora was one to two orders of magnitudes larger than the eruptions of Mount St. Helens in 1980 and Pinatubo in 1991 (respectively 100 times and 10 times).

The Yellowstone Caldera, a known supe-volcano that erupted some 600,000 years ago, is considered to have been 10 times more powerful than Tambora, and 1,000 times larger than Mount St. Helens. And everybody knows that some 2,000 years ago Mount Vesuvius destroyed Pompeii. Today a super-volcano near Vesuvius, the Phlegraean Fields caldera, lurks on the other side of Naples. An eruption of a *super-volcano in Southern Italy* (http://www.dailymail.co.uk/home/moslive/article-1342820/Vesuviuss-big-daddy-supervolcano-Campi-Flegrei-near-Naples-threatens-Europe.html#ixzz4sWMwPpXv) would have enormous repercussions for the entire European continent, the Mediterranean basin and beyond, due to interdependencies (Ventura et al. 2009).

Experts think major eruptions occur at a frequency rate of 1/100,000 event/years. That is at the threshold of credibility following industrial standards, as discussed earlier (Rulis 1986; Nehnevajsa 1984). Of course, there is great uncertainty in these predictions. For example, Professor Bill McGuire, director of the Benfield Hazard Research Centre at University College London and a member of Tony Blair's Natural Hazards working group stated:

> Approximately every 50,000 years the Earth experiences a super-volcano. A super-volcano is 12 times more likely than a large meteorite impact [we will discuss this hazard in Sect. 3.2.1 Places to watch now are those that have erupted in the past, such as Yellowstone in the US and Toba. But, even more worryingly, a super-volcano could also burst out from somewhere that has never erupted before, such as under the *Amazon rainforest* (https://www.theguardian.com/science/2005/apr/14/research.science2).

Thus, there is great uncertainty not only about the frequency, but also about the number of the super-volcanoes. Anecdote 3.2 tells the tale of a mystery super-volcano.

Anecdote 3.2: The Mystery Super-Volcano

In 1465 the sky in Naples turned deep azure. In the following months severe anomalies hit the European meteorological conditions: very heavy rains in Germany and flooding in Poland; the cellars of the castles of Teutonic knights flooded and villages disappeared. Four years later a mini-glaciation occurred. Trees did not blossom and canals froze in Italy, paralyzed by heavy snowfalls. Interestingly that century featured a series of colder periods called the Little Ice Age (Fagan 2019). The cause of all of the above was apparently the eruption of a super-volcano. Traces of its eruption—spikes in acidity in ice cores which date back to the fifteenth century—are visible from Antarctica to Greenland. Based on those findings one can safely assume the eruption was significantly bigger than the giant Tambora stratovolcano eruption in 1815, but the super-volcano remains unaccounted for. (Was it an asteroid impact instead? See Sect. 3.2.1

In fact, there is a volcanic crater one kilometre deep in the Pacific Ocean that could be the "mystery" super-volcano. The crater is known as the Kuwae caldera and layers of ashes testify to its eruption, including traditional lore about of "disappearing land". Cross-disciplinary studies date that eruption to 1453, that is, eight years earlier than the Naples event just mentioned. Was the Kuwae eruption the cause of a series of other mishaps across Europe, Scandinavia, Turkey, China, and the Americas (for example, the Atzec Empire famine). Unfortunately for researchers, there are no signs of such a large explosive event on the site, meaning there are not enough volcanic deposits. Thus the Kuwae is likely not the cause of the global impact.

The story is an unfinished one, as apparently inherent uncertainties cloud ice cores dating and cognitive biases may have led to connecting dots with biased lines.

Closing note. All of the above unfortunately leads to a worrisome conclusion: we may even be completely off with that window of a frequency rate between 1/50,000 and 1/100,000 years, and there isn't much we can do to correct that.

Of course we do not know the magnitude of a new eruption and the consequences could vary widely from local (Pompeii-like) to regional, and possibly multinational/continental scale. Just to give a perspective, the formation of the Phlegraean Fields caldera 39,000 years ago created the cliffs of Sorrento, on the other side of the bay of Naples.

Amazingly, we cannot reconstruct the events that provoked the global climate effects in 1465. No one can determine the culprit with any certainty, despite the fact that entire large mountains may have exploded and vaporized, new islands may have appeared, cities may have been destroyed. We do not even know if we should be looking for a super-volcano or a meteorite strike, which may itself have been responsible for volcanic eruptions.

The consequences of a major volcanic eruption range widely and we would be negligent if we did not contemplate the worst. In addition to the well-known consequences of lava flows and tsunamis, here are a few of the less obvious examples:

- **Climate Changes**. As the experience in 1465 in Naples shows, catastrophic volcanic eruptions can cause heavy rains and flooding, massive snowfalls and freezing, volcanic winters, and perhaps even little ice ages. According to UN estimates in 2012, *food reserves worldwide would last 74 days* (https://time.com/5216532/global-food-security-richard-deverell/) and updated information is not significantly different. Thus a volcanic winter caused by the eruption of a supervolcano could bring the planet to starvation.

- **Pumice Rafts**. The appearance of pumice rafts in seas and oceans is a well-documented consequence of large volcanic eruptions. During the Krakatoa eruption in 1883 pumice spewed into the ocean and clogged harbors in Indonesia. Pumice rafts also appeared near Fiji in 1979 and 1984 from eruptions around Tonga. In August 2012 a pumice raft 480 km long and 50 km wide floating 0.6 m above the ocean's surface appeared near New Zealand. *Floating pumice rafts can cause problems for ships at sea* (http://volcano.oregonstate.edu/floating-pumice-%E2%80%93-oceanic-hazard), including engine stalls. Pumice could also clog the cooling water intakes of nuclear power plants, even hundreds of miles away from the eruption. A research group at the University of Southampton, UK, has a computer model tracking the movement of pumice rafts to deliver alerts as needed.

- **Dust plumes**. As recently as 2010 the "small" eruption of the Icelandic volcano Eyjafjallajökull (Fig. 3.1) forced air traffic over Europe and transatlantic routes to halt for almost a week. Dust and ashes buried the local area (Thomas and Prata 2011). As for pumice rafts, Volcanic Ash Advisory Centers can track ash plumes for air traffic safety.

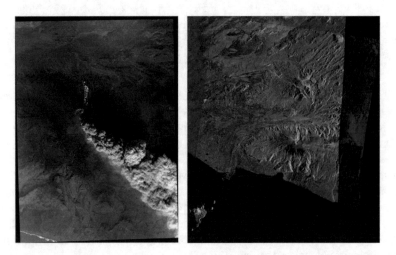

Fig. 3.1 Iceland volcano Eyjafjallajökul. Left: optical image of the eruption. Right: Overlay of two RADARSAT-2 images. The green before the eruption, and the red after the eruption. The combination highlights change. Left: Satellite image © 2021 Maxar Technologies. Right: RADARSAT-2 Data and Products © MDA Geospatial Services Inc. (2010) - All Rights Reserved. RADARSAT is an official mark of the Canadian Space Agency

However, the dangerous hazards posed by volcanic activity are no reason to blind ourselves and state there is nothing we can do to increase resilience of our systems and protect people. A large eruption of a climate-altering magnitude would be a large-scale challenge for humanity, our societies and possibly our survival. Preparedness should involve local/national, then international and planetary efforts. Here are some things that should be considered:

- **Evacuation plans**. These require careful planning and organization. Calderas are more complicated than volcanoes because the exit point of magma is not easy to spot in advance. However, we can monitor seismic swarms and deformations of the crust and win a bit of time to order evacuations. NASA and other groups are currently thinking about ways to *artificially cool down magma* (https://www.businessinsider.com/nasa-earth-supervolcano-apocalypse-2019-1?ir=t), while trying at the same time to extract beneficial energy. However, the implementation of such solutions is far away and many even question their feasibility.
- **Plans for taking care of survivors**. Efforts should include dealing with the flow of survivors, relocation, food shortage and travel difficulties, on top of possible epidemic and ruinous economic backlashes.
- **Steps for individuals and families**. Listening to the news and being ready to leave quickly are good rules in disaster-prone areas anywhere in the world. Have survival baggage ready; park the car with a full tank, ready to go. It does not cost much, but could save your life. Mormon leaders commanded their followers to maintain a full year of food and emergency supplies, but this has been relaxed more recently to keeping sufficient food and other necessities to last three months. Indeed, have a reserve of a couple of months' worth of non-perishable food could be a good idea for anyone. Self-aid kits similar to quake-preparedness ones, can also help each one of us to cope for a while.

3.2 Emerging Considerations on New/old Exposures

As of late, climate change risk started to being considered by the financial world, but we see very few industrialists considering them in their overall logistics. Let us look at the example of transportation accidents in a divergent environment.

Transportation related industrial accidents can be tricky due to their multiple aspects. In addition, they often occur outside of the operation perimeter. In general, transport infrastructure is exposed to natural hazards as well as man-made hazards. Indeed, in a recently published study entitled "*A global multi-hazard risk analysis of road and railway infrastructure assets*" (https://www.nature.com/articles/s41467-019-10442-3) produced interesting results: "~27% of all global road and railway assets are exposed to at least one hazard and ~ 7.5% of all assets are exposed to a 1/100 year flood event". 7.5% exposed to 1/100 year flood event? What if that 1/100 diverges, for example, to 3/10? Can anything be done to prepare, cope, or avoid?

Companies often take reliable transport infrastructure for granted, only to realize later how bad transportation-related accidents can be. That is because transport infrastructure remains the backbone of a prosperous economy as long as divergent exposures do not alter the equilibrium.

Climate change is already impacting industries' bottom lines. As it impacts logistic networks, it poses threats to global economy, hence to the banking system. In November 2019 the Federal Reserve Bank of San Francisco organized a conference, seeking to shed light on sought information on "*implications for monetary and prudential policy of climate change and its consequences*" (https://www.frbsf.org/economic-research/events/2019/november/economics-of-climate-change/) (Fig. 3.2).

Ten months later, on 21 September 2020, it was reported that a group of 560 of world's large companies, going by the name of "*Business for Nature*" (https://www.businessfornature.org/), with a combined revenue of approximately 4T USD, signed a *statement urging governmental actions on nature loss ahead of a UN biodiversity summit* (https://www.reuters.com/article/uk-climate-change-biodiversity-companies-idUSKCN26B0Y2). The statement declares that "Healthy societies, resilient economies and thriving businesses rely on nature. Governments must adopt policies now to reverse nature loss in this decade". This represents a significant and interesting evolution as we see companies asking governments for regulations rather than companies being regulated and struggling to comply.

We believe that the Federal Reserve conference, the UK actions in this field and the founding of "Business for Nature" as a tipping point in the awareness

---

Call for papers
The Economics of Climate Change
Federal Reserve Bank of San Francisco
November 8, 2019

The Federal Reserve Bank of San Francisco invites submissions for a research conference on "The Economics of Climate Change" to take place in San Francisco on **November 8, 2019**. This conference will bring together researchers interested in the economics of climate change to present new research that examines the economic and financial aspects of climate risk, mitigation, and adaptation. Theoretical, empirical, experimental, and policy-oriented submissions are all encouraged. Topics of interest include, but are not limited, to

- Quantifying the climate risk faced by households, firms, and the financial system
- Measures and models of the economic costs and consequences of climate change
- Improving the economic underpinnings of carbon emissions forecasts
- Incorporating climate change into macroeconomic and growth forecasting
- Implications for monetary and prudential policy of climate change and its consequences
- Green finance and financial innovation in climate change adaptation and mitigation.

Some of these topics were surveyed in "Climate Change and the Federal Reserve" (FRBSF Economic Letter 2019-09).

Fig. 3.2 Call for papers by the Federal Reserve Bank of San Francisco

of climate change. A new awareness that focuses on present, emerging, residual and divergent risks may lead to a consensus about the definition of an applied strategy. Interestingly, however, the Bank has not joined the *global bank initiative* (https://www.banque-france.fr/en/financial-stability/international-role/network-greening-financial-system/about-us).

Climate adaptation and risk assessment are also the main themes of a *recent ICMM report* (https://www.icmm.com/website/publications/pdfs/climate-change/191121_publication_climate_adaptation.pdf) (ICCM 2019), as well as of an IPCC 2001 report in nature (Kodra and Ganguly 2014) and a recent doctoral thesis (Aghdaie 2019). In particular, the ICMM report notes that physical climate-change risks and opportunities can impact mining companies in a multitude of ways, including as aspects of operation, production, finances, society and environment.

Let's look at California, now. While California is home to highly lucrative industries, high tech and IT, climate changes and natural hazards are very present and very significant, from "old style" hazards such as seismic events, wildfires and flooding and "new style" hazards such as *man-made contaminations* (https://www.latimes.com/politics/story/2019-10-10/california-finds-widespread-contamination-of-chemicals). Media reports are full of mesmerizing lucrative futures as AI, robots, the Internet of Things, *5G and driverless cars* (https://www.axios.com/coming-21-trillion-extreme-weather-bonanza-f6d7e2b3-ef87-44ca-939e-7cb743bf9114.html), but no one seems to pay attention yet to a new industry that results from the need to mitigate climate changes effects and other hazards. Indeed, old and new hazards are going to impact the new IT based world in ways that may seem difficult to grasp, as we pointed out at the end of Sect. 2.2.4. The ubiquitous character of the new and old hazards and the interconnected quality of the IT-based world require specific approaches. Indeed, it becomes paramount to enable the capacity to mitigate and act in rational, prioritized ways. That is particularly important when contemplating very large portfolios of potential targets.

In addition, AI is not good at predicting what it cannot learn from the past. Simply because it lacks imagination and builds relationships based on the past to project toward the future. AI works well when there are lots of data, such as in a monitored production process, but not when phenomena are rare (see Sect. 7.2.2).

Finally, the recent Carbon Disclosure Project (CDP) report shows how Pepsi, Allstate, Walmart, and dozens of other S&P 500 corporations face large financial impacts as climate change worsens (Luo et al. 2012). All of this makes it clear that there is no longer time for political diatribes over whether climate change is real; what is needed is action.

For some years now we have included climate change scenarios and their related costs in all our holistic risk assessments for a wide range of clients: from automotive producers and suppliers to mining, from insurers to critical infrastructure.

Declared impacts by various industries cover a wide spectrum of issues, such as:

- weather-related sourcing challenges;
- changes in consumer preference due to elevated temperatures;
- changes in crop yields due to changes in precipitation and temperature patterns;

- increases in production and transportation costs;
- increased supply-chain costs due to changes in crop locations;
- increased energy costs;
- disruptions due to flooding;
- volatility of hedging costs.

Weather-related insurance claims are, of course, also on the rise, with sometimes staggering single-event claims reaching and exceeding the billion dollar bar. Insurance payouts in 2020 hit record values due to climate change related events while COVID was causing severe economic distress as stressed by charities and *reinsurers like Zurich RE and Munich RE* (https://fortune.com/2021/01/02/2020-climate-cha nge-insurance-payouts-natural-disasters-covid-pandemic/).

Thus, a siloed approach to climate change risks will not work. Interconnected systems have to be analyzed though convergent approaches, that is, looking simultaneously at all hazards. Furthermore, a rational risk assessment deployment should be capable of:

- adapting to new situation within and outside the system;
- updating the values of probabilities and consequences as well as their respective uncertainties;
- retrieving data using various queries (Brown et al. 2019).

The requirement to allow an applied strategy definition for present, emerging, residual and latent climate change risks is simple. The risk register of a rational risk assessment should allow bottom-up aggregation of risks (see Chap. 11). As a result, it would deliver the risk information required for decision-making support and other tactical and strategic planning purposes (see Chap. 12). We have included climate change risks in our convergent, scalable, updatable and drillable deployments well before the concept became a buzzword.

The risk landscape of companies can change quite significantly due to the alternation of long-term normal patters by divergence, that is, shifts in probabilities of occurrence and consequences of climate-related events. Often that occurs with seemingly repeated extreme events (see Sect. 2.2, for example). The probabilities in the "New normal" may significantly alter the risk landscape around a project or a corporation. They may transform tolerable risks into intolerable ones, tactical risks into strategic ones (see Chaps. 10 and 12). As a result, framing probabilities of new normal patterns is often necessary. To ensure that decision-makers and management can keep optimizing tactical and strategic planning, a rational, dispassionate update of the probabilities is paramount.

Let us now review and discuss some divergent exposures that many might consider to be "exotic" but are way more present than we would hope.

3.2.1 Asteroid-Earth Collisions

The United Nations declared 30 June *"International Asteroid Day"* (https://astero idday.org/). The aim was to raise public awareness about what the event organizers describe as "humanity's greatest challenge".

Developing a risk assessment for collisions between Near-Earth Objects (NEOs, including asteroids and comets) and the Earth is not about knowing when the next big impact of an asteroid on Earth will be but is rather to understand, for example, if:

- The recent decision (December 2016) by European ministers to *decline funding part of the NEOShield project* (https://cordis.europa.eu/project/id/282703/rep orting) to intercept a space object is reasonable;
- The idea of a project such as NEOShield is reasonable in comparison, for example, to other space risks, for example, a repeat of the solar storm of 1859, known as the Carrington event (see Sect. 3.2.2).

Following an orderly approach, in order to develop a risk assessment for an asteroid- Earth collision we need to identify the hazard scenarios. The hazards, NEOs, range from objects that are very small (a few millimeters) to small and burning up in the atmosphere as shooting stars, to very large, say with a diameter of 10 km, likely the size that wiped out non-avian dinosaurs some 65 million years ago. There are agencies responsible for monitoring the asteroid hazard, including the European Space Agency's Space Situational Awareness project. However the monitoring is not perfect and is fraught with uncertainties: the 2013 Chelyabinsk impact, for example, caught everyone by surprise.

We can try to estimate the frequency and consequences of different scenarios.

1. **Worst case scenario**. The largest impacts may occur every 100 million years ($f = 1/10^8$), thus amply within the non-credible range of probabilities from a human point of view, but their impact could mean the end of mankind. The possible date is unknown, despite the efforts of scientists.

2. **Worst credible scenario**. There are millions of asteroids in the 15–140 m diameter range. 140 m is the threshold for regional damage at the scale of a country or a continent, while a 40 m object was the culprit in the largest impact in recent history, exploding over Tunguska, Siberia, on 30 June, 1908: 80 million trees were flattened over approximately 2,000 km^2 of a sparsely populated region. That area is the same covered by some modern mega-cities such as London, Tokyo and Sao Paolo. Experts consider the frequency of Tunguska-sized events in the 1/300 ($p = 3.3 * 10^{-3}$) range, which is comparable to that of large earthquakes and well in the realm of other credible events such as major catastrophic tailings dams failures (Azam and Li 2010).

3. **Most credible worst scenario**. More recently, the 2013 Chelyabinsk impact damaged approximately 5,000 buildings, and flying glass from blown out window injured over 1,200 people. That type of object impact may be in the 1/10 frequency ($p = 0.1$). It likely corresponds to an object 20 m in diameter"

exploding in the atmosphere with a force equal to multiple Hiroshima bombs. The consequences in 2013 were relatively benign.

If we now focus on the future and mitigations, we first have to realize that there is a significant difference between major earthquakes and space object impacts: earthquakes destroy buildings that are not built to withstand them, but there is little to do against a nuclear-blast-like impact coming from a colliding space object. Interestingly, where tsunamis, earthquakes and volcanoes remain widely unpredictable despite monitoring efforts, asteroid collision seems to be the only predictable large natural hazard that good monitoring and mathematics can predict. Thus the key may be alert systems.

Europe is setting up a network of telescopes to provide us with a heads-up before a collision. The network will scan systematically the sky, and is designed to be able to detect any asteroid with potential collision trajectory, allowing a preparation time of approximately two to three weeks. At the very least, it will make it possible to evacuate cities or to issue a shockwave warning. A different approach, namely an *asteroid deflection system* (https://www.france24.com/en/20170628-are-asteroids-humanitys-greatest-challenge) would require "something in the order of 300–400 million euros … a minuscule amount compared to the cost of disaster". However, there is great uncertainty about the cost, as another source quote €150 M EUR (Galvez et al. 2013).

3.2.2 Solar Storms

The Carrington event, one of the largest recorded geomagnetic storms due to solar activities, occurred on 1–2 September 1859. Beside anomalous and wide-spread occurrences of auroras, telegraph systems all over Europe and North America failed, telegraph operators were electrocuted, and *telegraph posts threw sparks* (https://www.history.com/news/a-perfect-solar-superstorm-the-1859-carrington-event#:~:text=The%20Carrington%20Event&text=Five%20minutes%20later%20the%20fireballs,operators%20and%20setting%20papers%20ablaze). Such a geomagnetic storm can be caused by solar flares, coronal mass ejections, and solar electromagnetic pulses (EMPs). For the sake of completeness, let us note there are modern researchers linking solar activity to the occurrence of *mega-quakes* (https://www.nature.com/articles/s41598-020-67860-3), but we will not enter in that interdependency discussion.

Let us look now at the likelihood and consequence scenarios of solar storm hazards and risks.

Frequency In July 2012 a Carrington-class solar superstorm missed the earth, producing no damages (Baker et al. 2013). It is likely that the average annual frequency of these events may lie in the range of 1/200 ($p = 5 * 10^{-3}$) to 1/500 ($p = 2 * 10^{-3}$) events/year. Less severe storms seem to occur with a frequency of approximately 1/50 ($p = 2 * 10^{-2}$) events/year with widespread radio disruption.

Consequences In March 1989 a large impulse in the Earth's geomagnetic field erupted along the US/Canada border. This started a chain of power systems disturbances that only 92 s later resulted in a complete collapse of the entire power grid in Quebec (Medford et al. 1989). In June 2013 experts from London and the United States used data from the Carrington Event to estimate the current cost of a similar event to the US alone at 0.6–2.6T USD. Beside electric lines, transformers are particularly vulnerable to such events and many will take months to replace and have no redundancy available. Indeed, the purchase placement of a single extra-high voltage transformer of the 300–400 MVA class has been quoted as taking up to 15 months to manufacture and test. Of course, manufacturing and testing the equipment does not mean the story ends there. The equipment will then need to be transported to site and commissioned before being put into service.

Two low-probability, high-impact events have recently stirred the attention of policymakers (Maynard et al. 2013). One is the potential for a *massive solar storm* (https://www.lloyds.com/~/media/lloyds/reports/emergingriskreports/solarstormrisktothenorthamericanelectricgrid.pdf), the other a human-caused EMP. The two have obviously different likelihoods and consequences but the mitigation would be the same because these low-probability, high-impact events constitute a threat to critical infrastructures. In the US, the White House released a National *Space Weather Strategy and Action Plan* (https://www.swpc.noaa.gov/news/national-space-weather-strategy-and-action-plan-released-0), and in 2019–2020 Congress considered bills on both subjects. Yet outside of defense-focused EMP research and hardening of certain military systems during the Cold War, efforts to assess and mitigate space weather and EMP threats to civilian infrastructure are relatively nascent. Policymakers are now pushing for enhanced research and preparedness efforts in this domain.

3.2.3 2020

We can still recall champagne popping on 31 December 2019. We had just come back from a mission to Brazil and had flown to Europe to give an MBA course at an Italian Business School and see our families. "Happy new Year! 2020 is going to be exceptional! Yes, indeed, and it is a bissextile year, thus a happy year for us! Cheeeeers!".

Two weeks later we flew to Brazil again for an interesting and pleasant trip despite some very *heavy localized precipitations in Belo Horizonte* (https://www.theguardian.com/world/2020/jan/27/heavy-rain-landslides-flooding-mass-evacuations-brazil). On our way back we returned to Italy and the driver at the airport greeted us by asking, "You weren't in China, by any chance?" Frankly, we had no clue, but in Italy they were already starting to fear the COVID-19 contagion. That was 23 January 2020.

A few days later we visited an automotive client near Turin and learned that they had already activated a crisis plan, due to the intense exchanges that they had with China.

On 20 February we had a meeting with a space observation specialist from Milan. We knew that something was going wrong, but we did not know yet how bad it would become. Indeed, the first "red zone" in northern Italy was declared only two days later, near Milan.

On 25 February we flew home to Vancouver, and due to mechanical problems with the plane we were rerouted on Icelandair via Keflavik. By only 24 h we missed crossing paths with 15 people coming from Ischgl, the Austrian skiing resort hotspot, who then brought the virus to Iceland. There were more near-misses in the following week, and for that reason we decided we had already spent all the 2020 good-luck tokens and to undertake a voluntary lockdown.

We had personal knowledge that *some countries had* (https://www.babs.admin.ch/fr/aufgabenbabs/gefaehrdrisiken/natgefaehrdanalyse.html), as we had done in some of our own assessments, evaluated the occurrence of a pandemic to a frequency rate of 1/30–1/50 event/years. Thus we knew a pandemic is a predictable event and even the generalized lockdown, with its share of economic pains was foreseeable. We had seen simulations (Petermann et al. 2011) of prolonged blackouts at national scale coming to even more catastrophic conclusions. However, we did not anticipate at that time that a concerted effort would be made, and that the situation would entail what we all now know: generalized quarantines, lockdowns and world-wide sorrow.

But 2020 had other surprises in store as well. Let's focus on the West Coast states and provinces of the US and Canada.

Wildfires Wildfires are common occurrence in this area. They are predictable and foreseeable. However, in 2020 the scale, duration and damages reached different levels (Witze 2020), dwarfing the 2017 Tubbs (Coen et al. 2018) and 2018 Camp fires (Fig. 3.3) (Spearing and Faust 2020). Smoke from the West Coast, after blanketing portions of Canada, *reached Europe* (https://edition.cnn.com/2020/09/16/weather/us-wildfires-smoke-europe-copernicus-intl/). We think the frequency rate of the 2020 fires is certainly lower than $f = 1/100$, and more likely closer to a $f = 1/500$ event, given the lack of historic reports of similar catastrophic events. We would not dare to go any lower, as we have no studies available on this subject. The final damage is very difficult to evaluate, given long-term issues (erosion, loss of topsoil, diseases, etc.) that will emerge in the future.

Social Unrest Riots and protests are rather common all over the world, but significant political polarization inhibiting swift actions and planning in a G8 country is a rather rare event. We would be tempted to say that this had not really occurred at the present scale since the American Civil War, so we would use $f = 1/160$ as a first estimate of the frequency. To account for other countries, we would use a rate of 1/100–1/200. Finally, reportedly the more than 1B USD billion plus (and counting) *riot damage* (https://www.axios.com/riots-cost-property-damage-276c9bcc-a455-4067-b06a-66f9db4cea9c.html) is the most expensive in US insurance history.

Fig. 3.3 Camp fire in California, Town of Paradise. Note the disparity of damage across the image. Satellite image © 2021 Maxar Technologies

Now comes the real surprise: as we can easily assume that these events are independent from each other (see Chap. 8) and are rather rare, the probability of seeing them occurring simultaneously is evaluated with Eq. 3.1 (maximum) and Eq. 3.2 (minimum).

$$p_{max} 1/30 * 1/100 * 1/100 = 3.3 * 10^{-6} \tag{3.1}$$

$$p_{min} 1/50 * 1/500 * 1/200 = 2.0 * 10^{-7} \tag{3.2}$$

These estimates are clearly at the limit of credibility or in the realm of act of God.

As for the consequences, beyond the direct impacts (loss of lives, businesses, living space, way of life, etc.), we have to consider the amplifying effect of smoke inhalation during a respiratory disease pandemic, the long-term consequences of loss of vegetation cover, etc. Altogether, if the impacts may be somewhat foreseeable singly, their cumulated effect is indeed unforeseeable.

2020 has brought to us a very good example of a real black swan!

3.3   Reporting Divergent Risks

One modern trend in our societies is the increased pressure on industrialists, project promoters to report hazards and risks to investors, lenders and comply with the duty to consult the population (see Chap. 5). Reporting risks means publishing in various forms what risks businesses face in their life stemming from due diligence to annual financial reports or new project being presented to the public. Newspapers, journals, magazines and of course social media are full of information and disinformation on scary scenarios like the ones we just discussed in Sects. 3.1 and 3.2.

Let us paraphrase the ICMM report (2019) cited in Sect. 3.2:

- Building climate resilience does not require reinvention of the wheel. Simple concepts and solutions can produce great effects (Johnson and Blackburn 2014; Gencer 2012).
- Climate change has to be integrated in ERM approaches.
- Climate change resilience has to consider all components of the productive system, thus avoiding a siloed approach to reality.

In order to build sustainable and rational resilience in a complex system, multi-hazard, convergent quantitative risk assessment (QRA) approaches are necessary (de Ruiter 2020).

Climate change is defined by the United Nations Framework Convention on Climate Change (UNFCCC) as, "a change of climate … which is in addition to natural climate variability observed over comparable time periods" (UNFCCC 2007).

The Panel on Climate Change (IPCC) argues that climate change leads to changes in extreme weather and climate events, potentially leading to disasters (IPCC 2012). In other words, climate change creates divergent hazardous events which can turn catastrophic if their consequences become very significant in any or all potential consequences dimensions (social, physical, environmental, economic, etc., see Chap. 9).

Indeed climate-related events have the potential to hit the:

- supply chain at various levels (ingress/egress of key materials, products, personnel, contractors, etc.);
- operations of any type;
- waste management (dumps, ponds, waste storage facilities).

These could ultimately result in the closure of the system.

Thus, as stated above, risk assessments have to deliver enough granularity to allow prioritization. In addition, they need to support decision-making on multiple simultaneous fronts. As a result, they will make it possible to decide where and when to allot resilience enhancement funds, and thus, in some cases, allow the pursuit of new opportunities. If unmanaged, climate-related risks will inevitably weaken company's finances and damage their CSR (Burke and Logsdon 1996) and SLO (Prno and Slocombe 2012) (see Sect. 6.1.2). Ill-conditioned resilience enhancements, such as those resulting from siloed approaches, will bear their legacy of unpleasant side effects toward the future.

The ICMM report (2019) suggests a stepped approach to identify climate risks and opportunities. At the same time it recognizes that each company may apply a different philosophy to ERM and resilience building. The stepped approach emphasizes the need to use monitoring (see Sects. 7.1 and 7.2) and forecasting techniques while integrating existing ERMs to study the complex issues related to climate change.

Due to the considerations above, reporting climate and other risks will become routine while climate risk reporting might remain voluntary. Indeed, the G20 Task force launched its voluntary framework in 2017. It calls on companies to provide climate-related financial disclosures in their public annual financial filings. Almost 800 companies and organizations with a combined market capitalization of more than 9.2T USD, have committed to support the framework. The *group includes* (https://fr.reuters.com/article/uk-climatechange-financial-disclo sure-idUKKBN19K0JF) insurance groups AXA and Aviva, oil companies Royal Dutch Shell and Total and mining companies Anglo American and BHP. More than 200 of the world's largest listed companies have forecast that climate change could cost them a combined total of almost one trillion dollars, as shown by a report by charity group *Carbon Disclosure Project* (https://www.cdp.net/en/info/about-us).

But the future may also go in the direction of compulsory reporting, perhaps in accordance to specific specifications, or with ad hoc documentation. Below is a sample of reporting requirements from various international jurisdictions for reference:

- **Canada**. National Instrument *NI43-101* (https://en.wikipedia.org/wiki/National_ Instrument_43-101) (CSA 2012)
- **Australasia**. The JORC Code, compulsory for listed companies in Australia and New Zealand (Stephenson 2001).
- **South Africa**. South African Code for the Reporting of Mineral Resources and Mineral Reserves (SAMREC 2000).
- **Hong Kong**. The Hong Kong Stock Exchange accepts reports prepared in accordance with NI 43-101, SAMREC or JORC.

The United Nations Framework Classification for Resources (*UNFC*) (https:// unece.org/sustainable-energy/unfc-and-sustainable-resource-management; https:// www.oecd.org/environment/outreach/UNECEandresponsiblemining6.6.17rev.pdf) (Falcone et al., 2017) is an international scheme for the classification, management and reporting of energy, mineral, and raw material resources. UNFC currently applies to minerals petroleum, renewable energy (Geothermal, Bioenergy, Solar Energy, Wind Energy, Hydropower), nuclear fuel resources (uranium and thorium resources), injection projects for geological storage (CO_2), and anthropogenic resources. Application of UNFC to groundwater resources is being evaluated.

Notably, Britain is the first G7 country to sign into law a requirement to reach net-zero emissions by 2050. Additionally, the green finance strategy sets out plans to foster investments in sustainable projects and infrastructure. These actions build on disclosure of climate risks set out by the G20 Task Force on Climate-Related Financial Disclosures. The addition of climate-related risks may become compulsory.

3.4   Goal of Convergent Leadership in a Divergent Risk World: The Example of Digital Transformation

First of all let's put any daunting and catastrophic thoughts aside: this is not the time to be frightened but rather time to plan tactically and strategically for a sustainable future. Allowing for sustainable and rational decision-making and considering uncertainties, divergence is the goal of convergent leadership. That is the same as turning on the headlights in your car when you drive at night rather than relying only on the instruments on the dashboard. NB: If you don't believe it, don't try it… just use your imagination, it's way safer. We do not want our readers to sue us for damages!

The first step should be reducing cognitive bias to a minimum (see Sects. 1.3, 2.3, Chap. 4, and Sect. 5.5.2). Of course, at this juncture we need to consider digital transformation, IoT and AI as these seem to permeate corporate discussions and planning (see Sect. 7.3). We regularly meet people that believe these technologies will save us from all uncertainties, risks and divergences. Of course, digital transformation brings significant benefits but also critical new risks. Like medicine, digital transformation has good primary effects and not-so-good side effects. Many consider digital transformation as the way to address all productivity and margin challenges, in other words, the way to deliver significant bottom-line value. They are likely right because in that area there are constant data being generated that AI can crunch in order to learn. However, in the realm of divergent hazards and risks we contend that these techniques may be less applicable, especially as the new data do not exist yet (see Sect. 7.2.2).

Furthermore, a recent EY's global report '*The digital disconnect* (https://assets. ey.com/content/dam/ey-sites/ey-com/en_gl/topics/digital/EY-digital-disconnect-in-mining-and-metals.pdf?download): problem or pathway?' points to some significant disconnect, in that many are talking, but not many actually embrace the new ways of digital transformation. The same report details some common pitfalls that are slowing the adoption, such as perception of high costs, lack of education and understanding, and lack of detail about the implementation pathway. That's probably why, as in many projects, people rush into implementation without developing an overall, non-siloed risk assessment that looks at the business as a whole, and shows what could happen once the digital transformation takes root in the heart of their production and logistic systems, and divergent exposures occur. Again, "new toy" enthusiasm causes a push to fast-track implementation while forgetting important details related, for example, to system interdependencies, etc. (see Sect. 8.2.2).

It is vital to clearly understand the existing system, including all its likely non-optimized partial digital solutions before slamming in a brand-new digital transformation program which will end up in trouble from day one, starting an endless line of costly patches.

Another fundamental step lies in defining, from the beginning and very clearly, the success and failure criteria. System definition is then paramount (see Chap. 6). Old-style vertical approaches, i.e., department-by-department siloed approaches, bring limited benefits and leave exposure gaps. This is because modern systems, including many digital transformation applications, also develop transversely, thus the *effects of malfunctions, cyber-attacks* (https://www.fmglobal-touchpoints.co. uk/mitigating-physical-losses-from-cyber-attacks.htm/), etc., become complex and highly interdependent.

The risk management approach must be holistic, cover the whole system and converge all hazards (natural, man-made/human error, technological, cyber, etc.) into one approach. Data will generally be missing. It becomes important to be able to integrate various sources while understanding their different levels of content details and their uncertainties. This is particularly important in face of climate change and other sudden alterations and divergencies. Well-developed risk assessments using all possible pre-existing information should support any decision-making.

Digital transformation, AI, IoT, etc., are not the panacea. They will not save us from divergent exposures, but will help us foster efficiency and streamline day to day work and planning under business-as-usual conditions.

Appendix

Links to more information about the Key terms from the Authors	
A, B	*Act of God* (https://www.riskope.com/2020/ 12/09/act-of-god-in-probabilistic-risk-assess ment/) *Black swan* (https://www.riskope.com/ 2011/06/14/black-swan-mania-using-buzzwo rds-can-be-a-dangerous-habit/) *Business-as-usual* (https://www.riskope.com/ 2021/01/13/business-as-usual-definition-in-risk-assessment/)
C, D	*Convergent* (https://www.riskope.com/2021/ 01/20/convergent-risk-assessments/) *Divergent* (https://www.riskope.com/2020/11/18/tactical-and-strategic-planning-to-mitigate-divergent-events/) *Drillable* (https://www.riskope.com/ 2020/01/15/probability-impact-graphs-do-not-fly/)

(continued)

(continued)

Links to more information about the Key terms from the Authors	
F	*Foreseeability/foreseeable* (https://www.ris kope.com/2021/01/06/foreseeability-and-pre dictability-in-risk-assessments/) *Fragile/fragility* (https://www.riskope.com/ 2020/04/01/antifragile-resilient-solutions-for-tactical-and-strategic-planning/)
P, R	*Predictability/predictable* (https://www.ris kope.com/2021/01/06/foreseeability-and-pre dictability-in-risk-assessments/) *Resilient, Resilience* (https://www.riskope.com/2016/11/ 23/resilience-cannot-based-instinctual-dec ision-making/)
S	*Scalable* (https://www.riskope.com/2015/04/ 16/how-system-definition-and-interdepende ncies-allow-transparent-and-scalable-risk-ass essments/) *Societal risk acceptability* (https:// www.riskope.com/2014/01/09/aspects-of-risk-tolerance-manageable-vs-unmanageable-risks-in-relation-to-critical-decisions-perpetuity-pro jects-public-opposition/) *Sustainability/sustainable* (https://www.ris kope.com/2019/01/16/improving-sustainab ility-through-reasonable-risk-and-crisis-man agement/) *Survivability* (https://www.riskope. com/2011/03/17/ale-fmea-fmeca-qualitative-methods-is-it-really-what-we-need/) *System* (https://www.riskope.com/2017/07/26/three-ways-to-enhancing-your-risk-registers/)
T, U	*Tolerance* (https://www.riskope.com/2020/04/ 29/risk-tolerance-thresholds/) *Uncertainty/uncertainties* (https://www.ris kope.com/2015/12/10/3-decision-making-tru ths-derived-from-uncertainty-taxonomy-sch eme-of-classification-and-a-road-sign/) *Updatable* (https://www.riskope.com/2020/01/ 07/climate-adaptation-and-risk-assessment/)

Other linked information (https://www.riskope.com/blog-news/) search Riskope blog and use the search box

Third Parties links in this section	
Colorado tainted waters	https://time.com/3991302/colorado-waste-water-spill/

(continued)

(continued)

Third Parties links in this section	
Locust invasion Crimea	https://www.aol.com/video/view/thousands-of-loc usts-invade-field-in-crimea/596792eb9efa89505f d948e5/?guccounter=1&guce_referrer=aHR0cH M6Ly93d3cuYmluZy5jb20v&guce_referrer_sig= AQAAAHi3gT0hNLRnUdNKXHkQkyXyXmiAX EIcEzWJmbwDIyHT28vvZRARcNZ3wZ9V56pAY qcBJytFcn1XIuovJySoSxBzzHr0BE7U3kSAb_Yuo HYRUzF8fefF7XK4jP3lcMZj5ll5DCwZgAOVg 6D4mGKNEQfxSVuVlEV1UrF0HoIkpCgF
Locusts Namibia and South Africa	https://www.bloomberg.com/news/articles/2020-02- 24/south-africa-confirms-locust-outbreak-in-nam ibia-karoo-region#:~:text=A%20locust%20outb reak%20in%20Namibia%20and%20South%20A frica%E2%80%99s,in%20a%20newsletter%20publ ished%20on%20its%20website%20Monday
Locusts Pakistan	https://www.dw.com/en/pakistan-declares-national- emergency-over-locust-swarms/a-52224762
Locusts Somalia	https://www.bbc.com/news/world-africa-513 48517#:~:text=Somalia%20has%20declared%20a% 20national%20emergency%20as%20large,the%20i nsects%2C%20which%20consume%20large%20a mounts%20of%20vegetation%2C
Bubonic plague Mongolia	https://edition.cnn.com/2020/08/07/asia/china-mon golia-bubonic-plague-death-intl-hnk-scli-scn/index. html
Volkswagen scandal	https://www.bbc.com/news/business-34324772
Diesel versus hybrid	https://www.marketwatch.com/story/volkswagen- loses-14-billion-in-value-as-scandal-related-to-emi ssions-tests-deepens-2015-09-21
Phlegraean caldera	https://www.regione.campania.it/regione/it/temati che/magazine-protezione-civile/rischio-vulcanico- campi-flegrei?page=1
Vesuvius super volcano	https://www.dailymail.co.uk/home/moslive/article- 1342820/Vesuviuss-big-daddy-supervolcano- Campi-Flegrei-near-Naples-threatens-Europe.html# ixzz4sWMwPpXv
Amazon rain forest super volcano	https://www.theguardian.com/science/2005/apr/14/ research.science2
Food security crisis	https://time.com/5216532/global-food-security-ric hard-deverell/
Floating pumice raft	https://volcano.oregonstate.edu/floating-pumice-% E2%80%93-oceanic-hazard
Artificially cooling magma	https://www.businessinsider.com/nasa-earth-superv olcano-apocalypse-2019-1?IR=T

(continued)

(continued)

Third Parties links in this section	
Multi hazard studies	https://www.nature.com/articles/s41467-019-104 42-3
Monetary policy and climate change	https://www.frbsf.org/economic-research/events/ 2019/november/economics-of-climate-change/
Business for nature	https://www.businessfornature.org/
Urging action on climate change	https://www.reuters.com/article/uk-climate-change-biodiversity-companies-idUSKCN26B0Y2
Global bank initiative	https://www.banque-france.fr/en/financial-stability/ international-role/network-greening-financial-sys tem/about-us
ICMM 2019 report on climate change	https://www.icmm.com/website/publications/pdfs/ climate-change/191121_publication_climate_adap tation.pdf
Extreme weather bonanza	https://www.axios.com/coming-21-trillion-extreme-weather-bonanza-f6d7e2b3-ef87-44ca-939e-7cb743 bf9114.html
Man-made contaminations	https://www.latimes.com/politics/story/2019-10-10/ california-finds-widespread-contamination-of-che micals
Reinsurers and climate change	https://fortune.com/2021/01/02/2020-climate-cha nge-insurance-payouts-natural-disasters-covid-pan demic/
International asteroid day	https://asteroidday.org/
NeoShield funding	https://cordis.europa.eu/project/id/282703/reporting
Asteroid deflection	https://www.france24.com/en/20170628-are-astero ids-humanitys-greatest-challenge
Carrington solar event	https://www.history.com/news/a-perfect-solar-sup erstorm-the-1859-carrington-event#:~:text=The% 20Carrington%20Event&text=Five%20minutes% 20later%20the%20fireballs,operators%20and%20s etting%20papers%20ablaze
Solar activity and mega quakes	https://www.nature.com/articles/s41598-020-678 60-3
Massive solar storm	https://www.lloyds.com/~/media/lloyds/reports/eme rging%20risk%20reports/solar%20storm%20risk% 20to%20the%20north%20american%20electric% 20grid.pdf
Space weather strategy	https://www.swpc.noaa.gov/news/national-space-weather-strategy-and-action-plan-released-0
Extreme rains in Brazil	https://www.theguardian.com/world/2020/jan/27/ heavy-rain-landslides-flooding-mass-evacuations-brazil
Pandemic risk assessment	https://www.babs.admin.ch/fr/aufgabenbabs/gefaeh rdrisiken/natgefaehrdanalyse.html

(continued)

(continued)

Third Parties links in this section	
Smoke from wildfires	https://edition.cnn.com/2020/09/16/weather/us-wildfires-smoke-europe-copernicus-intl/
US riots damage	https://www.axios.com/riots-cost-property-damage-276c9bcc-a455-4067-b06a-66f9db4cea9c.html
Financial disclosure	https://fr.reuters.com/article/uk-climatechange-financial-disclosure-idUKKBN19K0JF
Carbon disclosure group	https://www.cdp.net/en/info/about-us
NI 43–101	https://en.wikipedia.org/wiki/National_Instrument_43-101
UNFC	https://unece.org/sustainable-energy/unfc-and-sustainable-resource-management; https://www.oecd.org/environment/outreach/UNECE%20and%20responsible%20mining%206.6.17%20rev.pdf
The digital disconnect	https://assets.ey.com/content/dam/ey-sites/ey-com/en_gl/topics/digital/EY-digital-disconnect-in-mining-and-metals.pdf?download
Effects of malfunctions and cyber attacks	https://www.fmglobal-touchpoints.co.uk/mitigating-physical-losses-from-cyber-attacks.htm/

References

Aghdaie MH (2019) Robust water resource planning at river basins. Ph.D. dissertation. Concordia University. https://www.researchgate.net/publication/331213798_Robust_Water_Resource_Planning_at_River_Basins

Azam S, Li Q (2010) Tailings dam failures: a review of the last one hundred years. Geotech News 28(4):50–54

Baker DN, Li X, Pulkkinen A, Ngwira CM, Mays ML, Galvin AB, Simunac KDC (2013) A major solar eruptive event in July 2012: defining extreme space weather scenarios. Space Weather 11(10):585–591

Brown D, Oboni F, Oboni C, Parente S, Pritchard C, Sidorenko A, Stevens E, Dutra Villarroel I, Williams JA (2019) Global hot spots: how project and enterprise risk management practices drive business results around the world, simmer system book 4, ISBN 9781071277690

Burke L, Logsdon JM (1996) How corporate social responsibility pays off. Long Range Plan 29(4):495–502

Canadian Securities Administrators [CSA] (2012) National Instrument 43-101. Standards of disclosure for mineral projects. https://www.osc.gov.on.ca/documents/en/Securities-Category4/ni_20110624_43-101_mineral-projects.pdf

Coen JL, Schroeder W, Quayle B (2018) The generation and forecast of extreme winds during the origin and progression of the 2017 Tubbs Fire. Atmosphere 9(12):462

Copeland AH (1945) John von Neumann and Oskar Morgenstern, theory of games and economic behavior. Bull Am Math Soc 51(7):498–504

Cournot A (1838) Recherches sur les Principes Mathematiques de la Theorie de Richessess, Hachette, Paris; Eng. trans. Researches into the mathematical principle of the theory of wealth. Kelly, New York, 1960 de Ruiter MC (2020) Dynamics of Vulnerability: from single to multi-hazard risk across spatial scales. Ph.D. thesis, Vrije Universiteit Amsterdam, The Netherlands. ISBN: 978-94-028-2111-6

Ekeland I (1988) Mathematics and the unexpected. University of Chicago Press, Chicago

Fagan B (2019) The little ice age: how climate made history 1300–1850. Hachette, UK

Falcone G, Antics M, Baria R, Bayrante L, Conti P, Grant M, Hogarth R, Juiliusson E, Mijnlieff H, Nador A, Ussher G (2017) Application of the United Nations Framework Classification for Resources (UNFC) to geothermal energy resources-selected case studies, vol. 51. United Nations Publication

Galvez A, Carnelli I, Michel P, Cheng AF, Reed C, Ulamec S, Biele J, Zbrll P, Landis R (2013) AIDA: the asteroid impact and deflection assessment mission. In: European planetary science congress, EPSC abstracts, vol 8, 2013-1043. https://meetingorganizer.copernicus.org/EPSC2013/EPSC2013-1043.pdf

Gencer EA (2012) How to make cities more resilient. In: A handbook for local government leaders. United Nations Office for Disaster Risk Reduction

Gronewold N (2020) "Unprecedented" locust invasion approaches full-blown crisis. Scientific American, E&E News, 31 Jan 2020

International Council on Mining & Metals [ICMM] (2019) Adapting to a changing climate: implications for the mining and metals industry. Climate Change 2019. https://www.icmm.com/website/publications/pdfs/climate-change/191121_publication_climate_adaptation.pdf

International Panel on Climate Change [IPCC] (2012) Managing the risks of extreme events and disasters to advance climate change adaptation: special report of the Intergovernmental Panel on Climate Change. In: Field CB et al (eds) Cambridge University Press

Johnson C, Blackburn S (2014) Advocacy for urban resilience: UNISDR's making cities resilient campaign. Environ Urban 26(1):29

Kodra E, Ganguly AR (2014) Asymmetry of projected increases in extreme temperature distributions. Scientific reports 4:5884. https://www.nature.com/articles/srep05884

Luo L, Lan YC, Tang Q (2012) Corporate incentives to disclose carbon information: evidence from the CDP Global 500 report. J Int Fin Manage Acc 23(2):93–120

Maunder R, Hunter J, Vincent L, Bennett J, Peladeau N, Leszcz M, Sadavoy J, Verhaeghe LM, Steinberg R, Mazzulli T (2003) The immediate psychological and occupational impact of the 2003 SARS outbreak in a teaching hospital. CMAJ 168(10):1245–1251

Maupertuis PL (1744) Accord de différentes loix de la nature qui avoient jusqu'ici paru incompatibles. Académie Internationale d'Histoire des Sciences, Paris

Maynard T, Smith N, Gonzalez S (2013) Solar storm risk to the North American electric grid. Lloyd's

Medford LV, Lanzerotti LJ, Kraus JS, Maclennan CG (1989) Transatlantic earth potential variations during the March 1989 magnetic storms. Geophys Res Lett 16(10):1145–1148

Nash J (1950) Equilibrium points in n-person games. Proc Natl Acad Sci 36(1):48–49

Nash J (1951) Non-cooperative games. Ann Math Second Ser 54(2):286–295

Nehnevajsa J (1984) Low-probability/high-consequence risks: Issues in credibility and acceptance. In: Low-probability high-consequence risk analysis. Springer, pp 521–529

Petermann T, Bradke H, Lüllmann A, Poetzsch M, Riehm U (2011) Was bei einem Blackout geschieht. Folgen eines langandauernden und großräumigen Stromausfalls (Studien des Büros für Technikfolgen-Abschätzung beim Deutschen Bundestag, 33), Edition Sigma

Prno J, Slocombe DS (2012) Exploring the origins of 'social license to operate' in the mining sector: perspectives from governance and sustainability theories. Resources Policy 37(3):346–357

Rulis AM (1986) De minimis and the threshold of regulation. "De minimis and the threshold of regulation," Food protection technology. In: Felix C (ed) Proceedings of the 1986 conference for food protection. Lewis Publications, pp 29–37

South African Mineral Resources Committee [SAMREC] (2000) South African code for reporting of mineral resources and mineral reserves (the SAMREC code). South African Institute of Mining and Metallurgy

Skowronski DM, Astell C, Brunham RC, Low DE, Petric M, Roper RL, Talbot PJ, Tam T, Babiuk L (2005) Severe acute respiratory syndrome (SARS): a year in review. Annu Rev Med 56:357–381

Spearing LA, Faust KM (2020) Cascading system impacts of the 2018 Camp Fire in California: the interdependent provision of infrastructure services to displaced populations. Int J Disaster Risk Reduct 50:101822

Stephenson PR (2001) The JORC code. Appl Earth Sci 110(3):121–125

Thomas HE, Prata A (2011) Sulphur dioxide as a volcanic ash proxy during the April–May 2010 eruption of Eyjafjallajokull Volcano, Iceland. Atmos Chem Phys 11:6871–6880

United Nations Framework Convention on Climate Change [UNFCCC] (2007) Climate change: impacts, vulnerabilities and adaptation in developing countries. Bonn, Germany

Ventura G, Vilardo G, Sepe V (2009) Monitoring and structural significance of ground deformations at Campi Flegrei supervolcano (Italy) from the combined 2D and 3D analysis of PS-InSAR, geophysical, geological and structural data. In: 6th International symposium on digital earth

Witze A (2020) The Arctic is burning like never before—and that's bad news for climate change. Nature 585(7825):336–337

Part II
Divergent Exposures, the Public and Ethics

It is difficult to deal with divergence if there is no clarity on what differs between business-as-usual and divergence.

It is equally difficult, and unethical to discuss divergence without understanding what the public wants and what ethical decisions may be, how they relate to convergent risk assessments in a divergent environment.

This Part II aims at covering these points first by delving into the differences between business-as-usual and divergent events (Chap. 4), credible events and black swans and their relationship with standard levels of industrial mitigation (Sect. 4.1).

Attention is then shifted toward metaphoric description of risks (Sect. 4.2) that are useful when considering public reactions and a taxonomic description of uncertainties. Notions such as return period and force majeure are discussed (Sects. 4.3 and 4.4) to show how misleading, or plain erroneous, these notions can be in the face of divergent exposures and public expectations.

Chapter 5 discusses health, well-being and resiliency of business and for people (Sect. 5.2), followed by a review of what the contemporary public seems to want (Sect. 5.3), based on recent conferences and public hearings, and to a closure consisting of requirements of communication and transparency as well as considerations of ethics (Sect. 5.4).

Chapter 4
Business-as-Usual Versus Divergent Hazards

We start this chapter with a case history from Lao PDR (Anecdote 4.1) which illustrate how adverse conditions, but not necessarily divergent hazards, may lead to failure. The example also shows in its closing point why many countries have adopted the concepts of ALARA (as low as reasonably achievable), ALARP (as low as reasonably practicable), and BACT (best available control technology) (Wilson and Crouch 1982; see Sect. 4.1), in particular for hydroelectric dams and other infrastructural projects. We then discuss an interesting taxonomy of risk metaphors by German researchers based on uncertainties of likelihood and consequences (Sect. 4.2) and why using return periods is a disservice to the public (Sect. 4.3). We conclude with a discussion on force majeure contractual clauses (Sect. 4.4). A box with links to key terms is included in the references at the end of this chapter to facilitate the read.

Anecdote 4.1: Laotian hydroelectric dam collapse

A series of events starting on Sunday 22 July 2018 led a Laotian hydroelectric dam to collapse with disastrous consequences for the population downstream. The collapsed dam was part of the Xe Pian-Xe Namnoy hydropower project, still under construction. The project, which is located in the central Laotian Attapeu Province, involves Laotian, Thai and South Korean firms. Lao PDR is particularly rich in vertical drops with hydroelectric-generating potential. As a matter of fact, in 2017 the country had 46 operational hydroelectric power plants with 54 more under construction.

The feasibility study for this particular project was completed in November 2008, construction began in February 2013 and commercial operations were expected to begin in 2019 (Kim et al. 2019; Gong 2018; Oboni and Oboni 2020). The project includes two main dams, Xe Pian and Xe Namnoy, and five subsidiary dams. The dam that collapsed, Saddle Dam D was one of the subsidiary dams. The structure was 8 m wide at the crown, 770 m long and 16 m high. The purpose of the subsidiary dams of this project is to "help divert

---

F. Oboni and C. H. Oboni, *Convergent Leadership—Divergent Exposures*, https://doi.org/10.1007/978-3-030-74930-9_4

water around a local reservoir", as reported by the owner and delivered by local media.

After the Oroville Dam problems in 2017 (see Sect. 2.2.4), this was a catastrophic and devastating failure of a "minor" hydro structure (France et al. 2018). Are we seeing the development of a new trend in risk thanks to climate change? In 2018, a dam on the Nam Ao River, also under construction, failed as well, although in that case with no casualties.

Precipitation in the region is the lowest in January, with an average of 3 mm while most precipitation falls in August, with an average of 566 mm. According to a public news broadcast in February 2018 the Philippines, Vietnam, Cambodia and Laos were expected to face the *greatest threats from tropical storms and typhoons* (https://www.yahoo.com/news/2018-asia-spring-forecast-tropical-131525402.html?guccounter=2), including flooding rainfall and/or damaging winds and seas. Indeed, Tropical Storm Son Tinh on 18–19 July 2018 caused flooding in various provinces. As a result, heavy rainfall in Attapeu Province resulted in dangerously *high river and dam levels* (https://reliefweb.int/report/lao-peoples-democratic-republic/lao-pdr-flooding-office-un-resident-coordinator-situation).

The main Thai stakeholder declared that the *dam "was fractured"* (https://www.rnz.co.nz/news/world/362571/laos-dam-collapse-many-feared-dead-as-floods-hit-villages) after "continuous rainstorm[s]" caused a "high volume of water to flow into the project's reservoir". The consequences were unfortunately of tragic proportion. Indeed, the accident resulted in the flooding of eight villages. The Government declared the affected areas as National Disaster Area. Response activities were underway for several days and UN agencies also continued to gather information and assess the impacts. The collapse of the dam affected nearly *7000 people and displaced more than a thousand* (https://www.nytimes.com/2018/07/24/world/asia/laos-dam-collapse-hundreds-missing.html). Indeed, the failure damaged more than hundred houses, forcing people to seek shelter in local government buildings and schools. Red Cross teams in the Attepeu branch distributed *clothing, food and drinking water* (https://reliefweb.int/report/lao-peoples-democratic-rep ublic/lao-dam-collapse-floods-nearby-towns-and-affects-thousands) to households in the affected area.

Our task here is not to discuss responsibilities or faults, but rather to point out a few risk-related concepts. Our intention is to foster discussions that may prevent "the next one".

As stated above, Lao PDR sees frequent tropical storms and typhoons. At this time, we do not know if there is a climate change trend developing under the form of a true divergence, but these events are far from a surprise. NB: the difference between tropical storm and typhoons lies in the wind speed. Typhoons in Lao PDR are not direct hazard, because their force is normally diminished (reaching tropical storm level) once they have reached Lao PDR from the South China Sea. However, they can produce flooding as a consequence of heavy rainfall. *Up to three typhoons hit the country annually* (https://www.unisdr.org/files/33988_countryassessmentreportlaopdr%5B1%5D.pdf).

We do not know how many failures there might have been in the country over its history, or how many dam-years the country has. However, if we consider a round number of 50 hydroelectric projects under construction, this time we had a failure rate of 1/50. Combining that, for example, with the frequency rate of a major tropical storm of, let's say, 1/10, we have a probability of failure during construction $1/10 * 1/50 = 2 \times 10^{-3}$. If we look at the historical failure rate of hydroelectric dams, 10^{-4} to 10^{-5}, the gap is very significant!

Closing point. What seems clear is that the chances of failure of hydro projects under construction in Lao PDR were way higher than world-wide averages. A meteorological alert system and planning would have helped and maintaining emergency access to the operation would have had a positive impact. Will we see a significant policy change in Lao PDR in the aftermath of this catastrophe? That seem to be the case in most countries hit by nefarious events of this magnitude, leading to the promotion of the concepts of ALARA, ALARP and BACT.

4.1 Credible Events and Standard Levels of Mitigation

As discussed earlier, credible events have probabilities of occurrence larger than Acts of God, significantly larger if they are business-as-usual. Depending on the value of their probability credible events border with the limiting value of force majeure. We can also say that lack of preparation (contingency plans, maintenance, mitigation) for a credible event may be a fault, even a negligence.

Three standard levels of mitigation for credible events are commonly defined in our world: ALARA, ALARP, and BACT (Fig. 4.1). Unfortunately, these three levels of risk mitigation also represent a convenient way to avoid explicit tackling of risk tolerance (see Chap. 10), especially when dealing with the delicate theme of human life. However, we can also see these standardized levels of risk reduction as a definition of state-of-the-art practice. What is the risk in using these standardized levels of mitigation?

ALARA requires immediate recalibration after the occurrence of a divergent exposure. It requires a re-evaluation if hazards diverge from usual patterns: is this a new normal that requires the limit to be updated? ALARP and BACT focus on the practicality of the defenses and the control technology, so they are neutral with regard to possible divergent exposures unless new practical solutions and control technologies come to light. It is vital to understand which concept to implement if divergent exposures are possible. In recent years public opinion tends to consider negligent

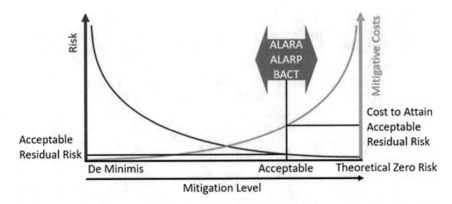

Fig. 4.1 Decrease of risk against increasing mitigative costs. The vertical line is set in the ALARP/ALARP/BACT zone. The position of the line, whose theoretical optimum would be at the intersection of the green and red curves, tends to slide toward the right, away from the optimum, due to the pressure of public opinion

even mitigative levels below these limits. The result is severe criticism even when corporations, governments follow reasonable mitigative behavior. Is there a defense against accusations of negligence? A partial answer may be using the so-called negligence test (Kumamoto and Henley 1996). But before explaining it let's go back to the definition of risk. In its simplest form (this formulation is a verbal extension of Eqs. 1.1 and 2.1):

$$\text{Risk} = \text{p of a hazard occurring} \times \text{C of consequences of the hazard hitting} \quad (4.1)$$

A company-wide risk can be expressed as the aggregation of individual risks. However, such an aggregation may result in unfortunate misjudgments in terms of selecting where to invest mitigative funds (NIST 2012). Unfortunately, the formulation in Eq. 4.1 (and similarly Eqs. 1.1 and 2.1) has indeed the disadvantage of being unable to distinguish between high-frequency, low-impact events and low-frequency, high-impact events. Later (Chap. 12) we will show how to solve this problem, as in many situations the former may be tolerable while the latter may be catastrophic. Ignoring the distinction above may indeed also put the company in a hazardous legal condition because tort law often uses the somewhat vague "reasonable man" standard to judge liability of negligence. The negligence of the injurer (company) is determined by relations between the probability of injurious event (p), the consequences of the resulting injury (C), and the burden, or cost, of adequate precautions or mitigations (M); that means the injurer is only negligent if M is less that the product of p and C (Eq. 4.2):

$$\text{Negligence} = M < p * C \quad (4.2)$$

In other words, a judge may deem a company negligent only if mitigative moneys M spent (per annum) are less than the annualized risks. If the distinction discussed above is ignored the negligence test may be in contradiction with good sense and not reasonable. Thus the legal negligence test is not a critical test for an operation while, clearly, transparency and rationality constitute a strong a priori defense in case something would go wrong. Let's look at two summarized examples to show how limited the concept of legal negligence test is, when confronted to public reactions.

In the first example, an industrial process system in Canada was assessed with a likely annualized probability of failure $p = 10^{-5}$. After an interview with the personnel, we defined a cost of consequences within a 95% confidence level to $C = 10$ M CAD. By strict application of the negligence criteria, the company would be negligent if and only if they spend less than $p * C = 10^7 \times 10^{-5} = 100$ CAD per annum for mitigative measures! Of course, the company would be the object of intense media and regulatory scrutiny should an accident occur. Even if the company were to spend way above the threshold value, it is difficult to imagine it would emerge unscathed from such a failure, proving the compliance with the legal negligence test is not sufficient.

In the second example, in the Andes a bus fleet shuttles the employees of a natural resources company to various locations. After ten years of regular services, a tragic accident took two lives. As management grew afraid of a crisis potentially ending in massive strikes, they asked us to study possible alternatives to the system.

The review of existing road safety measures revealed that 1 M CAD had already been invested in additional guard rails and buses were preceded by a patrol car, requiring an annual operational budget of 150,000 CAD per year. For ten years of operation it can then be approximatively evaluated that the company was spending $M = 1,000,000/10 + 150,000 = 250$ K CAD per year in mitigative measures.

Now, let's assume that the annualized probability of an accident is $p = 10^{-3}$, an extremely high value due to the local hazardous conditions. For comparison, in France or Switzerland, where strong data sets are available, the value would be at least two orders of magnitude lower. Having M and p, we can evaluate $C = M/p = 250,000/10^{-3} = 250$ M CAD. The company would therefore be considered negligent if and only if the cost of consequences associated with an accident would overcome 250 M CAD. As any accident scenario involving the shuttle buses would not reach that staggering value the company would not be negligent in the eye of the law. In the eye of the public and employees, however, it would again most likely be a completely different story.

The negligence test is not an end, but only the start of a continuous process of improvement for an operation's safety, health and risk and crisis management.

4.2 German Metaphors for Risks

In the mid-2000, German researchers in the field of theoretical risk management developed a series of metaphors to describe public perception of risks (Renn and

Table 4.1 This Table is identical to Table 2.1, copied here to facilitate reference

Knowledge level	Indicative examples of uncertainty sources	Typical approaches or considerations for parameters evaluations
Unpredictability	Projections of human behavior not easily amenable to prediction (e.g. evolution of political systems). Chaotic components of complex systems	Use of scenarios spanning a plausible range, clearly stating assumptions, limits considered, and subjective judgments. Ranges derived from ensembles of model runs
Structural uncertainty	Inadequate models, incomplete or competing conceptual frameworks, lack of agreement on model structure, ambiguous system boundaries or definitions, significant processes or relationships wrongly specified or not considered	Specify assumptions and system definitions clearly, compare models with observations for a range of conditions, assess maturity of the underlying science and degree to which understanding is based on fundamental concepts tested in other areas
Value uncertainty	Missing, inaccurate or non-representative data, inappropriate spatial or temporal resolution, poorly known or changing model parameters	Analysis of statistical properties of sets of values (observations, model ensemble results, etc.); bootstrap and hierarchical statistical tests; comparison of models with observations

Knowledge level versus sources of uncertainty and typical approaches or considerations for parameters evaluations

Klinke 2004). They are summarized below, with examples showing their relation to knowledge level, public fears and possibly panic. We have sorted the metaphors according to increasing levels of predictability and foreseeability and in addition we have used the levels of knowledge defined in Table 4.1 to compare them.

Pandora's Box risks are characterized by simultaneous uncertainty in the probability of occurrence of the generating event and the resulting extent of damage (only presumptions), coupled with high persistency. Persistent releases of organic pollutants and biosystem-changing endocrine disruptors can be cited as examples of Pandora's Box. Great uncertainty and presumptions indicate low predictability (Table 4.1: unpredictable) and foreseeability. A Pandora's Box can become a black swan if the damages are significant.

Cyclops risks are those where it is only possible to ascertain either the probability of occurrence or the extent of damage while the other side remains uncertain. In this risk class, the probability of occurrence is largely uncertain whereas the maximum damage can be estimated. A number of natural events such as volcanic eruptions (Vesuvius) and earthquakes (various "large ones" such as those in San Francisco, Tokyo, etc.) belong to this category. Predictability may be limited in some cases, but damages are foreseeable (Table 4.1: generally structural uncertainty), depending on the situation. Divergence is possible, but generally not black swans.

Pythias includes risks associated with the possibility of sudden non-linear climate changes, such as the risk of self-reinforcing global warming or the instability of the West Antarctic ice sheet, with far more disastrous consequences than those of gradual climate change. The extent of damage is unknown and the probability of occurrence cannot be ascertained with any accuracy (Table 4.1: unpredictable). Although accuracy may in general be low, models with a certain degree of predictability and foreseeability do exist and thus divergent exposures, not black swans, are expected.

Cassandra risks are characterized by a relatively lengthy delay between the triggering event (for example, nuclear radiation exposure below a certain critical threshold) and the occurrence of damage. This case is naturally only of interest if both the probability and magnitude of damage are relatively high. If the time interval were shorter, pertinent regulatory authorities would certainly intervene, because the risks would be perceived as intolerable. However, the time gap between trigger and consequence creates the fallacious impression of safety. No divergence, no black swans. Difficult to relate to Table 4.1.

Sword of Damocles risks indicate an allusion to the existence of imminent and ever-present peril. They have very high disaster potential, although the probability that this potential manifests itself is theoretically extremely low. Accidents involving nuclear power plants (Fukushima), large-scale chemical facilities (Bophal, Seveso), hydroelectric dams and catastrophic meteorite impacts are typical examples. Relatively good foreseeability and predictability (Table 4.1: value uncertainty), no divergence and generally no black swans.

Medusa risks refer to the potential for public mobilization. This criterion expresses the extent of individual aversion to risk and the political protest potentially fueled by this aversion. Both of these are triggered among the lay public when certain risks are taken without public consent or unsuccessful communication. This risk class is only of interest if there is a particularly large gap between lay risk perceptions and expert risk analysis findings. Some innovations are rejected although they are hardly assessed scientifically as threat (cell phones, high voltage lines, etc.). No divergence, no black swans.

In Table 4.2 we have attempted to relate knowledge level, degree of uncertainty, main criteria, the German metaphors, and the presence of divergence or black swans. Foreseeability and predictability generally increase from top to bottom (when applicable; for example, the Medusa metaphor, which looks at public mobilization, does not really apply).

Now, in our societies we often see different accident types with identical or very similar single-accident direct consequences but very different global impacts ending up generating surprisingly different public reactions. We think the German metaphors help explaining why as discussed below.

Although quite theoretical and apparently abstract, the Renn and Klinke classification has the merit of explicitly using the level of uncertainties on probability and/or consequences for the definition of risks as well as the concept of latency. The concept of latency is very important when dealing with complex systems, where errors and omissions and decisions made at the project's inception may indeed generate failures decades into the system's service life.

Table 4.2 Relationship between knowledge level, uncertainty, criteria, metaphor, and the possible existence of divergence or black swan (B-S)

Knowledge level	Degree of uncertainty	Main criteria	Metaphors	Divergence/B-S
Minimal	Ignorance	Probability of occurrence and extent of damage are highly unknown to science	Pandora's box	Possible B-S
Fair	Uncertainty	Probability of occurrence or extent of damage or both are uncertain (because of natural variations or genuine stochastic relationships)	Cyclops	Div. generally not B-S
			Pythias	Div. not B-S
			Cassandra	No div. no B-S
High	Known distribution of probabilities and corresponding damages	Probability of occurrence and extent of damage are known	Sword of Damocles	No div. generally not B-S
			Medusa	No div no B-S

Let's now work through examples presented in earlier chapters to see if we can find some explanations and use for the German metaphoric definitions. Table 4.3 summarizes these selected examples.

As Table 4.3 shows, all selected examples can be identified with a German metaphor. No case applies to the Pandora Box classification, as not a single example studied in this book met the condition of ignorance attached to it. All the cases, had they been discussed and developed as possible scenarios in a risk approach, would have led to the classification with one of the metaphors, which would then have determined, at very early stage, the roadmap to increase knowledge and reduce the existing uncertainties as far as possible and reasonable. Yes, the German metaphoric definitions are useful, both before and after the fact!

4.3   Talking About Return Period is a Disservice to the Public

Talking about return period is a disservice to the public, even if flood experts usually use the term when communicating to the public. Indeed, let's not forget returns periods are based on long term observations of business-as-usual.

Table 4.3 Selected examples with their predictability and foreseeability, possible divergence, black swan and their German metaphors (Renn and Klinke 2004)

Example case	Predictable likelihood	Foreseeable losses	Divergent	Black Swan	Systemic risks classification (Renn and Klinke 2004)
Houston, Toronto (Sect. 2.2.3)	Yes	Yes	YES	NO	Pythia
Commuter trains NY (Sect. 2.2.1)	Yes	Yes	NO	NO	Medusa
Oroville dam (Sect. 2.2.4)	Yes	Yes	NO	NO	Cyclop or Cassandra
Fukushima (Sect. 1.1)	Yes	Yes	NO	NO	Cassandra
Present Pandemic (Sect. 3.2.3)	Yes	No, due to interdependent reactions at planetary scale	NO	NO	Pythia
Planetary volcanic winter (Sect. 3.1.2)	Yes, but probabilities in the act of God range	Yes, catastrophic at planetary scale	NO	YES	Damocles
Swarm of equally strong seismic events (Technical note 3.1)	No and dependent on the primary quake	No, due to interdependent failures, wounded systems, etc.	YES	YES	Damocles

Experts use terms such as 1000-year flood or 100-year flood when designing building and zoning codes. Generally, they also indicate specifications for structures to withstand certain events (see the Houston example, Sect. 2.2). What does that really mean? Of course, we can compute the likelihood of a certain event per year and then talk about return period using a Poisson probability (see Sect. 8.1.3).

Often we hear people saying that since we saw a 100-year flood last year we are good for 99 years to come. This statement is completely absurd. From a probabilities point of view a 100-year flood magnitude event can happen two years in a row (or even twice in the same year, as we have repeatedly shown with examples!). Indeed, the frequency—1/100—is by definition a long-term average that indicates a central tendency but not the whole truth about a phenomenon.

In business-as-usual conditions, the likelihood of seeing a 1/100 event happen three years in a row exists as well, even though it is low, almost beyond the credible threshold. But it can happen with climate change disruptions. Above we indicated that based on long-term average occurrences (which is derived from gathered data) we can compute the probability of occurrence linked to an event. But what does it mean if the sample size is small or when the future diverges from the past? Climate change may be an example of this, but it is not the only one. Officials are starting to understand that what once was true might never be true again. The federal government of Canada warned that this type of extreme flooding, together with other costly phenomena is the new reality of climate change, and that Canadians should prepare to see it more often. Environment Minister Catherine McKenna said that the "one in 100-year flood" is indeed happening much more frequently: "This flooding is happening here in Quebec, it's happening in Ontario, it's happening in New Brunswick. And really sadly, what we thought was one-in-100-year floods are now *happening every five years, in this case, every two years*" (https://globalnews.ca/tag/100-year-flood-inc reasing/). Recent reports by *CICC* (https://climatechoices.ca/climate-impacts-are-getting-worse-canada-must-adapt/) (2020) describe how the combined losses per weather-related disaster have ballooned from an average of 8.3 M CAD million per event in the 1970s to an average of 112 M in the 2010s, including government and some private costs, a 13-fold increase.

Talking about probabilities per year as opposed to return period is the easy first step in presenting the probabilities of events to the public in a more transparent way. A single-number estimate will always be wrong whereas a range of annualized probabilities is more ethical and correct (see Sect. 8.1.4).

4.4   The Force Majeure Myths

In many cases, in the aftermath of a catastrophic event, all parties may want to resume operations and rebound. Finding a win/win situation rather than losing time in force majeure litigation will speed up the rebound. In addition, many of our clients are fully aware that no business interruption (BI) insurance will cover the market share they lose if they are not servicing their clientele around the world. The most difficult cases arise precisely when one party to the contract claims performance is impossible and seeks to apply force majeure clauses, while the opposing party generally maintains that performance is possible but just at a greater expense to the first party.

It is of critical importance for business management and survival to spell out what force majeure is. That requires the definition of what is predictable versus what is unpredictable. That goes through the definition of the threshold values of probability and magnitude. It is not chatting about it and putting together some ad hoc decision-making committees that will take decision based on past experience and gut feelings!

Why are we talking about force majeure now? Simply because, due to divergences, force majeure will become a very critical issue in the short and medium terms for any business entity.

At first sight it may seem like FM clauses are written for the benefit of one side only. They may, for example, allow insurers to avoid paying. Indeed, some homeowner policies state that wind shear is an act of nature but a tornado is an act of God and is therefore not covered. FM clauses may be used as a way to exit a contract. An audacious yet inventive attempt at utilizing the inherent ambiguity of FM occurred in the US, when Donald Trump argued that the 2008 financial crisis qualified as a force majeure event, thus excusing him from making payments on a real estate loan (Trump, Donald et al. vs. Deutsche Bank Trust et al., New York Supreme Court, Queens, 026841/2008).

A brief parenthesis on germs, shipping and force majeure seems appropriate at this point, given the evolution of the COVID-19 pandemic and our experience. Indeed, clients have repeatedly asked us to perform risk assessments and BI quantifications of harbors, wharves, and ship loaders-unloaders (see Part V). These facilities are historically the first to be hit by quarantine orders. Indeed, at the very beginning of the COVID-19 pandemic we read that Chinese ports were shut down before anything else.

If we are looking to quantify the BI of a shipping facility, we have to break the informational silo, because looking at categories of hazards one by one limits the resulting understanding of the risk landscape. Indeed, cyber hazards and natural and man-made hazards of various types (including sabotage and terrorism), which can hit infrastructure, logistic networks and, of course, ships, require convergent evaluations.

Let's discuss at what point, if any, the BI of the shipping facility will become a force majeure scenario. In addition, we will also look at the difference between business-as-usual hazard scenarios and runaway divergent ones. The scenario of an epidemic creating BI is way within the credibility realm insofar as humanity has experienced numerous epidemics and pandemics. In addition, closing access to afflicted areas is a common and useful strategy.

Cordoning off dangerous areas and groups of people began centuries ago to contain leprosy and plague, for example in Lyon in 1583 and in China around the year 600. In 1348, Venice, it became a well-structured process to protect coastal cities from plague: ships arriving from infected ports waited at anchor for 40 days before landing (the practice is called "quarantine" from the Italian *quaranta giorni*, literally forty 40 days). Over time, thanks to antibiotics and routine vaccinations, the likelihood of large-scale quarantines has been reduced significantly. However, today the potential threats of bioterrorism and diseases like SARS and COVID have resurrected this custom. In the case of COVID-19, we have seen quarantines deployed at scales not seen for quite a long time.

Actually, framing the likelihood of a quarantine is rather easy. Although epidemics of a certain scale are quite common, of a frequency rate between 1/20 and 1/50, we can estimate major ones with widespread quarantines as follows. From a *site gathering historic data* (https://web.archive.org/web/20200303184702/https://www.his tory.com/topics/middle-ages/pandemics-timeline) we count ten occurrences worldwide in a period of roughly 400 years, leading to an estimate of f = 1/40, mostly in harbor cities and logistic/commerce/business hubs (we have excluded from our count two blatantly politically or racially motivated cases of quarantine).

So, now the question: Is quarantine a force majeure case? Amazingly analysts oftentimes disregard quarantine, a rather predictable hazard as shown above, because of the information silo syndrome. Are the consequences of the predictable event of a large-scale quarantine foreseeable? Can we foresee the damage generated by a (predicted) hazard hit based on present or future mitigation and policies/actions? The answer is yes. We can foresee what a quarantine will generate by looking at BI and the interdependencies within and outside the system, as well as to the potential ripple effects (see Technical note 9.2). The evaluated consequences will of course depend on the entity requiring the evaluation: is it a shipping company, the government of the afflicted area, etc.? As usual, it would be misleading to evaluate consequences as "one number". Multiple scenarios are necessary. So, in conclusion: quarantine is not a force majeure case, unless it becomes extremely long and the consequences can be considered as unforeseeable.

Business-as-usual hazards, including their usual extremes, are predictable and foreseeable within a relative narrow margin of uncertainty. That is, of course, provided interdependencies and systemic amplification are carefully considered. Temperature variations and rains belong to this category including centennial, bi-centennial, and one-in-a-thousand extremes. Divergent hazards are unpredictable insofar their frequency (at a given magnitude level) explodes: for example, say *two 1000-year floods in two years* (https://www.forbes.com/sites/ericmack/2019/09/19/ imelda-brings-texas-its-second-1000-year-flood-in-two-years/?sh=1c04b5e77f1e), or *three 500-year floods in three years* (https://www.washingtonpost.com/gdpr-con sent/?next_url=https%3a%2f%2fwww.washingtonpost.com%2fnews%2fwonk% 2fwp%2f2017%2f08%2f29%2fhouston-is-experiencing-its-third-500-year-flood- in-3-years-how-is-that-possible%2f). An example of this are the storms in Houston we discussed in Sect. 2.2 (see Technical note 2.2).

Foreseeability may also reduce, as wounded systems become gradually less robust and consequences may be amplified in unforeseeable ways. Think, for example, about massive wildfires leading to large scale erosion, slope failures, etc., in case of rain.

Our risk assessments routinely include usual and divergent scenarios, often with multiple sub-scenarios to deal with climate change, political changes (tariffs), etc.

Force majeure clauses in contracts should always be optimized to reduce costs and the potential of litigation. Indeed, any time spent in the aftermath of a mishap discussing whether the event was caused by force majeure, negligence or other causes severely impacts operations. In addition, it can significantly increase the costs of consequences. Thus, it is necessary to optimize the force majeure formulation. If it is possible to renegotiate contracts in the future, or for any new contract, a proper force majeure formulation would constitute an important proactive mitigative measure, one with a very large Return On Investment (ROI).

There are numerous areas where optimization could take place, for example in the form of:

- a more detailed explanation of terms;
- definition of threshold values;

- enhanced definition of the mitigative levels that industry considers as common practices or best practices;
- a definition of negligence;
- the establishment of a limiting value of force majeure.

The risk analysts working in this area should provide the lawyers drafting the contract with specialized technical support based on risk-informed decision making (RIDM) concepts (see Sect. 12.2). This allows for the transparent definition of terms such as reasonable precautions/alternatives, and due diligence in running an operation, maintaining it, being proactive to ensure that contractual performance criteria are met, etc.

Again, we recommend the use of a register of potential hazards which possibly lead to a force majeure event. The register may be either generic or detailed, but should be homogeneous and balanced. Furthermore, an optimized force majeure clause formulation should encompass likelihood and/or probability and/or frequency rates versus magnitude limits for ordinary versus act-of-God events. We need a very clear definition of what constitutes an act of God in order to optimize the force majeure clause. We can illustrate this with the example of a commercial contract for a facility in Salt Lake City. Before 1999, one could have said that a tornado in Salt Lake City was an act of God. Indeed, scientific consensus was that a tornado was a non-credible, impossible event there. That consensus broke up when one tornado occurred on 11 August 1999. From that point on, a tornado in Salt Lake City became a rare, but credible event.

Of course, at this point terms such as credible have to be clarified (see Sect. 4.1). Indeed, there is a definite link between their definition and the limiting value of force majeure. We can also say that lack of preparation (contingency plans, maintenance, mitigation) for a credible event is a fault, even a negligence, as we discussed in Sect. 4.1.

Appendix

Links to more information about the Key terms from the Authors	
A, B	*Act of God* (https://www.riskope.com/2020/12/09/act-of-god-in-probabilistic-risk-assessment/) *Black swan* (https://www.riskope.com/2011/06/14/black-swan-mania-using-buzzwords-can-be-a-dangerous-habit/) *Business-as-usual* (https://www.riskope.com/2021/01/13/business-as-usual-definition-in-risk-assessment/)

(continued)

(continued)

Links to more information about the Key terms from the Authors	
C, D	*Convergent* (https://www.riskope.com/2021/01/20/convergent-risk-assessments/) *Divergent* (https://www.riskope.com/2020/11/18/tactical-and-strategic-planning-to-mitigate-divergent-events/) *Drillable* (https://www.riskope.com/2020/01/15/probability-impact-graphs-do-not-fly/)
F	*Foreseeability/foreseeable* (https://www.riskope.com/2021/01/06/foreseeability-and-predictability-in-risk-assessments/) *Fragile/fragility* (https://www.riskope.com/2020/04/01/antifragile-resilient-solutions-for-tactical-and-strategic-planning/)
P, R	*Predictability/predictable* (https://www.riskope.com/2021/01/06/foreseeability-and-predictability-in-risk-assessments/) *Resilient, Resilience* (https://www.riskope.com/2016/11/23/resilience-cannot-based-instinctual-decision-making/)
S	*Scalable* (https://www.riskope.com/2015/04/16/how-system-definition-and-interdependencies-allow-transparent-and-scalable-risk-assessments/) *Societal risk acceptability* (https://www.riskope.com/2014/01/09/aspects-of-risk-tolerance-manageable-vs-unmanageable-risks-in-relation-to-critical-decisions-perpetuity-projects-public-opposition/) *Sustainability/sustainable* (https://www.riskope.com/2019/01/16/improving-sustainability-through-reasonable-risk-and-crisis-management/) *Survivability* (https://www.riskope.com/2011/03/17/ale-fmea-fmeca-qualitative-methods-is-it-really-what-we-need/) *System* (https://www.riskope.com/2017/07/26/three-ways-to-enhancing-your-risk-registers/)
T, U	*Tolerance* (https://www.riskope.com/2020/04/29/risk-tolerance-thresholds/) *Uncertainty/uncertainties* (https://www.riskope.com/2015/12/10/3-decision-making-truths-derived-from-uncertainty-taxonomy-scheme-of-classification-and-a-road-sign/) *Updatable* (https://www.riskope.com/2020/01/07/climate-adaptation-and-risk-assessment/)

Other linked information (https://www.riskope.com/blog-news/) search Riskope blog and use the search box

Third Parties links in this section:

February 2018 Asia typhoon	https://www.yahoo.com/news/2018-asia-spring-forecast-tropical-131525402.html?guccounter=2
Attapeu floods	https://reliefweb.int/report/lao-peoples-democratic-republic/lao-pdr-flooding-office-un-resident-coordinator-situation
Lao PDR dam collapse	https://www.rnz.co.nz/news/world/362571/laos-dam-collapse-many-feared-dead-as-floods-hit-villages
Lao PDR victims of flooding	https://www.nytimes.com/2018/07/24/world/asia/laos-dam-collapse-hundreds-missing.html
Lao PDR human impact	https://reliefweb.int/report/lao-peoples-democratic-republic/lao-dam-collapse-floods-nearby-towns-and-affects-thousands
Lao PDR country assessment	https://www.unisdr.org/files/33988_countryassesmentreportlaopdr%5B1%5D.pdf
100 years flood increasing	https://globalnews.ca/tag/100-year-flood-increasing/
Climate impacts in Canada	https://climatechoices.ca/climate-impacts-are-getting-worse-canada-must-adapt/
Historic data on pandemics	https://web.archive.org/web/20200303184702/ https://www.history.com/topics/middle-ages/pandemics-timeline
Texas 1000 years floods in two years	https://www.forbes.com/sites/ericmack/2019/09/19/imelda-brings-texas-its-second-1000-year-flood-in-two-years/?sh=1c04b5e77f1e
Houston 3 500 years floods in three years	https://www.washingtonpost.com/gdpr-consent/?next_url=https%3a%2f%2fwww.washingtonpost.com%2fnews%2fwonk%2fwp%2f2017%2f08%2f29%2fhouston-is-experiencing-its-third-500-year-flood-in-3-years-how-is-that-possible%2f

References

[CICC] Canadian Institute for Climate Choices (2020) Tip of the Iceberg: navigating the known and unknown costs of climate change for Canada. https://climatechoices.ca/wp-content/uploads/2020/12/Tip-of-the-Iceberg-_-CoCC_-Institute_-Full.pdf

France JW, Alvi I, Dickson P, Falvey H, Rigbey S, Trojanowski J (2018) Independent Forensic Team Report: Oroville Dam Spillway Incident. California Institution of Water Resources, Riverside, CA. https://cawaterlibrary.net/wp-content/uploads/2018/03/Independent-Forensic-Team-Report-Final-01-05-18.pdf

Gong L (2018) Laos Dam Collapse: The Regional Response. RISI Commentary. https://think-asia.org/bitstream/handle/11540/8699/CO18139.pdf?sequence=1

Kim Y, Lee M, Lee S (2019) Detection of change in water system due to collapse of Laos Xe pian-Xe namnoy dam using KOMPSAT-5 satellites. Korean J Remote Sens 35(6_4):1417–1424

Kumamoto H, Henley EJ (1996) Probabilistic risk assessment and management for engineers and scientists, 2nd edn. IEEE Press

[NIST] Initiative, J.T.F.T. (2012) Guide for conducting risk assessments. National Institute of Standards and Technology, Gaithersburg

Oboni F, Oboni C (2020) Tailings dam management for the twenty-first century. Springer

Renn O, Klinke A (2004) Systemic risks: a new challenge for risk management: as risk analysis and risk management get increasingly caught up in political debates, a new way of looking at and defining the risks of modern technologies becomes necessary. EMBO Rep 5(S1):S41–S46

Wilson AC, Crouch E (1982) Risk/benefit analysis. Ballinger Publishing Company

Chapter 5
Corporate Risks and Exposures Versus the Public's Wants and Reactions

This chapter starts by reviewing corporate risks and exposures with examples from the world of insurance, then delves into health and well-being of business and people, and what the public wants when confronted with risks. Communication of risk to the public, transparency and ethics are discussed.

A box with links to key terms is included in the references at the end of this chapter to facilitate the read.

5.1 Corporate Risks and Exposures

Every medal has two faces. In insurance there is the insured and the insurer. The insured is an entity that wants their risks transferred to the insurer, who in his turn agrees to bear that risk for a fee. An old rule of thumb says that if an insured event has a probability around 1% or higher, it should not be insured because the insurer will lose money. In order to avoid losing money on existing contracts insurers have one way: deny insurance. Anecdote 5.1 shows an example of insurance denial. Quantitative Risk Assessments (QRA) can help solving some conundrums and finding alternative ways to deal with corporate and "private" risks.

> **Anecdote 5.1: Fires in Australia and Canada**
> We explore how insurance denial amid climate change will impact assets and other aspects of business. In particular, we think at fires like the *Australian bushfires* (https://www.cnbc.com/2020/01/06/australian-bush-fire-could-aff ect-consumer-confidence-says-economist.html) and *Fort McMurray (Canada) fire* (https://globalnews.ca/news/3138183/fort-mcmurray-wildfire-named-can adas-news-story-of-2016/) (Fig. 5.1) in addition to increasing *rains leading to flooding* (https://www.bbc.com/news/world-latin-america-51254669). These

F. Oboni and C. H. Oboni, *Convergent Leadership—Divergent Exposures*,
https://doi.org/10.1007/978-3-030-74930-9_5

Fig. 5.1 Fort Mc Murray Fire, Hilltop Estates. Notice the disparity of fire damage across the property. Satellite image © 2021 Maxar Technologies

will undeniably be a 'big drag' on growth and have already generated invaluable *losses to biodiversity* (https://www.sciencedirect.com/science/article/pii/S0960982215003942). Furthermore, insurance denial will become a problem as we *discussed back in 2016* (https://www.riskope.com/2016/07/07/geo hazards-probabilities-frequencies-and-insurance-denial/) and even *as far as 2009* (https://www.riskope.com/2009/09/08/denial-of-insurance-coverage-plagues-mining-industry-developments-world-wide/). Climate change will likely create areas of uninsurable assets in many locations, or even regions around the world. Indeed, some part of Australia will likely share the same fate (Thyagarajan 2014; van Oldenborgh et al. 2020) as Fort MacMurray from an insurance point of view.

Lately Canadians are starting to understand the deep implications of insurance denial amid climate change. For example, in Fort MacMurray Canada fires in May 2016 consumed ten per cent of the buildings and forced 88,000 people from their homes for at least a month. As a result many *buildings were denied renewal* (https://www.cbc.ca/news/canada/edmonton/fort-mcm urray-condominium-insurance-1.5318750) by their insurers, and hundreds of claims remained unresolved for quite a long time. Climate change will likely create areas of uninsurable assets in many locations, or even regions around the world. Indeed, some part of Australia will likely share the same fate as Fort MacMurray from an insurance point of view.

Flood maps that are outdated and therefore likely misleading, compounded by rising sea levels and more powerful storms, will create similar problems for many cities and regions. Indeed, they fail to alert home and business owners about their true exposure to hazards as climate change heightens, for example, threats to floodplains and coastal areas. For instance *2019 saw a record-breaking 12 months of precipitation* (https://www.noaa.gov/news/us-has-its-wettest-12-months-on-record-again) over the continental U.S: three times, i.e. in April, May and June 2019 the precipitation record hit an all-time high. Because of these extremes insurance premiums may fail to reflect true risks to properties. Thus, insurance companies might run out of funds or even go under. Insurers may be tempted to deny insurance due to these extreme events. However, insurance denial is not a good solution when your business is to insure assets! Thus, insurers need to find balanced solutions to limit insurance denial amid climate change. Quantitative risk assessment will guide insurers to make risk informed decision. Quantitative risk assessment on assets will bring insight that government risk map will fail to provide. They will help insurers as well as their clients to find reasonable negotiated solutions.

Closing point. If you are an owner or an insurer and need to understand what your risk profile looks like, quantitative risk assessment will shed light on your owned assets exposures. This will help you to make better decisions, risk informed decisions (see Chap. 12).

When we read consultancies' reports entitled, for example, *"Tracking the risk trends, the top 10 issues companies will face in the coming year"* (https://www2.deloitte.com/content/dam/Deloitte/global/Documents/Energy-and-Resources/gx-er-tracking-the-trends-2019.pdf), we are always amazed by the fact that these reflect opinions from executives, with no justification; at best, they deliver a view on their perception of emerging exposures.

Table 5.1 summarized the issues highlighted in one of these reports together with our comments and the German metaphoric interpretation (see Sect. 4.2).

However, the list above seems to forget how complicated relationships with insurers may be. For instance, Swiss Re Ltd. will no longer provide insurance or reinsurance to businesses with more than 30% *thermal coal exposure* (https://www.swissre.com/dam/jcr:6697586a-4fb9-4d58-a7f4-5d50f2c64 86f/nr-20180702-swiss-re-establishes-thermal-coal-policy-en.pdf). The new policy applies to existing and new thermal coal mines and power plants, and covers all lines of business and Swiss Re's global operations. We have also seen cases of insurance denial in cases based on excessive environmental liabilities. In Technical note 5.1 we delve more deeply into the details of insurance denial, also seen from a societal point of view.

Table 5.1 Issues outlined by executives and German metaphoric interpretations

Issues outlined by executives	German metaphor
1. Financing	Cyclops, Cassandra
2. Volatility	Cyclops, Cassandra
3. Stakeholders engagement	Medusa
4. Taxes, regulations and governments	Cyclops
5. How to invest more strategically	Cassandra
6. Hiring and retaining talented workers	Cassandra
7. Business development (in hazardous areas from a geological, climatic, geographic, and political points of view)	Cassandra
8. Climate change and other hazards (including regulatory hazards)	Cassandra Pandora's box
9. Infrastructure gap in the countries of operation	Pythia
10. IP protection	Pythia
11. Exploring new revenues opportunities	Cassandra

Technical note 5.1: Insurance denial

Seen from a societal risk management point of view the problem is quite intriguing:

- Governments (at all levels, from central to municipalities) have allowed residential development of areas that were already, or may become (because of climate change), hazardous (at higher frequency than "calm" areas). Reasons for those decisions vary from demographic pressure (need to find a place for people to settle) to business friendliness (more people means a larger tax base, more employment, etc.)
- People have gone into hazardous areas unknowingly, or because they were affordable.
- Insurers were happy to insure, as it meant more business. In some cases, they established what amounted to agreements with the government, as they were providing what could be considered a public service.

Whatever the reasons and the history, insurers are now finding themselves overexposed. In fact, governments cannot offer that public service (people in non hazardous areas would not agree to pay for the others) and people in hazardous areas demand safety, rescue and support.

Three actors, three sets of historic reasons, three sets of diverging interests.

In the Anthropocene Epoch (see Chap. 2) insurers are facing new challenges when insuring against geohazards. That is true especially for those caused by human activity. Indeed, geohazards probabilities, frequencies and insurance denial are intimately related. Insurers have realized that, because of the dynamic evolution, the usual actuarial point of view may be inadequate and can be misleading. The indiscriminate use of force majeure and insurance denial to protect themselves is actually detrimental to their business and their clients.

What an epiphany! Looking only in the rear-view mirror while driving is indeed going to complicate the steering of the vehicle! Insurers have traditionally worked like that. In fact, using past data (statistics) to evaluate their business opportunities, they have already had their share of misery from climate changes and other divergent events. Geohazards caused by human activity fast-track evolution is typically an arena where using actuarial data and statistics can only be wrong and expose everyone, including the insurers, to enormous liabilities. Unfortunately, insurers have asked hazard specialists (geologists, meteorologists) rather than risk specialists help in solving their conundrum, a mistake we often see occurring in various business arenas as well.

Closing point. Obviously, hazard specialists want to measure what they know, but they often confuse hazard with risks. By managing hazards instead of risks, they end up being ineffective or inefficient, with the undesirable results of squandering money, not getting results, or identifying targets for unjustified insurance denial.

International polices may trigger insurance denial as well. For instance, the Paris agreement confirmed a goal of limiting the increase of global temperature to below 2 °C, and countries that signed the agreement are committed to developing plans to reduce greenhouse gas emissions and regularly report on their progress. As a result, the implementation of Swiss Re's new thermal coal policy, mentioned earlier, comes, for example, from their strong commitment to uphold the principles of the Paris climate agreement.

Obviously, there are chances that an asset may become stranded due to regulatory changes, shifts in customs and fashions, obsolescence, etc., and in some cases the definition of exclusion zones, etc. For instance, candle companies found themselves with stranded assets when gas and electricity became mainstream (some were converted into chocolate making units as the wax mixers were perfect for the fats used in chocolate blending). Traditionally, stranded asset were assets with little or no residual value in the aftermath of regulatory and market changes. In this day and time, however, stranded assets are studied and evaluated before their stranding. For example, the California Insurance Commissioner has focused on the issue within the *Climate Risk Carbon Initiative* (https://www.businessinsurance.com/article/00010101/NEWS06/912311473/California-tallies-insurer-investments-in-traditional-energy-industry) of January 2016, an initiative that requires California insurers with 100M USD in annual premiums to disclose investments in fossil fuels. Furthermore, it asks all insurers operating in the state to divest investments in thermal coal. The commissioner launched the initiative precisely because of the potential for fossil fuel investments to become stranded assets on the books of insurers. As said before, in this book we will not address issues of stranded assets.

5.2 Health, Well-Being and Resiliency of Business and for People

In this section we explore an integral and sensitive risk management topic for the health, well-being and resiliency of business and people. We discuss trends and critical issues and lay out a future vision of managing corporate risks and exposures in relation to what people who are associated with or affected by projects need and deserve.

To set the stage for this discussion it is important to recognize that people who are affected by or associated with projects and major business sectors continue to have little influence over risk exposures, tolerances and relevant risk mitigation/management planning (Kemp et al. 2020). This fact dovetails with diverse business sector catastrophes that have occurred over the last decade, including the Fukushima disaster in 2012 (see Sects. 1.1 and 4.2), and the Mount Polley tailings dam failure in British Columbia in 2014 (see Sect. 1.3), which economically displaced and deeply impacted cultural norms and instigated conflict within and amongst numerous Indigenous communities (Shandro et al. 2018). Other examples of catastrophic occurrences that have had significant public impact are:

- 2015, the port explosion in Tinjian, People's Republic of China, which killed 173 people and injured hundreds more (Zhao 2016);
- 2015, the Samarco tailings dam failure (see Sects. 1.3 and 5.3.1) in Brazil, which killed 17 people and resulted in significant environmental damage and human displacement (Freitas 2016);
- 2016, the release of untreated wastewater along coastal Viet Nam, which resulted in mass fish death and adverse impacts to the livelihoods and well-being of multiple coastal communities (Fan et al. 2020);
- 2018, the catastrophic hydroelectric dam failure in Lao PDR (see Anecdote 4.1), which resulted in both fatalities and physical displacement of thousands of individuals (Latrubesse et al. 2020);
- Beirut 2020 port explosion in Lebanon leaving almost 200 people dead with thousands injured, brutally exasperating an already fragile state (Abouzeid et al. 2020), and, more recently,
- 2020, the destruction of the Juukan Gorge Caves in Western Australia by mining major Rio Tinto, which violated archeological and cultural sites dating back tens of thousands of years, to the detriment of Aboriginal peoples.

As Janis Shandro and her colleagues have noted, business sector responses and the appearance of cluster cases during the COVID-19 global pandemic has also been shown to have had a deep impact on projects (Todt de Azevedo et al. 2020), as well as on local communities supporting such development, as observed, for example in *LNG Canada cases* (https://www.lngcanada.ca/news/new-positive-cases-of-covid-19-at-the-lng-canada-project-site/).

The state of affairs presents a serious challenge for risk management professionals and demands a critical review and re-evaluation of current approaches to risks, namely to probability of occurrence and consequences of adverse phenomena, which is a main theme across this book. Many of the disasters listed above are associated with some levels of international risk management standards and safety codes, and yet still have failed to satisfy basic needs for people: to be safe, to feel safe, to have their health and well-being and that of their family attended to, and to be taken seriously. With over fifty years having passed since the mercury poisoning disaster in Minamata, Japan, as a result of waste dumped by the Nippon Steel factory, we still find ourselves in a precarious era of risk mismanagement across business sectors, especially as it relates to health and safety.

5.2.1 Current International Standards for Mitigating Risk Associated with People

The primary source of applicable standards strategies and guidance related to the identification mitigation and management of risks associated with people are the International Finance Corporation (IFC) performance standards on environmental and social sustainability (IFC 2012a). First developed in 2006 with revisions in 2012, IFC now has eight performance standards that focus on overarching risk management, labor and working conditions, resource efficiency, community health safety and security, land resettlement, biodiversity, Indigenous people and cultural heritage.

Performance standard one (PS1) (IFC 2012b) is focused on the assessment and management of environmental and social risks and impacts and encourages projects to identify and manage those through the adoption of the mitigation hierarchy. This hierarchy encourages avoidance as the optimal strategy and mitigation when avoidance is impossible; as a last resort it articulates cases where compensation may be the only option. PS1 promotes improved environmental and social performance of a project through the use of risk management systems, which are covered in Sect. 5.2.2 more detail below. PS1 also requires the development of a systematic tool known as a grievance mechanism to be able to capture concerns, queries and impacts from people affected by a project and to ensure that issues are responded to and managed appropriately. Finally, an integral facet of this standard focuses on the importance of engagement with affected communities throughout the project cycle on issues that could potentially affect them and to ensure that relevant environmental and social information is disclosed and disseminated.

To meet PS1 a management system addressing risks to people requires, at minimum, five core components/features (Green and Shandro 2014). We note that the complexity of the management system will be directly related to the complexity of the project and its operational context and that the system has to be dynamic, as social, environmental, and operational conditions will change over time.

The core components of the management system include:

- assessments (such as environment/social/health impact assessments) and planning (feasibility studies/health sector evaluations);

- the development of associated management plans focused on risk mitigation measures and implementation requirements to make the plan actionable;
- the articulation and development of associated resources for both project staff and external institutions/affected communities (with a focus on the organization of people associated with this work and raising awareness on the significance of particular topics mitigation strategy required);
- a dedicated team and associated resources to handle grievances from workers and community members;
- a robust monitoring and response initiative to inform further refinement/development of all the above core components.

Such a management system does require human and financial resources. The integration of and collaboration with people who are affected by projects is also required to ensure the system for managing risk is both culturally, socially and politically appropriate, i.e. effective at meeting the needs of people.

Performance standard 4 (IFC 2012c) is also very relevant to the topics addressed in this present chapter. This standard on community health safety and security seeks to avoid and anticipate adverse impacts on the health and safety of affected communities during the life cycle of the project. These impacts and risks could occur from both routine and non-routine circumstances, thus the standard supports the use of additionally refined risk management procedures, including the use of health impact assessments, security assessments, or human rights assessments to mitigate risk to affected people. The standard requires a list of topics to be explored, including:

- infrastructure design and equipment;
- design safety;
- hazardous materials management and safety;
- community exposure to the disease, traffic;
- emergency preparedness and response;
- ecosystem services;
- use of private or public security forces.

This standard also relies on specific guidelines for the environmental, health and safety sectors, also developed by IFC (2007).

These standards are held within a larger group of guidance material and resources known as the *Equator Principles* (https://equator-principles.com/wp-con tent/uploads/2020/05/The-Equator-Principles-July-2020-v2.pdf). As extensions of these standards, major business sectors, and financial institutions such as ICMM (International Council on Mining and Metal) and IPIECA (International Petroleum Industry Environmental Conservation Association) have all developed and made public robust guidance for managing a variety of risk topics. We see major firms publicly agreeing to uphold and conform to such standards and guidance. We see such projects supported by industry associations. And yet when non-conformance occurs, there often is little retribution. We continue to fail.

5.2.2 Risk Management for a Healthier Society

Today we no longer need to make the business case for health. Research has shown the importance *of investing in health prevention* (https://2017.mediainprevention. org/en/archive/return-on-prevention-cost-benefit-analysis-of-prevention-measures. html?pid=415). A cost–benefit analysis confirms that every 1 USD invested in upstream preventative health care for international assignees (workers) brought financial return benefits between 1.60 and 2.50 USD; compare this to the *cost of a failed international project assignment* (https://www.prevent.be/sites/default/files/ return_on_prevention_study.pdf) that has been estimated to be between 570,000 USD and 950,000 USD for one worker, depending on the scenario.

We also recognize that today 80% of an individual's health is determined by behaviors and the social and environmental conditions in which they live and work (see Technical note 9.1), and that *upstream interventions can reduce our downstream business and societal costs* (https://www.pwc.com/us/en/health-industries/ health-services/assets/the-case-for-intervening-upstream.pdf). We need to focus now more than ever on preventative actions. When we look more deeply at projects, especially those within the extractive sector or those that are complex in nature, we see that health concerns often underpin the social acceptance of a project. We see that health and safety conditions in the local environment can also be a driver of corporate performance and impact a company's ability to meet commitments and targets. Local conditions can also be a driver of occupational health risk and can therefore have a direct and significant impact on the bottom line. In looking at these topics through this lens, as Janis Shandro and her colleagues have done (Todt de Azevedo et al. 2020), we can see that they are often underappreciated and uncharacterized.

Today we need to bring sectors and people together and remove the informational silos. We need to ensure multisectoral development provides maximum benefits for people, and we need to adopt a new, more robust process for health risk management within convergent approaches. We should not be checking the box and meeting the often insufficient requirements legislated by various national governments. We need to remind ourselves that when addressing topics that deal with people the conditions are always changing. Our assessments, risk prioritizations and associated management plans are not documents intended to sit stagnant on the shelf and be used solely for obtaining permits. Our implementation, our actions, our processes need to be refined as conditions change. We need a monitoring program that doesn't just look at the numbers but triggers response, action and innovation especially when facing divergent exposures. And we need to strive for the Japanese term *Kaizen* (改善), meaning continuous improvement and change for the better. We need a business philosophy encouraging continuous modification of strategy and actions to reach the optimum outcome supported by adequate risk information.

When we talk about risk management involving health and safety issues along with other topics associated with people, we need to collaborate both internally and externally. We also need to be practical and intuitive, as often the unsafe that is very obvious is neglected. It is time to look at those unsafe conditions head on, whether they be inside or outside the fence, and have a procedure to address the very serious risks that stem from them. We also need to incorporate in our assessment procedures an approach to understand how people and communities experience, perceive and respond to risks and impacts. This approach is fundamental, as no two locals will respond in the same way.

Over the course of the last two years we have started to see inspirational change. We have seen health risk management guidance be developed for one of the largest industrial sectors in Asia and successfully adopted into legislative requirements within at least two countries of the Greater Mekong Subregion (Shandro et al. 2019). We have seen Anglo American, a major mining company, update their corporate guidance on managing risk. They have expanded current topics and are to our knowledge the first to have a dedicated section on community health and safety and emergency management that *focuses on the people they affect* (https://social way.angloamerican.com/toolkit/impact-and-risk-prevention-and-management/eme rgency-preparedness-and-response-planning/introduction/about-this-section). This guidance is more in-depth and current than associated international standards which were developed almost a decade ago. We've also seen an entire nation focus their efforts and attention on the health and well-being of their people as compared to previous decades of rampant economic development. The Healthy China 2030 plan (Chen et al. 2019) lays out a platform for all nations to consider. We have seen a development bank that will, over the course of the next five years, make an investment of over 23T USD on infrastructure projects release updated guidance on identifying and managing health risks associated with their projects (ADB 2018).

Most recently, Janis Shandro and her colleagues, on behalf of a major development bank serving the Latin American and Caribbean region, supported a call for action "towards a healthy future where civil society, governments, businesses, and institutions can achieve enhanced resilience to current and potential future health risks". Let us quote their roadmap (Todt de Azevedo et al. 2020, p. 5):

1. **Business leadership embraces and acts on a broader definition of health**—fully integrating worker health, community health and the link between the two.
2. **Community-worker health issues are well-integrated in management systems**—in a systematic way and incorporate a continuous improvement process as part of environmental, health and social management systems.
3. **Management gaps in labor and working conditions are addressed**—these gaps place the health of workers and communities at risk and present vulnerabilities to business continuity.
4. **Engagement, collaboration, and alliances are established with the public health sector**—proactive relationship-building and partnerships will facilitate the planning and collaboration required ahead of future health crisis.
5. **Strategic investment to support local health initiatives are promoted**—focusing on investment in areas where the private sector can play a key support role and areas of mutual benefit.

6. **Business plays a leadership role on global health issues**—business has a strategic role to play in global health issues and needs to be integrated as a key player in global health and climate change discussions and action.

We need to collectively embrace this future vision as the divergent risks are many, and we still have yet to hone our skill at managing the risks we can.

5.3 What People Want

As we discussed in the previous section, the public is exposed to risks associated with projects and operations through two main pathways. The first occurs through consultation and engagement activities, where proponents discuss a project with potentially affected people as part of early planning stages (see Sect. 5.3.3). The second, more traumatic way follows a negative impact, such as the experience of injury, death, dislocation, displacement, or the loss of something of value including sacred or cultural heritage sites, traditional territory, etc. during or in the aftermath of a disaster. Community-Based Disaster Risk Management (CBDRM) focuses on community participation at various levels, which is seen as a central component of CBDRM initiatives (Zwi et al. 2013).

The Asian Disaster Preparedness Centre (ADPC) defines CBDRM as "a process in which at-risk communities are actively engaged in the identification, analysis, treatment, monitoring and evaluation of disaster risks in order to reduce their vulnerabilities and enhance their capacities. This means that people are at the centre of decision making and implementation" (ADPC 2003; Abarquez Murshed 2004). As a matter of fact some CBDRM programs are entirely driven by the community while others may be led by a partnership with various types of agencies.

The objective of CBDRM is to "reduce vulnerabilities and to increase the capacities of vulnerable groups to prevent or minimize loss and damage to life, property, livelihoods and the environment, and to minimize human suffering and hasten recovery" (Abarquez Murshed 2004). CBDRM may incorporate disaster risk management (DRM) and Disaster Risk Reduction (DRR). These may also include climate change adaptation. In other words, the subject of this book supports the creation of rational, sustainable and ethical CBDRM.

Memories of risks linger for a long time and such legacies can potentially impact future relationships for other projects/operations for decades/generations. Furthermore, defining what people want can be a daunting task for socio-economic, cultural, and even religious or spiritual reasons. In this section we present a few examples where the theme was publicly discussed among technical/engineering experts and/or the public. We start with a workshop in Montreal which took place after a major accident at a tailings dam, then we look at what came out from a general discussion held in London on mining and oil and gas. Next we report on a public hearing which took place in Yellowknife (NWT, Canada), where the impacts on a major arsenic contamination case and related possible mitigation project were debated. Finally, we describe the results of a Global Platform by UNISDR and the Aashukan Declaration.

5.3.1   The CIM 2015 Conference Workshop

At the Canadian Institute of Mining, Metallurgy and Petroleum convention 2015 (May 10–13 2015, Montréal), shortly after the 2014 Mount Polley dam disaster, the worst tailings dam failure in Canadian history, one question was asked by both specialists and the public: Why is there a gap between "risk management" and "safety" approaches? The reason is that safety management strives for "zero fatalities" whereas risk management is about making sensible decisions and boosting sustainability while creating value. We stress the fact that while zero fatalities is certainly a honorable goal and works well as a slogan, it is a simplification of a complex reality. It keeps key personnel accountable and alert but is certainly not a scientifically acceptable statement, as zero and nil are merely theoretical values in practical world.

We again heard participants using the term "risk" as a synonym for a number of other technical words, and thus we again stressed the need for a proper glossary that specialists should use to avoid misunderstandings.

Finally, there was a specific request for clarity in defining tolerance to risk and enhancing understanding of complex consequences. It is obvious that this question cannot be answered as long as common practice relies on probability impact graphs (PIGs), risk matrices and other fallacious methodologies are used in mining and other industries (Cresswell 2015; Hubbard 2009; Cox 2008, see also Appendix B).

Only a few months later the world was gain shaken by a mining tragedy when the Samarco dam disaster occurred. To our knowledge that one surpassed anything reported that far in terms of volume making it evident that the issues addressed at the convention were of paramount importance.

5.3.2   The London "Managing Risks" Conference

On 10–12 July 2017, as part of the Geological Society's "Year of Risk" Imperial College London in cooperation with the Institute of Risk Management hosted a conference entitled "Managing Risks Across the Mining and Oil & Gas Lifecycle". The conference's aim was to bring together mining and oil & gas professionals to discuss their respective and common risks, including how the industries manage their wastes for the long term and communicate risks to various audiences. We chaired a session entitled "Managing our wastes for the long term", during which we again observed the frustration felt by both technical people and representatives of the public.

Despite almost unanimous agreement that changes and enhancements from common practices are badly needed, everyone seemed to prefer status quo. Recurring phrases heard at the conference were: "we would like to… but for this reason…we can't" and "we would like to…but we do not know how". Both of these difficulties can be overcome by abandoning obsolete recipes, leaving behind common risk assessment practices, using scientific methods to evaluate holistic risks, and including social and environmental consequences, as will be discussed.

A case history was presented and industry specialists were asked to come up with the preliminary steps of a risk assessment, i.e., performance criteria, system definition and hazard identification. The outcome was extremely interesting and clearly showed the asymmetry between the requirements of the two groups. It became apparent to all participants that reaching a consensus early on in the risk assessment procedure represents a viable step towards mitigating later chances of confrontation and dismissal of the results by the public.

The common thread among the presentations was that systems have to be described and defined before sensible risk assessment and decision-making can be undertaken. To start with, managing risks across the mining and oil and gas lifecycle requires holistic approaches and a well-defined glossary. Other points then followed:

- Both industries need approaches and solutions that are simple but not simplistic, because over-simplification can have serious consequences at the levels of both modelling and mitigation.
- There is major confusion even about basic definitions of risk and uncertainty, even within a single industry, or, even worse, across industries. It is time to see a single, robust glossary used world-wide.
- Terms such as "never", "nil" or "zero" should never be used, especially when talking about hazards and risks.
- Transparent credibility ranges (e.g., 10^{-5} to 10^{-6}, see Sect. 1.1) should be used.
- What is generally considered "long term" may be very short, especially in the face of events driven by climate change. And perpetuity may be an awfully long time.
- Censoring and biasing reality are lurking "mortal" sins (see Sect. 2.1).
- CSR and SLO (see Sect. 5.4) and ESG (see Sect. 6.1.2) dictate ethical behavior and correct evaluation and presentation of risks.

If a risk assessment procedure bypasses all the above pitfalls and follows the recommended standard procedure that we discuss in this book, then it can be considered a rational risk assessment.

5.3.3 Public Hearing for the Giant Mine

To be sustainable, holistic and ethical, a rational (technical) risk-based approach has to include all possible hazards and their consequences. Thus rational sustainability requires holistic and convergent risk assessments in order to reduce the chances of misallocation of mitigative resources, which will likely lead to an increase of exposure to the public and possibly casualties. However, it is paramount to consider the psychological sustainability of any mitigation plan (Kasperson et al. 1988). A perfectly rational plan that keeps the public in a state of perceived risk will prove to be psychologically unsustainable and will fail its objectives (Waksberg et al. 2009). It is indeed necessary to integrate rational risk assessments and risk perception (Renn 1998). For both, a single, but often complex scenario (Sects. 4.1 and 4.2), or a portfolio of scenarios, technical and ethical issues, communication and perception management require transparent and unbiased mitigation prioritization to ensure integrated sustainability. In Part IV and Part V we show why risk-triaging is an aid to tactical and strategic mitigation planning especially when influenced by climate change.

The integration of rational risk assessment, communication and risk perception issues goes well beyond "good old", sound engineering practice. Risk assessment will indeed guide decision-making when adaptation (Khakzad et al. 2012) is needed (for example, in times of changing climate), foster communication, and address issues related to risk perception.

As an illustration of the above we summarize below a result from the MacKenzie Valley Review Board's "Report of Environmental Assessment and Reasons for Decision" (MVRB 2013: Appendix D, Specific Risk Assessment Requirements), which lays out what a rational and societally acceptable risk assessment should include. Our comments are shown in parentheses.

- The compilation of a proper glossary containing a description of all the terms used in the Project and its development, especially those that might have a common use which differs from the technical meaning (such as "risk", "crisis", "hazard") in compliance with ISO 31000.
- The definition of the project context in compliance with ISO 31000, including all the assumptions on the Project environment, chronology, etc.
- A properly defined hazard and risk register.
- A clearly defined system of macro and subsystems/elements and their functional and interdependent links, describing for each of them:

  - expected performances;
  - possible failure modes and quantification of the related ranges (to include uncertainties) of probabilities evaluated as numbers in the range 0–1 (mathematical characterization) with a clear explanation of the assumptions underlying their determination; for the determination of the probabilities the assessors will use a selection of methods taken from ISO 31010 international code;
  - associated magnitude of the hazards and related scenarios.

- An independent analysis of failure/success objectives (see, for example, NASA's *Fault Tree Handbook with Aerospace Applications* (http://everyspec.com/NASA/NASA-General/download.php?spec=NASA_FTA_1.1.68.pdf), which should model success and failure with various pre-selected criteria).

- A holistic consequence function integrating all health and safety (including inference of casualties and pathologies deducted from health studies), environmental, economic and financial direct and indirect effects.
- Applicable published correlations (For example, using Holmes and Rae empirical correlation between "life changing units" and the likelihood to become ill due to external changes, stress, Societal Willingness to Pay (See Technical note 9.1, Sects. 10.1.2 and 10.2.1), etc.) and information.
- It is expected the risk assessment will use a unified metric showing consequence as a function of all health and safety, environmental, economic and financial direct and indirect effects/dimensions. This will be done in a manner that allows transparent comparison of holistic risks with the selected tolerance threshold.
- Consequences will be expressed as ranges, to include uncertainties. When evaluating the consequences, the risk assessment will explicitly define risk acceptability/tolerance thresholds, in compliance with ISO 31000 international code. These will be determined in consultation with potentially affected communities, using a unified metric compatible with the one described above for consequences.

The last points above can be summarized as follows: in Chap. 1, Eq. 1.1 (and later, similarly, Eqs. 2.1, 4.1) we defined the calculation of risk R, in its simplest form, as the product of the probability of occurrence of an undesirable event (probability of failure, or p_f) and the damages it potentially causes D. For our purposes now we extend that equation to:

$$R = p_f(\text{range}) \cdot D(\text{range}) \tag{5.1}$$

In keeping with this formulation, risks and tolerance or acceptability (see Chap. 10) will be developed separately, in such a way as to not influence or bias the judgment of the assessors or evaluators. Risks will then be grouped into "tolerable" and "intolerable" classes (see Chap. 12). The risks in the intolerable group will be ranked as a function of their intolerable part. Mitigation efforts will be allotted proportionally according to that ranking.

5.3.4 The Sendai Framework

The so-called Global Platform was established for implementing the Sendai Framework for Disaster Risk Reduction (UNISDR 2015) adopted in Sendai, Japan, in 2015. In the same year international humanitarian legal researchers (da Costa and Pospieszna 2015) reviewed four country case studies and their main regulations on disaster risk reduction. They took into account the extent to which these countries adopt a human rights-based approach and argued that where community engagement is fostered greater community empowerment is generated. In parallel, where it is possible to hold states accountable for their disaster management performance greater levels of accountability are generated. Thus based on their analyses the authors consider there is a synergy between empowerment and accountability in disaster risk reduction, while human rights based approaches may contribute to foster progress in these areas.

The *2017 Global Platform conference* (https://www.unisdr.org/conferences/2017/globalplatform/en) stated that there continues to be real risks (they used this wording) deriving from political attitudes (McClain et al. 2017). For example:

- indicators may continue to measure disasters costs in monetary terms;
- there may be a failure to take into full account the implications on the health, culture, environment, customs and ways of life of affected social groups.
- best technologies may be used to identify the arrival of drought or the poisoning of water, earth and air, but without the powers necessary to mitigate them.

We were very pleased to see that health, culture, environment, customs and ways of life of affected social groups are all considered to be valid failure criteria that should be taken into account in the multidimensional consequences of potential accidents (see Chap. 9).

As a matter of fact, in 2013 we wrote that the public distrust towards, for example, the mining industry originated in the fact that consequences are oftentimes poorly defined (Oboni et al. 2013). We also wrote that we should consider indirect and life-changing effects on population and other social aspects (See Technical note 9.1). These can be grasped by modern techniques using simplified methods and considering the wide uncertainties that surrounds the driving parameters. Among these parameters we can cite:

- human and social impacts including violence. Indeed, projects, accidents and catastrophic events cause social and environmental impacts that constitute examples of endemic (slow) and sudden (catastrophic) violence (Baird 2020). The development of endemic violence, generated by projects, and sudden violence, generated by accidents and catastrophic events, are temporally related and may be strongly interdependent. An example (Baird 2020) was the 2018 Xe Pian Xe Namnoy dam accident (see Anecdote 4.1). Reportedly endemic violence started during the construction of the Xe Pian Xe Namnoy dam, under the form of neglect, loss of livelihood, increased poverty, etc.—associated with the dam. The catastrophic violence exploded in the aftermath of the accident. Media contribute to biasing perception of the significance of the two kind of violence because quite obviously the catastrophic violence is more vivid and media-genic than the endemic one.
- Fish, fauna and top-soil/vegetation consequences (See Chap. 9);
- long term economic and development consequences (See Chap. 9).

From the Sendai Framework we infer that it is likely that peoples and communities will recover confidence in institutions if there is clear evidence of the willingness of nations to guarantee the right to life. This means that national governments would have to increase their efforts towards effective regulation and protection of people, and that industry will have to adapt and respect international agreements and pacts.

We can note here the similarity between the Sendai Framework and the *Aashukan Declaration* (https://aashukandotcom.files.wordpress.com/2017/04/the-aashukan-declaration.pdf#_blankhttps://aashukandotcom.files.wordpress.com/2017/04/the-aashukan-declaration.pdf#_blank) of April 2017.

It seems again that preserving SLO and showing leadership in CSR (See Sect. 6.1.2) require risk assessments to become transparent, analyse the complexities of consequences and allow transparent dialogue between stakeholders.

5.4   A Note on Communication and Transparency

Communicating and fostering transparency are often perceived as hazardous by projects proponents. Indeed, the proponent of a project might feel vulnerable if risk information becomes public. However, communicating and anticipating objections to risk assessments is paramount to foster social license to operate (SLO) and corporate social responsibility (CSR). The first step is to avoid any suspicion of conflict of interest. Indeed, risk assessments should be performed by an independent entity (Brehaut 2017; Roche et al. 2017) in order to avoid conflict of interest (Oboni et al. 2013) and assuage public concerns.

Overcoming possible objections and fostering SLO and CSR are possible by applying a mix of technical and soft concepts and skills to the communication plan. We are strong promoters of the idea that the glossary and failure metrics need clear definitions (see Sect. 1.3) in order to avoid plunging into a state of confusion at the first contact with the public. For example, if a toxic dump survives a given hazard but the water source for the local population is contaminated, does this constitute a success or a failure? Too often we have seen a failure considered only when major breaches occur, generating distrust among the public. Other times engineers looked only at stability, disregarding other issues and forgetting to define the system's success metric, again resulting in panic. If predictability and foreseeability are weak due to poor information and communication, then the public will react vehemently. This point became obvious with the *Oroville Dam* (https://www.riskope.com/2018/04/18/independent-forensic-team-report-oroville-dam-spillway-analysis/) forensic analysis (see Sect. 2.2.4), where the Independent Forensic Team:

> formed the impression that most DWR staff and those involved in the Potential Failure Modes Analysis (PFMA) studies considered the use of the emergency spillway in terms of only an "extreme" flood event. The Independent Forensic Team notes that a "1 in 100" year storm might be considered an "extreme" event in an operational sense. However, from a dam safety viewpoint an "extreme" dam safety would be a much larger storm (France et al. 2018).

This is again an example of unclear glossary and definitions which led to misleading evaluations of risks. Indeed, as the Oroville report states, failure modes can vanish from detailed considerations simply because of the specific definition of failure. Once media and the public got hold of the narrative, it was too late.

Oftentimes when explaining risks to management or the public, the "demand for zero risk" objection arise. Repeatedly, stakeholders may point out that any non-zero risk is unacceptable. Indeed, in the aftermaths of all recent accident, public opinion, regulators, law enforcement agencies and the media vehemently embraced that vision. However, the first reaction should be to declare that such a goal is not realistic, as any endeavor has intrinsic risks. Indeed, we have been exposed to hazards and resulting risks since the dawn of humanity, but of course we are more ready to accept voluntary risks than involuntary ones. Nevertheless, it is vain to expect zero risk in any human endeavor.

More realistic goals may be those comprised in the concept of ALARP or the implementation of BACT (see Chap. 4). Both these notions are of course open for discussion, as even their definitions are debatable and clouded with uncertainty, especially in view of divergent exposures. In fact, the response to such demands depends on the assessment of their source and the quality of the request.

The demands for zero risk may be sincere but ill-informed. This is when risk education becomes paramount. Thus, the main themes to address may be, for example:

- the misconception of zero risk;
- tolerance and acceptability;
- the definitions of ALARA and BACT (see Sect. 4.1);
- planned actions.

In some cases, political motivations may also come into play. Opposing groups often will appeal to the community health awareness and base their approach on "risks". As a result, experts may be recruited to support this endeavor. We thus see that attacks based on "risks" seem to have gained momentum in recent years, thanks also to the relative impunity warranted by social media and the bad examples offered by politicians of all flavors and in many countries. Let's remember that in any case, attacking the legitimacy and sincerity of opposing groups, and questioning their integrity is a poor approach and will very likely backfire.

Finally, the demand for zero risk may be the result of distress stemming from outrage, anger, and distrust. That generally occurs together with negative judgments about the project promoter who becomes the enemy, allegedly arrogant, dishonest, and focused solely on profits. At that point risks are not the central issue anymore. In fact, the root cause of the hostile behavior may be a serious communication problem. Mending it is a labor of patience and tact: the goal is understanding what "broke the bridge", certainly not trying to make a convincing case that risks are bearable.

5.4.1 Communication

As far as the public is concerned "Trust us" no longer works. As stated in Sect. 1.3, obtaining SLO, the social license to receive a permit to construct and operate any facility, depends nowadays on a company's willingness to engage in meaningful communication with the public with the objective of gaining their trust. This view is supported by the Australian Government, which in a document outlining leading practice in tailings management, stated:

> A key challenge for companies is to earn the trust of the communities in which they operate and to gain the support and approval of stakeholders to carry out the business. A 'social license to operate' can only be earned and preserved if mining projects are planned, implemented and operated by incorporating meaningful consultation with stakeholders, in particular with the host communities. The decision-making process, including where possible the technical design process, should involve relevant interest groups, from the initial stages of project conceptualization right through the mine's life and beyond (AG 2016).

Stakeholder consultation, information sharing and dialogue should occur throughout the system design, operation and closure phases, so viewpoints, concerns and expectations can be identified and considered. Regular, meaningful engagement between the company and affected communities is particularly important for developing trust and preventing conflict.

It should be noted that the term "consultation" is only one aspect of a meaningful communication program by a company. According to the International Association for Public Participation (IAP2), community engagement consists of a spectrum of actions:

- inform (provide information);
- consult (obtain feedback);
- involve (act on what we hear);
- collaborate (public participates in decision-making process but company makes the final decision);
- empower (public decides).

The fourth level, "collaboration", closely parallels one of the leadership level requirements of the Mining Association of Canada's Aboriginal and Community Outreach Protocol (MAC 2015), part of its "Towards Sustainable Mining" initiative. That requirement is that formal mechanisms are in place to ensure that the public "...can effectively participate in issues and influence decisions that may interest or affect them".

Meaningful communication also requires that a company demonstrate its commitment by making its assessment protocols and results publicly available. In addition to helping to drive internal improvement, this practice will go a long way towards earning public trust by showing the comprehensive nature of the standards of practice being used and the efforts being made to ensure that they provide ongoing protection for the public and the environment.

To foster meaningful communication and engagement, leading practice requires that the public is adequately informed of the nature of the risks relating to proposed and existing projects and can effectively participate, in a collaborative manner, in decisions that may interest or affect them. Leading practice also requires that a company make its assessment protocols and reports available to the public to foster transparency.

5.4.2 Transparency

To convince the government and the public beyond a reasonable doubt that a proposed site selection and system provides an acceptable level of risk protection, a company must fully disclose the nature of the risks and demonstrate that its risk management strategies and its commitment to a strong governance framework will adequately address their concerns (Oboni and Oboni 2018).

This will require that the results of a catastrophic failure simulation study be disclosed, and its risk mitigation measures be described. It will require that information be provided that supports the selection of the proposed alternative based on operating and end of service life requirements.

There are two main benefits of a meaningful communication process. The first is that by listening to and collaborating with the public regarding their concerns, a company will have a better appreciation of the risk mitigation measures it should adopt. The second is that a company will gain the opportunity to demonstrate its commitment to high governance and risk management standards in a constructive manner and, if done right, set the basis for it to earn the SLO.

For new alternatives to be credible they must be supported by a high level of design, operating and closure expertise. They must also be subject to the highest level of corporate governance as the new technologies will present their own challenges and require greater attention to design assumptions and operating controls.

5.5 A Note on Ethics and Risk Assessment

In this section we are going to change the point of observation and ask ourselves what qualities a risk assessment method should possess in order to support ethical projects and endeavors. As we will see, some authors propose a "silo-breaking" theory of ethics. In our professional roles, we are adamant in fostering silo-breaking attitudes in risk management. We are also strong believers in the need for geoethics, transparency and sensible risk assessments. Accordingly, in the next two sections we discuss first how general ethics theory can be linked to our practice and thus is synergistic to silo-breaking risk assessments, and then how geoethics can finally guide our actions.

5.5.1 General Ethics

In our earlier book, we discussed general ethics (Oboni and Oboni 2019) but we believe it is important to discuss it again in this present context. A very interesting book by Warwick Fox, entitled *A Theory of General Ethics* (Fox 2006), can help us understand how such a theory helps to link risk assessment and ethics. Fox starts his philosophical discourse by stating that at the core of ethics lies the definition of the values we should live by. Philosophers refer to this as "normative ethics". The aim of normative ethics is to define norms and standards that we should meet or attempt to meet in our conduct. Traditional thinking focused on humans and their relationships with each other. That is an inter-human ethics, i.e., an anthropocentric view of ethics, disregarding, for example, animal suffering.

Ethics really started transcending inter-human ethics in the 1970s. There was the desire for a holistic approach of ethics to replace silo-based approaches such as solely anthropocentric (humans), solely bio-centric (animals), or solely eco-centric (environment). That desire stemmed from the observation that, the actual environment in which we live consists of natural, self-organizing systems as well as the intentionally organized systems and built environments that have arisen during the Anthropocene epoch. As a result, Fox argues that general ethics is a single integrated approach covering inter-human, natural environment and human constructed environment ethics.

Thus, the theory of general ethics, which Fox also calls "the theory of responsive cohesion", attempts to address three questions:

- What values should we live by?
- Why should we live by those values?
- How should we live by those values?

Fox defines three possible kinds of cohesion: fixed cohesion, discohesion, and responsive cohesion. Of the three, responsive cohesion is the most important and valuable. It is normative of relational qualities and tells us what relational qualities ought to be respected. Let's see the relationship between the different types of cohesion and risk management.

- The characteristics of fixed cohesion are rigid, frozen, mechanical and prone to routine. Fixed cohesion is based on compliance, audits, codes and the application of fixed rules. It is a backward-looking cohesion, rooted in past habits. Fixed cohesion corresponds to systems where compliance is the tool of choice. Risk management is disregarded.
- Discohesion is chaotic, anarchic, all over the place, without logic. It corresponds to systems where knee-jerk reactions to hazard take place, with no preparation and planning. Again, risk management is disregarded.
- Responsive cohesion holds a system together. It is found when elements of the system respond to each other in deep, significant and meaningful ways. A responsively cohesive system is internally and contextually responsive, flexible, adaptive, and creative. Using Fox's terms, its salient elements and features "keep company with each other", mixing predictability and surprise. The system is genuinely complex, organic and "alive". With regard to risk management, it is based on transparent, updatable risk-informed decision making, capable of evaluating convergent and divergent exposures.

Now that we have defined a responsive cohesion system and its characteristics, that we know that responsive cohesion is the foundational value, we also know that an ethical system has to "live" by selecting responsive cohesion maximization.

It is possible to develop procedures that abide by the ethics of responsive cohesion by:

- understanding and managing risks (maintaining an internally and contextually responsive system);
- designing rational risk mitigation based on risk informed decisions (flexible solutions based on evolving uncertainties and hazards);
- updating on a regular basis system structure and hazards, risks (adapting mitigations);
- maintaining transparency of information, SLO and CSR (integrating inter-human ethics with the other aspects).

A scalable, drillable, updatable, convergent risk assessment methodology has the necessary qualities to support Ethics. Inversely, classic common practice 4 × 4 or 5 × 5 risk matrix approaches (FMEAs, PIGs, see Appendix B) do not have those qualities and contribute to create and maintain blind spots (see Technical Note 7.2).

The principle of cohesive response entails responding to ecological, social and built contexts in that order of priority. Then such a project will create sustainable value. This will require a significant reorganization of the decision-making and design workflow. Projects will start with risk assessments and environmental impacts evaluations and then investigate and solve technical aspects. To cite one example, Radford (2009) reported that a group of Australian architects (Williamson, Radford and Bennets, University of Adelaide) have shown how to apply the theory of responsive cohesion in the conception, design and construction of architectural systems.

We can envision the day where we will design large infrastructure projects, natural resources operations as responsive cohesive systems, thus fostering General Ethics.

5.5.2 Geoethics

As stated in the Introduction (see Chap. 1) for millennia humans have altered the earth's environment in various manners and to various degrees, in what many call now the Anthropocene epoch (see Chap. 2). Of course, there is no specific starting date for this epoch, and some place its beginning as late as the industrial era. Some voices have arisen stating that only now do we have the means to evaluate and understand the global impacts of man-made modifications. The era of this new understanding is called by some the Sapiezoic era. Before this, every industrial accident was a "fatality", and all slopes failures were "natural". Today we know that the root cause of many environmental damages may be linked to human activities or natural processes that have occurred since humans started altering their habitats because impacts on the geosphere have altered many natural processes, increasing the frequency and magnitude of failures, in some cases leading to divergence. Today only self-blinding humans and negationists refuse to evaluate their risks and act accordingly clearly act against geoethics.

The discussion related to causality can go on forever, but that is not within the scope of this book whereas prioritizing and mitigating risks are.

Sapiezoic geoethics demands new tools for managing any project that has the potential to alter the environment of the earth. Based on the previous section on general ethics we can see that geoethics is indeed part of the cohesive response and covers the environmental and social aspects of the issue (Oboni and Oboni 2020). There are four elements that characterize the Anthropocene and Sapiezoic eras and require the utmost attention in relation to risk management and geoethics:

1. unprecedented scale: the unprecedented scale of impact and alteration through *systemic inter-dependencies* (http://www.riskope.com/2015/04/23/ five-keys-to-healthy-project-management-that-every-manager-needs-but-only-successful-ones-know/) (see Sect. 6.2.2), with the related loss of resilience (Sanderson and Sharma 2016);
2. the need to understand of our role, including the capacity to transition from inadvertent global changes to thoughtful and deliberate control of our effects on the planet. Indeed we are now using by far the largest volumes ever of toxic and hazardous chemicals and radionuclides;
3. the need for evaluation skills: incomplete risk assessments do not help evaluate or control man-made or natural hazards and therefore should be proscribed. Our ability to live sustainably is linked to the capacity to evaluate voluntary and *involuntary risks* (https://www.riskope.com/2015/08/06/world-news-related-to-mining-and-health-and-safety/) and establish reasonable tolerances to risk (see Chap. 10), thus prioritize them and their mitigations in the best possible manner (see Chap. 11, , 12);
4. the need to end cognitive bias such as blind spots and therefore accepting that we have to end the usual societal condoning. Geoethics demand the ending of our societal condoning. We have to foster the widening of what we see and perceiv7.2e and what others see and perceive. That means fostering the "public arena" at the expense of our blind-spots—what we cannot see, but others see (see Technical note). These objectives outline a very *ethical and beneficial way* (https://www.riskope.com/2016/07/21/anthropocene-ethical-geoethical-issues/) of ending of our innocence, a way to ensure that practices in the Anthropocene epoch, the demands of the Sapiezoic area and risk management integrate seamlessly to deliver a more liveable, geoethical world within a responsive cohesion.

In particular our research discusses the management of large portfolios (e.g., regional, provincial or even national scale), and shows the difference between hazard management and risk management.

The above considerations will support the thesis that developing transparent discussions with all stakeholders and sensible mitigative programs ensures better allotment of mitigative funds while complying with the goals of responsive cohesion and geoethics.

Appendix

Links to more information about the Key terms from the Authors	
A, B	*Act of God* (https://www.riskope.com/2020/12/09/act-of-god-in-probabilistic-risk-assessment/) *Black swan*(https://www.riskope.com/2011/06/14/black-swan-mania-using-buzzwords-can-be-a-dangerous-habit/) *Business-as-usual* (https://www.riskope.com/2021/01/13/business-as-usual-definition-in-risk-assessment/)
C, D	*Convergent* (https://www.riskope.com/2021/01/20/convergent-risk-assessments/) *Divergent* (https://www.riskope.com/2020/11/18/tactical-and-strategic-planning-to-mitigate-divergent-events/) *Drillable* (https://www.riskope.com/2020/01/15/probability-impact-graphs-do-not-fly/)
F	*Foreseeability/foreseeable* (https://www.riskope.com/2021/01/06/foreseeability-and-predictability-in-risk-assessments/) *Fragile/fragility* (https://www.riskope.com/2020/04/01/antifragile-resilient-solutions-for-tactical-and-strategic-planning/)
P, R	*Predictability/predictable* (https://www.riskope.com/2021/01/06/foreseeability-and-predictability-in-risk-assessments/) *Resilient, Resilience* (https://www.riskope.com/2016/11/23/resilience-cannot-based-instinctual-decision-making/)

(continued)

(continued)

Links to more information about the Key terms from the Authors	
S	*Scalable* (https://www.riskope.com/2015/04/16/how-system-definition-and-interdependencies-allow-transparent-and-scalable-risk-assessments/) *Societal risk acceptability* (https://www.riskope.com/2014/01/09/aspects-of-risk-tolerance-manageable-vs-unmanageable-risks-in-relation-to-critical-decisions-perpetuity-projects-public-opposition/) *Sustainability/sustainable* (https://www.riskope.com/2019/01/16/improving-sustainability-through-reasonable-risk-and-crisis-management/) *Survivability* (https://www.riskope.com/2011/03/17/ale-fmea-fmeca-qualitative-methods-is-it-really-what-we-need/) *System* (https://www.riskope.com/2017/07/26/three-ways-to-enhancing-your-risk-registers/)
T, U	*Tolerance* (https://www.riskope.com/2020/04/29/risk-tolerance-thresholds/) *Uncertainty/uncertainties* (https://www.riskope.com/2015/12/10/3-decision-making-truths-derived-from-uncertainty-taxonomy-scheme-of-classification-and-a-road-sign/) *Updatable* (https://www.riskope.com/2020/01/07/climate-adaptation-and-risk-assessment/)

Other linked information (https://www.riskope.com/blog-news/) search Riskope blog and use the search box

Third Parties links in this section:	
Australian bushfires	https://www.cnbc.com/2020/01/06/australian-bush-fire-could-affect-consumer-confidence-says-economist.html
Fort McMurray fire	https://globalnews.ca/news/3138183/fort-mcmurray-wildfire-named-canadas-news-story-of-2016/
Flood in LATAM	https://www.bbc.com/news/world-latin-america-51254669
Losses to biodiversity	https://www.sciencedirect.com/science/article/pii/S0960982215003942
Insurance denial	https://www.riskope.com/2016/07/07/geohazards-probabilities-frequencies-and-insurance-denial/
Insurance denial in mining	https://www.riskope.com/2009/09/08/denial-of-insurance-coverage-plagues-mining-industry-developments-world-wide/
Building insurance denial	https://www.cbc.ca/news/canada/edmonton/fort-mcmurray-condominium-insurance-1.5318750
US 2019 wettest month	https://www.noaa.gov/news/us-has-its-wettest-12-months-on-record-again

(continued)

(continued)

Third Parties links in this section:	
Tracking risk trends	https://www2.deloitte.com/content/dam/Deloitte/global/Documents/Energy-and-Resources/gx-er-tracking-the-trends-2019.pdf
Insurance denial for coal exposures	https://www.swissre.com/dam/jcr:6697586a-4fb9-4d58-a7f4-5d50f2c6486f/nr-20180702-swiss-re-establishes-thermal-coal-policy-en.pdf
Climate risk carbon initiative	https://www.businessinsurance.com/article/00010101/NEWS06/912311473/California-tallies-insurer-investments-in-traditional-energy-industry
LNG Canada cases	https://www.lngcanada.ca/news/new-positive-cases-of-covid-19-at-the-lng-canada-project-site/
Equator principles	https://equator-principles.com/wp-content/uploads/2020/05/The-Equator-Principles-July-2020-v2.pdf
Investing in health prevention	https://2017.mediainprevention.org/en/archive/return-on-prevention-cost-benefit-analysis-of-prevention-measures.html?pid=415
Financial return of health prevention	https://www.prevent.be/sites/default/files/return_on_prevention_study.pdf
Upstream intervention	https://www.pwc.com/us/en/health-industries/health-services/assets/the-case-for-intervening-upstream.pdf
Focusing on the people one affects	https://socialway.angloamerican.com/toolkit/impact-and-risk-prevention-and-management/emergency-preparedness-and-response-planning/introduction/about-this-section
NASA Fault tress handbook	https://everyspec.com/NASA/NASA-General/download.php?spec=NASA_FTA_1.1.68.pdf
UNISDR 2017 conference	https://www.unisdr.org/conferences/2017/globalplatform/en
Asshukan declaration	https://aashukandotcom.files.wordpress.com/2017/04/the-aashukan-declaration.pdf#_blank
Oroville dam forensic report	https://www.riskope.com/2018/04/18/independent-forensic-team-report-oroville-dam-spillway-analysis/
Systemic interdependencies	https://www.riskope.com/2015/04/23/five-keys-to-healthy-project-management-that-every-manager-needs-but-only-successful-ones-know/
Involuntary risks	https://www.riskope.com/2015/08/06/world-news-related-to-mining-and-health-and-safety/
Ethical and beneficial	https://www.riskope.com/2016/07/21/anthropocene-ethical-geoethical-issues/

References

Abarquez I, Murshed Z (2004) Community-based disaster risk management: field practitioners' handbook. Asian Disaster Preparedness Center (ADPC), Bangkok

Abouzeid M, Habib RR, Jabbour S, Mokdad AH, Nuwayhid I (2020) Lebanon's humanitarian crisis escalates after the Beirut blast. The Lancet.

[ADPC] Asian Disaster Preparedness Center (2003) CBDRM 11 course materials. Bangkok

[AG 2016] Australian Government (2016) Tailings Management: Leading Practice Sustainable Development Program for the Mining Industry. https://www.industry.gov.au/sites/g/files/net3906/f/July%202018/document/pdf/tailings-management.pdf

[ADB] Asian Development Bank (2018) Health impact assessment: a good practice sourcebook. http://dx.doi.org/https://doi.org/10.22617/TIM189515-2

Baird IG (2020) Catastrophic and slow violence: thinking about the impacts of the Xe Pian Xe Namnoy dam in southern Laos. J Peasant Stud

Brehaut H (2017) Catastrophic dam failures path forward. Keynote lecture, Tailings and Mine Waste 2017, Banff, Nov 5–9, 2017. https://docs.google.com/viewer?a=v&pid=sites&srcid=dWFsYmVydGEuY2F8dG13LTE3fGd4OjE4MzRkN2UyYTYwMjUwNTU

Chen P, Li F, Harmer P (2019) Healthy China 2030: moving from blueprint to action with a new focus on public health. The Lancet: Public Health. https://doi.org/https://doi.org/10.1016/S2468-2667(19)30160-4

Cox LA Jr (2008) What's wrong with risk matrices? Risk Anal 28(2):497–512

Cresswell S (2015) Qualitative risk & probability impact graphs: time for a rethink. https://d9c8ca5f-9f12-4b1a-9d2f-29b8760500bc.filesusr.com/ugd/f61fa4_acb0c7d7d06a49d9b081ede847b2f370.pdf

da Costa K, Pospieszna P (2015), The relationship between human rights and disaster risk reduction revisited: bringing the legal perspective into the discussion. J Int Humanitarian Legal Stud 6:64–86

Fan MF, Chiu CM, Mabon L (2020) Environmental justice and the politics of pollution: the case of the Formosa Ha Tinh Steel pollution incident in Vietnam. Environment and Planning E: Nature and Space. https://doi.org/10.1177/2514848620973164

Fox WA (2006) Theory of general ethics: human relationships, nature, and the built environment. MA, MIT Press, Cambridge

Freitas CMD, Silva MAD, Menezes FCD (2016) The disaster at the Samarco mining barrage: exposed fracture of Brazil's limits in disaster risk reduction. Sci Cult 68(3):25–30

Green T, Shandro J (2014) Canadian institute of mining metallurgy and petroleum, training course on health, safety and security, Vancouver, Canada May 10, 2014. https://issuu.com/cim-icm/docs/van2014_preliminary_program

Hubbard D (2009) Worse than useless. The most popular risk assessment method and why it doesn't work. In: The failure of risk management: why it's broken and how to fix it, Chap. 7, Wiley

[IFC] International Finance Corporation (2007) Environment, health and safety guidelines. https://www.ifc.org/wps/wcm/connect/topics_ext_content/ifc_external_corporate_site/sustainability-at-ifc/policies-standards/ehs-guidelines

[IFC] International Finance Corporation (2012a) Performance standards on environmental and social sustainability. https://www.ifc.org/wps/wcm/connect/Topics_Ext_Content/IFC_External_Corporate_Site/Sustainability-At-IFC/Policies-Standards/Performance-Standards

[IFC] International Finance Corporation (2012b) Performance standard 1: assessment and management of environmental and social risks and impacts. https://www.ifc.org/wps/wcm/connect/topics_ext_content/ifc_external_corporate_site/sustainability-at-ifc/policies-standards/performance-standards/ps1

[IFC] International Finance Corporation (2012c) Performance standard 4: community health, safety and security. https://www.ifc.org/wps/wcm/connect/topics_ext_content/ifc_external_corporate_site/sustainability-at-ifc/policies-standards/performance-standards/ps4

France JW, Alvi I, Dickson P, Falvey H, Rigbey S, Trojanowski J (2018) Independent forensic team report: Oroville Dam spillway incident. California Institution of Water Resources,

Riverside. https://cawaterlibrary.net/wp-content/uploads/2018/03/Independent-Forensic-Team-Report-Final-01-05-18.pdf

Kasperson RE, Renn O, Slovic P, Brown HS, Emel J, Goble R, Kasperson JX, Ratick S (1988) The social amplification of risk: a conceptual framework. Risk Anal 8(2):177–187

Kemp D, Owen JR, Lebre E (2020) Tailings facility failures in the global mining industry: Will a 'transparency turn' drive change? Bus Strat Environ 30(1):122–134

Khakzad N, Khan F, Amyotte P (2012) Dynamic risk analysis using bow-tie approach. Reliab Eng Syst Saf 104:36–44

Latrubesse EM, Park E, Sieh K, Dang T, Lin YN, Yun S (2020) Dam failure and a catastrophic flood in the Mekong basin, southern Laos, 2018. Geomorphology 362(1): 107221

[MAC 2015] Mining Association of Canada (2015) Towards sustainable mining aboriginal and community outreach protocol. https://mining.ca/towards-sustainable-mining/protocols-framew orks/aboriginal-and-community-outreach

McClain S, Secchi S, Bruch C, Remo JW (2017) What does nature have to do with it? Reconsidering distinctions in international disaster response frameworks in the Danube basin. Nat Hazards Earth Syst Sci 17(12):2151–2162

[MVRB 2013] MacKenzie Valley Review Board (2013) Report of environmental assessment and reasons for decision giant mine remediation project https://reviewboard.ca/upload/project_docu ment/EA0809-001_Giant_Report_of_Environmental_Assessment_June_20_2013.PDF

Oboni C, Oboni F (2018) Geoethical consensus building through independent risk assessments. Resources for future generations 2018 (RFG2018), Vancouver BC, June 16–21 2018 https://gac. ca/wp-content/uploads/2019/10/RFG_Printed_Program_FINAL_low-res_05312018.pdf

Oboni F, Oboni C (2019) Tailings dam management for the twenty-first century: what mining companies need to know and do to thrive in our complex world. Springer.

Oboni F, Oboni C (2020) Holistic geoethical slopes' portfolio risk assessment. Geological Society, London, Special Publications 508. https://doi.org/https://doi.org/10.1144/SP508-2019-157

Oboni F, Oboni C, Zabolotniuk S (2013) Can we stop misrepresenting reality to the public? CIM 2013, Toronto. https://www.riskope.com/wp-content/uploads/Can-We-Stop-Misrepresent ing-Reality-to-the-Public.pdf

Radford A (2009) Responsive cohesion as the foundational value in architecture. J Architect 14(4):511–532. https://doi.org/10.1080/13602360903119553

Renn O (1998) The role of risk perception for risk management. Reliab Eng Syst Saf 59(1):49–62

Roche C, Thygesen K, Baker E (eds) (2017) Mine tailings storage: safety is no accident. A UNEP rapid response assessment. United Nations Environment Programme and GRID-Arendal, Nairobi and Arendal, ISBN: 978-82-7701-170-7 https://www.grida.no/publications/383

Sanderson D, Sharma A (eds) (2016) Resilience: saving lives today, investing for tomorrow. International Federation of Red Cross and Red Crescent Societies, World Disasters Report. ISBN 978-92-9139-240-7

Shandro J, Jokinen L, Stockwell A, Mazzei F, Winkler M (2018) Risks and impacts to First Nation health and the Mount Polley Mine tailings dam failure. Int J Indigenous Health. http://dx.doi.org/ https://doi.org/10.18357/ijih122201717786

Shandro J, Peralta G, Roth SA (2019) Health impact assessment framework for special economic zones in the greater Mekong subregion. Asian Development Bank. https://www.adb.org/public ations/health-impact-assessment-framework-economic-zones-gms

Thyagarajan R (2014) Constructing a negligence case under Australian law against statutory authorities in relation to climate change damages. Carbon Clim Law Rev 8(3):208–222

Todt de Azevedo LG, Barron T, Shandro J (2020) Health risk management and resiliency in private sector projects: IDB invest clients' response to the COVID-19 pandemic, early lessons and a road map for action. IDB Invest. https://www.idbinvest.org/en/download/11646

[UNISDR 2015] United Nations Office for Disaster Risk Reduction (2015) Sendai framework for disaster risk reduction 2015–2030. https://www.preventionweb.net/files/43291_sendaiframew orkfordrren.pdf

Van Oldenborgh GJ, Krikken F, Lewis S, Leach NJ, Lehner F, Saunders KR, van Weele M, Haustein K, Li S, Wallom D, Sparrow S (2020) Attribution of the Australian bushfire risk to anthropogenic climate change. Nat Hazards Earth Syst Sci (preprint). https://doi.org/10.5194/nhess-2020-69

Waksberg AJ, Smith AB, Burd M (2009) Can irrational behaviour maximise fitness? Behav Ecol Sociobiol 63(3):461–471

Zhao B (2016) Facts and lessons related to the explosion accident in Tianjin Port, China. Nat Hazards 84:707–713. https://doi.org/10.1007/s11069-016-2403-0

Zwi A, Spurway K, Ranmuthugala G, Marincowitz R, Thompson L, Hobday K (2013) Do community based disaster risk management (CBDRM) initiatives reduce the social and economic cost of disasters? Protocol. EPPI-Centre, Social Science Research Unit, Institute of Education, University of London, London

Part III
Convergent Assessment of Exposures

Convergent assessment is needed because we need to enhance our understanding of the influence of multiple, possibly divergent hazards acting simultaneously or in short sequence on our systems.

Given the potentially increasing frequency and severity of disaster hits, it is paramount to recognize the importance of evaluating risks comprehensively (convergent assessment), to understand the influence of multiple hazards on risk (holistic approach) and to account for consecutive events and their impact on vulnerabilities and extant mitigations (interdependencies and common cause failures).

Part III starts to describe how to define a system (Chap. 6) both in the social and legal dimensions and then in physical terms including possible interdependencies.

The next three chapters, 7 (hazard identification), 8 (evaluation of probabilities) and 9 (evaluation of consequences), deliver the information supporting the creation of better hazards and risk registers.

Chapter 6
System Definition in a Convergent Platform

The system definition in a convergent platform needs to cover all the dimensions in which the system exists. Please check the box with links to key terms in the references at the end of this chapter to facilitate the read.

A first macro taxonomy can be based on the "soft" system (Sect. 6.1), i.e. the social dimensions of the system, the legal environment, and the physical system (Sect. 6.2) including the various possible interdependencies.

6.1 Definition of the "Soft" System

6.1.1 Social Dimensions of the System

Incorporating social dimensions into risk management has been a practice for well over 20 years, albeit to varying degrees of implementation in various arenas. Indeed, the IFC performance standards were established in 2006 by the lending arm of the World Bank Group. Nevertheless, industry professionals face significant challenges when it comes to risk management for people especially in more complex projects. This is further challenged when multiple contractors and subcontractors are involved. This section highlights two topics that demand a more refined approach or attention as they are highly influential in implementing mitigation of risks that involve people. These topics include understanding the local context, contracts to build and the management of contractors and subcontractors, especially around risk mitigation measures.

Understanding the local context
Understanding the local context is paramount for a correct and sensible definition of the system. This allows for a 360-view of reality, including environmental, health and societal well-being in decision making processes. Well understood local context will

F. Oboni and C. H. Oboni, *Convergent Leadership—Divergent Exposures*,
https://doi.org/10.1007/978-3-030-74930-9_6

allow to reduce the chance to see violence (see Sect. 5.3.4) and unethical behaviors (see Sect. 5.5) from inception.

Contracting

Social risk associated with contracts emerges typically during construction. This sensitive phase of a project generally involves dramatic change to a landscape, the influx of workers supporting the building phase, the influx of opportunity seekers to a project site and often involves heavy industrial traffic on roadways and increased movement of people in an area. The reason why contracts are important to consider in identifying a project's potential risks for the local population revolves around the fact that businesses involved in the building of such projects are often contracted to do only that: to build, not to engage with local community members or monitor local conditions to ensure their workforce is not impacting the social environment. Normally, the owners of a project develop a range of management plans and mitigations, but these requirements do not often make it into the contractual obligations of a builder, who is the one who actually causes the bulk of risks to people during this phase. The focus tends to be on the performance while on the project site (Anecdote 6.1).

Anecdote 6.1: Poor implementation of mitigations by contractors

A client learned the lesson the hard way when a third-party evaluation related to the implementation of required mitigation measures across 29 management plans focused on environmental and social risk was undertaken. This evaluation was passed to the main builder who after one week returned it to the client. The evaluation involved a simple matrix of green, yellow and red colour scores which were reflective of the level of implementation achieved. A green score indicated implementation was complete. A yellow score identified implementation was partially under way. A red score indicated implementation had not yet occurred.

In this case, the returned evaluation identified that 95% of all mitigations required as part of this project across 29 management plans *were red*. The owner of the project convened an emergency meeting with the builder and went ballistic. "Why have you not implemented these management plans!" demanded the normally very quiet client. The representative from the EPC (Engineering, Procurement and Construction) company responded in a very calm manner, "Because we don't have the money to. Our contract was a contract to build".

Unfortunately, this situation isn't uncommon and demands a stronger review and integration of risk management topics into contractual obligations. The majority of these projects are based on lump-sum financial arrangements and the tendency is to award the contract to the least expensive bidder. This is a critical piece as conditions related to risks for people are always changing, and contractual arrangements such as lump sum to build are prohibitive of the much-needed flexible risk mitigation approaches. The characterization of those involved in builds also needs to be considered in relation to contracts.

Closing point. Many EPC companies have a very small profit margin as part of a project and are consistently pushed into the red by their clients. In our experience EPC executives have been very vocal on this point. This is especially the case for complex projects that involve multiple contracts.

Management of contractors and subcontractors

In projects that involve multiple contractors and subcontractors we often see a number of challenges emerge that can have a direct implication for risk management strategies required to safeguard the public and workers. Firstly, it is important to look at diversity amongst the workforces. In this case, we are not referring to gender diversity (although this is an important topic requiring further exploration and research), but rather the need to look at cultural diversity. This topic rarely is raised in impact assessments let alone considered in the implementation of mitigation measures but yields some of the most important stress-mediated risks associated with project development. Who the builders of such a project are and where they have come from are two important questions risk assessors should ask when undertaking a system definition aiming to cover in a convergent way potential social risks generated by a project. Indeed, cultural norms vary considerably across jurisdictions and countries, even if they are physically located in the same region. Tensions between ethnicities or races are common, and behavioral norms that differ from those of the people who are born and raised in certain jurisdictions, often can prompt conflict in others. An example of this is off-hours behavior: workers from countries where significant recreational consumption of alcohol is normal can pose elevated risks for conflict, accidents and injuries in a project area.

Time for a word of caution: it is basically impossible to identify all hazards (see Chap. 7) and predict all emerging risks from a complex project so the capacity to adapt is a must. Indeed, a systematic procedure is needed internally to address such a scenario, and it should be anticipated that issues will emerge that have not been articulated in an a priori risk assessment. These are the reasons for building convergent risk assessment that are easily updatable, drillable and scalable: they maximize the opportunities for adapting and making the best possible decisions.

6.1.2 Legal Dimensions of the System

Corporate Responsibility: CSR and ESG

The term Corporate Social Responsibility (CSR) is ubiquitous in this book, as it appears in Sects. 1.3, 3.3, 7.2 and in Chaps. 5 and 9, and thus it merits a specific discussion. CSR is a form of self-regulation that helps ensure that a company's actions have a positive impact on the environment, consumers, employees, and communities. Environment, Social and Governance (ESG) is a term used to evaluate corporate behavior using specific, measurable criteria. ESG performance indicators include sustainable, ethical and corporate governance issues such as managing the company's climate impacts and making sure there are systems in place to ensure accountability. Based on these considerations its relation to holistic risk assessments is immediately evident.

Recently, with its emphasis on more specific metrics, ESG has largely replaced CSR in discussions of corporate risk and crisis management. As we will explore, climate change increasingly drives crises management, and ESG can be implemented to account and plan for, manage and mitigate climate-related risk and crises with significant economic implications. Together, ESG and CSR require companies to take an active rather than passive approach to climate change and the risks and opportunities it presents. This active approach demands holistic convergent risk assessments for business-as-usual conditions as well as divergent exposures. It is indeed essential that companies view and demonstrate through concrete actions that ESG objectives are an integral part of their strategy. Indigenous communities and stakeholders such as shareholders, investors, lenders, regulators, and communities all expect this (see Sects. 5.3, 5.4, 5.5 and 6.1). The bar for what constitutes best practices is rising (see Sect. 4.1) and access and cost of capital depends on getting the systems and processes in place to demonstrate transparency, action and accountability.

Legal Framework for Corporate Decision-Making
Corporate decision-making takes place within various levels within a corporation. Importantly, the directors appointed to the board of directors of the corporation are generally responsible under the applicable legal framework for the fundamental obligation of managing the affairs of the corporation and the overall stewardship of it. Given the significance of the authority and responsibility of directors, the legal framework, which can include corporate statutes and the common law, imposes specific duties (and attendant liabilities) on directors with respect to the exercise of their responsibilities.

For example, in Canada, directors of corporations are entrusted with two principal duties that are relevant to decision making: (1) the duty to act honestly and in good faith with a view to the best interests of the corporation (often referred to as a fiduciary duty or duty of loyalty); (2) the duty to exercise the care, diligence and skill that a reasonably prudent person would exercise in similar circumstances. These duties are enshrined in corporate statutes, such as Section 122 of the Canada Business Corporations Act (*CBCA*) (https://laws-lois.justice.gc.ca/eng/acts/C-44/page-21.html#doc Cont). From a risk management point of view this corresponds to comparing risks to corporate and societal risk tolerances (see Chap. 10) to ensure that the best interests of the corporation are protected and that the public is not overexposed.

It has been long accepted by the courts in Canada that a director's duty to pursue the best interests of the corporation are not confined to short-term profits or immediate share value. In two frequently-cited decisions of the Supreme Court of Canada (SCC)—Peoples Department Stores Inc. (Trustee of) v. *Wise* (https://scc-csc.lexum.

com/scc-csc/scc-csc/en/item/2184/index.do) (2004) and BCE Inc. v. 1976 Debenture Holders (2008)—the Supreme Court held that the "best interests of the corporation" are not solely aligned with the "best interests of shareholders", that boards can look to other stakeholders, such as employees, suppliers, consumers, the government, the environment, and that directors have a fiduciary duty to consider the longer-term interests of the corporation (assuming that the corporation is an ongoing concern). A holistic quantitative risk assessment that looks at all dimensions of potential failures (see Chap. 9) and satisfies the principles described in Chap. 5 for business-as-usual as well as divergent hazards is in keeping with that alignment.

The first of the two cases cited, Peoples Department Stores (2004), involved two retail clothing companies—Peoples Department Store and Wise Stores Inc. Peoples was fully owned by Wise (making Peoples a subsidiary). The Wise brothers became the only shareholders of Wise and were the directors for both companies. When both companies ran into financial troubles in the 1990s, the Wise brothers devised a scheme whereby inventory would be purchased through Peoples and given to Wise on credit, such that Wise owed more than $18 million to Peoples. In 1995 both Wise and Peoples declared bankruptcy. The creditors for Peoples brought an action against the Wise brothers, claiming that the directors breached their fiduciary duty and duty of care under the CBCA by making decisions that were better for Wise but harmful to Peoples and its creditors.

In a unanimous decision, the SCC found that the directors did not owe a duty of care to the creditors, and when examining the fiduciary duty, the court found that the decisions made by the Wise brothers were made to try to resolve the problems that both corporations were facing, and that the decisions were found to be in the bests interests of the corporation. The SCC made it clear that the best interests of the corporation are not the best interests of the creditors or any other specific stakeholder. The SCC also affirmed the business judgement rule, that is, the role of the court is not to second-guess the business decisions of the directors (note that in this decision certain transactions leading up to the bankruptcy were found to be in violation of the Bankruptcy and Solvency Act (BIA) because they were not transacted at fair market value).

In the second of the two cases cited, BCE was under financial pressure and the directors decided to allow competing bids for its outstanding shares through an auction process. Three competing bids were made, each bid requiring Bell Canada, a wholly owned subsidiary of BCE Inc., to incur substantial debt. The BCE board accepted the offer that they believed was in the best interests of BCE and its shareholders. Although 97.93% of BCE's shareholders approved the offer, the plan of arrangement was opposed by a group that held debentures issued by Bell Canada. They argued that the actions of the board were oppressive, and that if the sale went ahead, the short-term training value of the debentures would decline. The debenture holder brought an oppression action under section 241 of the CBCA. The SCC determined that where there was a conflict (in this case, between the interests of the shareholders and the debenture holders), the directors must resolve the conflict in accordance with their fiduciary duty to act in the best interests of the corporation. The SCC reaffirmed its earlier decision in the Peoples Department Stores case that the fiduciary duty is owed to the corporation, and not to any particular constituency. Where there is a conflict between the interest of the corporation and the interests of any other stakeholder, the director's first duty is to the corporation. In considering the best interests of the corporation the directors may look to the interests of a range of stakeholders – shareholders, employees, creditors, governments and the environment – to inform their decision. The courts should give appropriate deference to the business judgement of the directors.

The SCC decision in BCE Inc. has been frequently cited for the business judgment rule, including the fact that the directors may look to the interests of the environment. Ultimately, BCE Inc. was codified in 2020 amendments to the CBCA at s. 122(1.1):

> (1.1) When acting with a view to the best interests of the corporation under paragraph (1)(a), the directors and officers of the corporation may consider, but are not limited to, the following factors:
>
> (a) the interests of (i) shareholders, (ii) employees, (iii) retirees and pensioners, (iv) creditors, (v) consumers, and (vi) governments; (b) the environment; and (c) the long-term interests of the corporation.

The CBCA had already been amended in 2019 to codify the court findings from these prior judicial decisions, and Section 122 of the CBCA expressly provides that when acting with a view to the best interests of the corporation, the directors and officers may consider a broader group of stakeholder interests, including the interests of shareholders, employees, retirees, consumers, government, the environment and long-term interests of the corporation (and this is not intended to be an exhaustive or exclusive list). Therefore, as a starting place, directors are required to consider the best interests of the corporation, which can include broader interests, such as the environment.

For example, for many corporations, climate issues present a significant risk and/or opportunity for the corporation, and therefore these issues must be considered as part of a director's duty to act in the best interests of the corporation. Secondly, directors must act with diligence, and therefore they must be sufficiently informed about these issues, seek advice and information and analysis so that that the directors are in a position to make informed decisions. The supporting role of risk-informed decision-making again becomes evident (see Part IV, Tactical and Strategic Planning for Convergent/Divergent Reality and Part V, Convergent Assessment for Divergent Exposures: Case Studies, which features risk-informed decision-making deployments examples on railroads, wharves, and chemical plants in isolation and as a system).

Although it is entirely appropriate and necessary for boards to obtain assistance from management as well as other internal and external advisors and experts on these issues, this is not a matter that can be delegated to management to address. The obligation of directors to consider the implications of material issues is grounded in the duties the directors owe to the corporation, and therefore boards must have ultimate oversight of these issues.

Boards must consider what governance should look like within their companies and what structures and processes are in place or need to be put in place to provide oversight of the material issues. This includes structures to ensure that there is appropriate assessment, tracking, analysis, etc., of these developments in the jurisdictions in which they do or want to do business. Boards require sufficient information and expertise, either within the board or reporting to the board, to make informed decisions. There are guidance documents with respect to how to set up effective climate governance on a corporate board. For example, the World Economic Forum published a report in 2019 on Guiding Principles and Questions on setting up effective climate governance, which was prepared in collaboration with PwC.

Corporate Responsibility—Additional Considerations

A. **Corporate Disclosure and Reporting**

There are a variety of ESG reporting frameworks created by investment firms, industry organizations and non-profit organizations. Governments, regulatory bodies, and companies, amongst other entities, are all assessing opportunities to standardize ESG reporting and disclosure in general and for specific industries. An increasing number of companies and investors are using ESG frameworks to support their investment analysis, including frameworks such as United Nations Sustainable Development Goals, Sustainability Accounting Standards Board, Task Force on Climate related Financial Disclosures, and Equator principles (see Sect. 5.2.1).

Apart from voluntary ESG disclosures or disclosures required by investors, lenders, or commercial partners, public companies may also be subject to mandatory disclosures related to certain ESG pillars, for example climate change (see Sect. 3.3) and board diversity requirements.

B. **Public Engagement and Indigenous Nations**

There are circumstances where corporations are required to engage in forms of public communications or engagement with respect to corporate initiatives (see Sects. 5.3 and 6.1).

- **Aboriginal Consultation**. In Canada and certain other countries, rights of indigenous peoples are recognized by the Constitution and legislation, including implementation in certain jurisdictions of the United Nations Declaration on the Rights of Indigenous Peoples (see Sect. 5.3.4) Such recognition, along with the ESG principles, requires procedural requirements being placed on the "Crown" and companies, including relating to engagement, consultation, and accommodation of indigenous peoples for activities that are of interest or affect indigenous peoples.
- **Environmental Assessment and Regulatory Processes**. Across Canada, environmental/impact assessment and regulatory processes require an assessment of the need to and the carrying out of consultation with indigenous peoples and public engagement, where required, as a precondition to a decision being rendered on the particular approval requested (see Sect. 5.3.3).

6.2 Physical System Definition

The key to a world-class risk assessment endeavor, be it an ERM, a project, or a processing facility, is the proper definition of the system. We all know that ISO 31000 and other international and national risk codes stress the fact that the context of the study—the environment in which the system operates—has to be described. However, so many times we have seen project teams and facilitators embarking in FMEAs (see Appendix B) or other risk assessments without taking the time to rigorously describe the system's anatomy and physiology. It may seem weird to use medical terms but let's see why they are appropriate.

6.2.1 The Emergence of Systems

Most common practice risk analysis tools date from the years of WWII and its aftermath in the 1950s.

At the beginning, only weapons and very "scary" systems were studied using those methodologies, covering business-as-usual risks, mostly looking at components. Divergent risks were not yet an issue. Industry was still employing the so-called "insurance gal" (an unfortunately derogatory term) to transfer risk, without any serious evaluations or considerations for the "systems", to insurance companies willing to take a bet on them.

A series of mishaps, public outcry and political pressure brought "risk" to become a buzzword. Risk assessment and risk management were nice words to say, and common practice percolated down to the minimum common denominator to provide a placebo to society, but accidents were still occurring, failures were still called unforeseeable, and potential consequences were still looked at cursorily and in a siloed way. No one was carefully describing systems' anatomy and physiology. It was the time of open risk workshops gaining the status of "instant risk assessment". Actually, most of the time participants were able to voice concerns and fears, without having dissected the system under consideration, pretty much as we used to do in medicine before understanding anatomy and physiology.

Then large-scale terrorist acts such as the September 11 attacks occurred on US soil, and in 2008 there was a global recession. All of a sudden new words were coined to describe what we humans already knew very well: poorly made risk assessments, without systemic understanding, do not have any value.

There was a spate of magic revivals, obscurantism, and denial of bad habits, all just to hide one simple fact: unless we take the time and effort to properly define our systems, we cannot perform any serious analysis on them! Voices were raised about systemic risk, non-functioning models, black swans (legitimate ones and mostly silly ones), the Olympus menagerie we cited in Sect. 1.1, fragility, complexity, etc., but no mention was made of the blatant gap of divergent exposures.

The parallel to medicine is striking: if we do not know the anatomy and physiology of the human body, any surgery or drug will have a very poor rate of success and can even make matters worse.

So, getting back to risk assessments:

- Is it true that our systems are complex? Yes.
- Are they fragile because of their complexity and other reasons? Yes.
- Do rare, extreme events occur? Yes.
- Do we have systemic risks in our systems? Yes.
- Are we capable of hiding our heads in the sand, saying that there is nothing we humans can do to evaluate the above, and merrily keep making the same mistakes? Yes.
- Is it reasonable, socially acceptable, or good for humanity to do so? No.

(If you want to have fun for a moment, you can set up the same list of questions, replacing "system" with "human body" and "events" by "diseases". Enjoy!).

By undertaking a systematic analysis of a system's anatomy and physiology, most, if not all, of those pitfalls can be avoided. That preliminary effort brings rationality, clarity and transparency to the table. It makes risk studies scalable, flexible, adaptable to new conditions. It yields a holistic understanding of the risk landscape surrounding operations/projects.

6.2.2 How to Dissect Your System

In prehistorical and early historical times human health (the system of interest in medical science) was in the hand of shamans and other healers who were using empirically selected remedies (herbs and roots, for example) or ceremonies and rituals (including inducing mental alterations of various kind) to cure and protect their people. We are not passing judgment on these techniques, especially since, at the time, there were no alternatives to select from and we know by now that some of those traditional remedies actually worked very well.

Clearly humans were neither really happy with the understanding they had of human body nor with the overall rate of survival. They needed to understand more. Hence, for example, Leonardo da Vinci started to perform anatomical studies, even though dissection was prohibited by the Church and against the law in those times and recorded his acute observations in the famous sketches that delivered a first understanding of human anatomy. A few more centuries of research have brought us to be able to detect genetic mutations, hereditary diseases and much more. The development of this understanding was not always easy, as religion, obscurantism and other agents were not always open to the enhancement of science, to say the least.

Similarly, when performing a risk assessment, it is necessary to understand the system under consideration. Its architecture must be carefully studied by people who intimately know it. Risk assessment experts can only offer support in this phase as they do not know the system's intricacies. However, in specific cases they may help customers to create a model that will reflect reality while remaining as simple as possible.

The battery limits of the system define who and what is within and outside of the system (Table 6.1). In fact, the battery limits help characterizing threats-to and threats-from among system's elements as we will see in Chaps. 7 and 8. Further, the definition of the project general "context" in compliance with ISO 31000 (Chap. 11), including all the assumptions on the project environment, chronology etc., also helps to define the system's battery limits.

Using the definitions in Table 6.1 it is possible to dissect the elements of the system within the physical perimeter or any other battery line. Usually after defining the battery lines of the system, first the types of primary nodes are listed (Fig. 6.1), followed by the secondary nodes. Defining the secondary nodes may result in a

Table 6.1 Defining the project/operation's battery lines: physical perimeter, environment, personnel and cyberspace

Battery lines	Comments
Physical perimeter	The area inside the physical perimeter, where operations and personnel are under corporate direct supervision and responsibility. Hazards to the system can occur within the perimeter and bear consequences inside and/or in the environment
Environment	The area outside the perimeter, where external environmental damages can occur, public is present. Hazards to the system can occur in the environment to a distance to be determined during the study
Personnel	Any person working directly (employees or contractors) or indirectly (subcontractors) for the corporation within the perimeter. During off-hours personnel become residents of the environment. Based on the frequency and duration of visits, visitors may be assimilated to personnel
Cyber space	The realm of information: the web, data, including all project's/operation's IT systems, SCADA, sensors, etc. as well as the links to cyberspace. It also includes private cell phones and other devices connected, temporarily or permanently, to corporate's systems

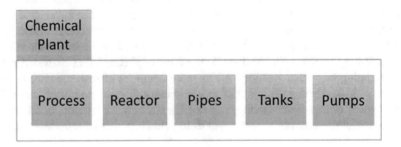

Fig. 6.1 Initial chemical plant system dissection with main nodes listed

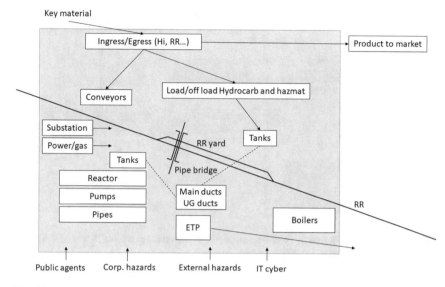

Fig. 6.2 Detailed chemical plant system dissection after discussion with personnel, with added intricacies

revision of the primary nodes as a systemic view emerges from the initial process-centric one. This procedure continues, depending on the required level of detail or "granularity".

"Granularity" indicates scalability. The scalability of the model will thereafter make it possible to zoom in on one or another (or all) of the nodes to examine details, depending on the needs. Any system, no matter how complicated, can be described by as a system of nodes whose granularity is appropriately selected. If needed, the chemical plant node shown in Fig. 6.1 can be subdivided into increasingly finer levels of granularity, as in Fig. 6.2. These may go all the way down, for example, to a micro- or pico-node called, say, "pump A". The level of granularity is dictated by the stage of development (at pre-feasibility level only macro-nodes may be necessary) and the purpose of the risk assessment (part of the system may be modelled with macro-nodes, others be more detailed, see Case Study 2, 3, Chaps. 15 and 16)). The advantage of scalability becomes evident as the phases of development progress.

The system description is completed when the incoming resources, produced, processed, transported and the outgoing ones are listed in each node (Fig. 6.2). As Anecdote 6.2 shows, an operation-centric view may censor reality. Therefore in this phase it is necessary to use engineering good sense and modelling tact in order to prepare lists compatible with the level of detail required by the client. The scalability of the system will eventually allow refining the descriptions as needs arise in the development of the system's life.

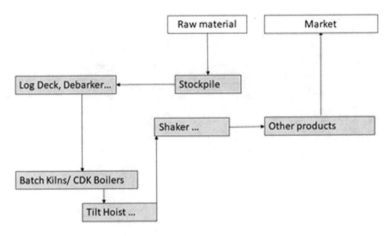

Fig. 6.3 The system description we received from the management of a sawmill. Each element has sub-elements it is necessary to zoom in to see

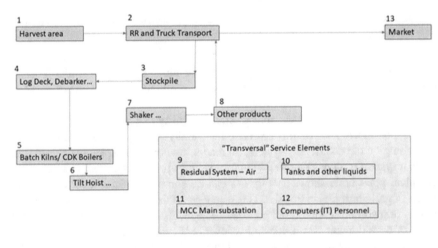

Fig. 6.4 Completed sawmill system, including transversal service elements and peripheral elements (in orange)

Anecdote 6.2: Industrial sawmill system

We were asked by a client to assess industrial sawmills. Figure 6.3 shows the system description of such an operation that we received back from management.

Undoubtedly, Fig. 6.3 corresponds to an operation-centric view. It is censoring reality, as numerous elements of the operation's "environment" are missing. These can be separated in two types: the transversal service elements and peripheral elements. Transversal services are those such as power, IT/cyber, office building, tanks and residual systems (air/dust), which cover all the other systems to a varied degree; for instance, power will be fed through a substation to all the sawmills machines and buildings. Peripheral elements, such as the harvest area, ingress/egress logistics and end users (market), are those that are mostly outside the physical perimeter of the operation but nevertheless necessary to the performance. Included among the peripheral elements is, of course, the market, with all the hazards (geopolitical, tariffs, etc.) that may be linked to it.

Figure 6.4 completes the system description begun in Fig. 6.3 by adding those two categories and their related elements.

Closing point. It is this system definition that we used through all sawmills of the client.

Once the system is well defined it is easier to identify hazards and pair them to nodes and resources (see Chap. 7), thus setting up the first steps of a clean, well-balanced risk and hazard register. For instance, in Fig. 6.2, the railroad crossing the plant is a hazard source (threat-to the plant), as is the pipe bridge over the railroad yard (threat-to the railroad yard and then to the plant). Thus, proper definition of the system you are analyzing is the key to meaningful risk assessment.

As shown above, the system definition also helps to identify threats-to and threats-from among system's elements (see Chaps. 7 and 8). Threat-to and threat-from analysis is used to link identified external or internal hazards to nodes and resources and in particular:

- to targets (elements of the system) or
- from elements to targets lying outside of the system (population, environment, third parties, etc.) or
- inside the system (which then become interdependent).

As a result, each couple is qualified in terms of possible dire outcomes (consequences).

It is important to note that elements within the system and elements in the environment can involve threats-to (for example, a tank breach can damage another infrastructure) and threats-from (for example, an act of vandalism on a pipe by angry residents).

Oftentimes highly respected professionals maintain that highest risk to safe operation is a certain element or a certain hazard. They may justify their judgement with parameters and data. One has to be wary of this type of judgement because we humans are not so good at simultaneously considering two parameters, in this case, likelihood and consequences. In order to develop a study compatible with the budgeted effort, assumptions and simplifications will generally be necessary, but a simplified, well-thought-out assessment is better than an arbitrary judgment.

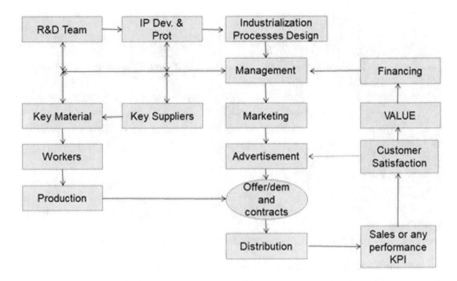

Fig. 6.5 System definition for a startup, including functional links between elements. NB: such a scheme facilitates value chain discussions

A well-defined system will avoid many blunders and confusions typical of common practice risk registers. Technical note 6.1 shows an example for a business system that we define as non-physical because of the preponderance of non-physical elements, i.e. human, management administrative, etc.

Technical note 6.1: A system definition of a non-physical system

The technique discussed in the previous section can also be used for physical (tanks, machines, etc.), non-physical (administrative/commercial) and hybrid systems. Figure 6.5 shows a preliminary system definition of a start-up that we use to foster discussions with CEOs and during our MBA courses on risk. It makes it possible to understand that literally any system can be analyzed regardless of its specific characteristics. Incidentally, note how this scheme bears similarities to the map you would draw to generate a value chain discussion for the same company.

Closing point. Dissecting the system in elements at various levels of granularity is paramount to prepare rational and logical risk assessments.

6.3 A Note on Interdependencies

The definition of the source of the resources and client nodes makes it possible to reasonably identify the system's interdependencies (internal–external). Of course, interdependencies between nodes (of given levels) have to be identified in a simple, but effective way, in order to avoid "paralysis by analysis". In fact, the system definition should include all pertinent interdependencies (physical, geographical, logical, informational) necessary to its operation as well as a clear delimitation of selected battery lines assumptions (Table 6.1).

In Chap. 1 we gave the formula for calculating risk based on probability of failure and potential damages. We repeat it here:

$$R = p_f * D \qquad (6.1)$$

But sometimes local probability p_f and local damages D of an element do not help to capture a fair representation of the risk to a system. For example, if pressure relief valves are installed on a system at two different locations, they will not have fully independent probabilities of failure: if one fails, the other one has a greater chance of also failing. This leads to the instinctive reaction oftentimes seen in the aftermath of a element's failure, when all the similar elements are immediately checked. The logic is that:

- if the elements are from the same manufacturer there may be a potential common defect or weakness;
- they may be installed in the same location, that is, they share the same physical environment (for example, both underwent the same lateral movement due to an earthquake, etc.);
- they all are maintained in the same way (the same operator, etc.).

What we just described is known as the Common Cause Failure (CCF) phenomenon (Rausand 2011), which emerged as an important aspect in aerospace applications. However, it is not limited to those and we generally study it in any industrial or infrastructural endeavor as a special case of interdependency. We will discuss CCF again starting in Sect. 8.2.2.

6.3.1 Internal Interdependency

Interdependencies are often present even within a single machine or device or administrative office. In this case they are called "internal interdependencies" or "domino effects". Internal interdependencies, including CCFs, mean we have to assume a greater probability of failure of the system despite redundancies (see Sect. 8.2.1) installed for the purposes of enhancing reliability (mitigation of the probability of failure, and sometimes consequences reduction).

If we consider, for example, a chemical process which uses a Water Treatment PLant (WTPL) to recirculate water through its productive circuits, we may be in a situation where the process has a certain probability of "failure", while the WTPL has its own probability of failure. But the process and the WTPL are interdependent. If the process has an upset, the WTPL may have to stop treatment, or even be damaged. If the WTPL has a problem, the process does not get a vital resource.

Below we quote a series of examples of internal inter-dependencies:

- At Point Comfort, TX, a forklift towing a trailer collided with a line containing highly flammable liquid propylene, causing a release and a vapor cloud explosion. Sixteen workers were injured, the process unit was heavily damaged, and a nearby school was evacuated. If a risk assessment had been prepared by an inattentive analyst, the forklift may have been assessed, including possible "traffic collisions", and the line might have been evaluated in terms of age, corrosion, maintenance. But the domino scenario of the forklift (or its trailer, supposing that was legal to do) hitting a line with highly flammable fluid would likely have been forgotten or dismissed as non-credible. Now, many accidents follow this pattern!
- A similar case was the explosion and fire hitting a chemical plant in Bakersfield where a raging fire engulfed a chemical storage facility but no one was injured. The blaze at the chemical plant burned throughout a 9600-square-foot building. The fire began when a 250-gallon container of methanol came in contact with an electrical pump, and the flames then engulfed storages.
- Changing from industrial to military: the US Pentagon mistakenly sent live anthrax to at least nine labs in the US and to a US military base in South Korea using Fedex. This is a classic example of logical internal interdependency. That's the classification, although the fact in itself seems quite unbelievable!
- Thomé et al. (2015) give an example of physical interdependency from Fukushima where off-site power failed during the earthquake and the on-site power supply redundant systems (emergency diesel generators and batteries) were lost due to the tsunami induced flooding. As a result, the reactors monitoring instrumentation and remote controls became unavailable except in Unit 3. The company recognized the solution to this physical interdependency was to install diesel generators and batteries at different levels at the Fukushima nuclear site, reducing the vulnerability of this system to flooding and ensuring its survivability and service.
- Logical interdependencies were also present at Fukushima (Koshizuka 2012). Indeed, controls rooms were shared by couples of adjacent reactors. Each control room had a shift supervisor with decision-making and emergency control responsibilities. Reportedly there were communication failures between the shift staff and the on-site Emergency Response Organization (ERO) and wrong assumptions causing nefarious delays (Thomé et al. 2015). We unfortunately all know the rest of the story.

6.3.2 External Interdependencies

A classic case of external interdependencies may be the following: an external hazard impacts operations and other similar elements not within the considered system. For example, a major earthquake may disrupt the road network giving access to an operation within a certain perimeter. As a result, if we are looking at a business interruption (BI) evaluation of the operation it would be wrong to assume that repairs to operate and run can be made within the normal time frame, as the road network will impede access for quite a while. The probability we should investigate is the probability of a quake provoking a damage D1 on the operation and the probability of a quake generating damages D2 on the network. Those probabilities should then be combined, and the consequences adeptly evaluated.

One example comes to mind: when in 1999 the tunnel under Mont Blanc, at the border between Italy and France, was closed due to a severe fire, harbors and wharves across Europe felt the shockwave. It was also evaluated that traffic patterns changed on a radius of several hundred kilometers, creating havoc for all classes of users. Here again, as an operator of a wharf, we would be interested in the probability of "some accident across Europe" to create havoc, with a probability p and consequence C for the wharf itself. The consequence could be negative: less traffic, thus a need to fire employees and store more perishable goods than normal, etc., but also perhaps, and rather unexpectedly, an excess of business boosting profits but creating congestion. Obviously, the accident would not have a direct impact on the wharf, but the interdependence could create riots, strikes and violence issues of perishable goods and accidents due to the excessive congestion that would normally not occur with the same probability and consequences.

Finally, an accidental spill from a third-party vessel in a harbor may result in a severe restriction of docking and loading/unloading operations for an innocent, but interdependent operation. Indeed, these consequences can impact all the present vessels. This is an example of geographical interdependency for the owners of the other vessels or the wharves.

Other examples of interdependencies with patterns similar to these are the 2015 collapse of the Tex Wash bridge of interstate I10 in Arizona, which caused difficulties in the traffic for quite a long time with increased costs and travel times for all kinds of third parties, and the 2020 explosion at a refinery in California, which stressed the gasoline supply.

In our global world, interdependencies may also intervene at planetary scale. One sad example came with the *medical isotopes production shortage* (https://www. nature.com/news/reactor-shutdown-threatens-world-s-medical-isotope-supply-1. 20577#:~:text=Canada's%20Chalk%20River%20reactor%2C%20which,will% 20end%20production%20next%20month) (Tolzefson 2016) in 2016. While in itself a reactor shutdown due to technical safety reasons could be considered business-as-usual as regards the reactor itself, the world experienced a shortage of vital medical isotopes. A risk assessment of a medical system in Europe, for instance, would see this shortage as an interdependent effect of an incident experienced by a foreign producer.

Let close this chapter by noting that in addition to natural and classic man-made hazards, severe interdependent scenarios may be triggered by Forex, tariffs, changes in regulations, geopolitical changes, climate-linked hazards, pandemics, etc., especially when they become divergent.

Appendix

Links to more information about the Key terms from the Authors	
A, B	*Act of God* (https://www.riskope.com/2020/12/09/act-of-god-in-probabilistic-risk-assessment/) *Black swan* (https://www.riskope.com/2011/06/14/black-swan-mania-using-buzzwords-can-be-a-dangerous-habit/) *Business-as-usual* (https://www.riskope.com/2021/01/13/business-as-usual-definition-in-risk-assessment/)
C, D	*Convergent* (https://www.riskope.com/2021/01/20/convergent-risk-assessments/) *Divergent* (https://www.riskope.com/2020/11/18/tactical-and-strategic-planning-to-mitigate-divergent-events/) *Drillable* (https://www.riskope.com/2020/01/15/probability-impact-graphs-do-not-fly/)
F	*Foreseeability/foreseeable* (https://www.riskope.com/2021/01/06/foreseeability-and-predictability-in-risk-assessments/) *Fragile/fragility* (https://www.riskope.com/2020/04/01/antifragile-resilient-solutions-for-tactical-and-strategic-planning/)
P, R	*Predictability/predictable* (https://www.riskope.com/2021/01/06/foreseeability-and-predictability-in-risk-assessments/) *Resilient, Resilience* (https://www.riskope.com/2016/11/23/resilience-cannot-based-instinctual-decision-making/)

(continued)

(continued)

Links to more information about the Key terms from the Authors	
S	*Scalable* (https://www.riskope.com/2015/04/16/how-system-definition-and-interdependencies-allow-transparent-and-scalable-risk-assessments/) *Societal risk acceptability* (https://www.riskope.com/2014/01/09/aspects-of-risk-tolerance-manageable-vs-unmanageable-risks-in-relation-to-critical-decisions-perpetuity-projects-public-opposition/) *Sustainability/sustainable* (https://www.riskope.com/2019/01/16/improving-sustainability-through-reasonable-risk-and-crisis-management/) *Survivability* (https://www.riskope.com/2011/03/17/ale-fmea-fmeca-qualitative-methods-is-it-really-what-we-need/) *System* (https://www.riskope.com/2017/07/26/three-ways-to-enhancing-your-risk-registers/)
T, U	*Tolerance* (https://www.riskope.com/2020/04/29/risk-tolerance-thresholds/) *Uncertainty/uncertainties* (https://www.riskope.com/2015/12/10/3-decision-making-truths-derived-from-uncertainty-taxonomy-scheme-of-classification-and-a-road-sign/) *Updatable* (https://www.riskope.com/2020/01/07/climate-adaptation-and-risk-assessment/)

Other linked information (https://www.riskope.com/blog-news/) search Riskope blog and use the search box

Third Parties links in this section:

Canada Business Corporation Act	https://laws-lois.justice.gc.ca/eng/acts/C-44/page-21.html#docCont
Peoples v.Wise	https://scc-csc.lexum.com/scc-csc/scc-csc/en/item/2184/index.do
BCE v. Debenture	https://scc-csc.lexum.com/scc-csc/scc-csc/en/item/6238/index.do
Medical isotopes interdependency	https://www.nature.com/news/reactor-shutdown-threatens-world-s-medical-isotope-supply-1.20577#:~:text=Canada's%20Chalk%20River%20reactor%2C%20which,will%20end%20production%20next%20month

References

[ISO 31000:2009] International Organization for Standardization (2009) ISO 31000, Risk manage-
    ment—principles and guidelines. International Standards Organization. https://www.iso.org/iso-
    31000-risk-management.html
Koshizuka S (2012) Report from investigation committee on the accident at the Fukushima Nuclear
    Power Stations of Tokyo Electric Power Company. Nippon Genshiryoku Gakkai-Shi 54(10):642–
    646. https://inis.iaea.org/search/search.aspx?orig_q=RN:44022364
Rausand M (2011) Risk assessment: theory, methods, and applications. Wiley
Thomé ZD dos Santos Gomes R, da Silva FC, de Oliveira Vellozo S (2015) The Fukushima nuclear
    accident: insights on the safety aspects. World J Nucl Sci Technol 5(03):169. https://inis.iaea.org/
    collection/NCLCollectionStore/_Public/46/006/46006744.pdf
Tolzefson J (2016) Reactor shutdown threatens world's medical-isotope supply. Nature 12. https://
    doi.org/10.1038/nature.2016.20577

Chapter 7
Comprehensive Hazard Identification

Hazards are sometimes blatant, sometimes scary, and sometimes difficult to identify. Hazard-based prioritizations generally lead to poor decisions because what is scary or big does not necessarily generate large consequences hence may not lead to greater risks. Furthermore, hazards do not act alone. That is one main reason why convergent risk assessments are needed, as they deliver a 360-view of potential exposures on a system.

In this chapter we will review a selection of hazard identification approaches and techniques. A box with links to key terms is included in the references at the end of this chapter to facilitate the read.

7.1 Standard Methods for Hazard Identification

Hazards related to natural and human causes are of primary concern through all phases of the design and system lifecycle of any operation, project.

When tackling a risk assessment, we define hazard scenarios as any malfunctioning or deviations from the intended level of performance of the system or any of its elements "as is". System "as is" means with the level of mitigation and controls present at the moment of the study and with the quality of investigations, design and maintenance which becomes apparent during the preparation of the study, such as components failure, repair or other phenomena, including not only those generating harm to people, properties and the environment, but also business interruption (BI) and other consequences (see Chap. 9). Thus, design choices and considerations performed by the designers/project engineers, such as factors of Safety (FoS) in mechanical or other analyses, climatic and seismic environment, quality of existing documents and reporting, social and geopolitical environment, etc. are to be included in the probability of failure of each element but do not constitute a hazard as defined above. The same is true for maintenance and monitoring. However, deviations from the intended level of care in maintenance and monitoring are considered as hazards.

Table 7.1 Specific example from a real-life study

Threat from	Vectors
Geosphere (including climate change)	Earthquake
	High wind (windstorm and hurricane)
	Lightning
	Snowstorms (snowmelts are included in G6)
	Extreme cold, freezing rain
	Flooding, extreme rains/extreme drought
Electrical	System of communication, IT
	Power, electric
	Power, hydrocarbons
Fixed and moving equipment	Equipment failure from natural causes such as wear
	Fire, explosion due to any hazard EXCEPT terrorism, pandemic, outages and natural disasters
Personnel	Human error
	Succession planning of key personnel
	Pandemic
	Employee dishonesty
Public and public agents	Riots
	Arson
	Cyber attacks

Table 7.1 defines—as a specific example from a real-life study, not as a general list—a series of threats-from families and vectors potentially lurking on the system's elements, and thus potentially generating hazard scenarios.

The threat-from and threat-to concepts are essential to risk assessment and risk management. Oftentimes engineers and technical people have a "facility-centric" approach, that is, they see the facility they are studying as a potential source of hazard exposures and forget about external exposures. This means looking, for example, at a process or sub-process as the sources of hazards to the environment but not looking at what the environment can do to the process.

Likewise the analysts oftentimes see the facility as generating threats to the population, environment, etc. but forget to think about what hostile population could do to their process, logistics, etc. Routine uses of threats-from are seismicity, climate and perhaps terrorism but many other threats exist. In reality, *taxonomies clearly stating the two classes* (https://croninprojects.org/Ethics-RFG2018/GeoEthics-Acr oss-Property-Lines/Cronin-RFG2018-PropLines-LowRes.pdf) are rare. This leads to confusion of terms similar to the one between hazard consequences and probabilities. As we have noted many times, this is a common pitfall of many risk assessments. A recent paper states, "The mining industry's approach to disaster risk reduction (DRR) focuses on a narrow set of external vulnerability factors in understanding the cause of dam disasters" (Owen et al. 2020), confirming at least for that industry our point of view.

In order to better the information level to all stakeholders we believe the taxonomy requires some examples. However, before delivering these examples, we need to define potential hazard on operations and corporate level (not to be misunderstood with operational hazards or strategic hazards, see Chap. 12):

- Operation hazards consist in what happens within the operation's system (that is, within the physical and informational battery lines, see Chap. 6). As a matter of fact, they include external hazards hitting the system. Resources needed to run the operation, such as raw and key materials and key suppliers, are considered to be operation's hazards even thus they may lie outside of the operation's physical perimeter. In like manner protests by local groups (see Sect. 5.3) are also operation hazards, unless they escalate to the regional or national level.
- Corporate hazards are external hazards impacting operations indirectly (outside of the operation's perimeter), such as strikes at commercial wharves, logistics difficulties, Forex and tariffs, loss of IP (due to cyber attacks, theft, etc.), and finally, changes in regulations. In addition, personnel salaries levitation due to governmental policies or other geo-economic causes are also corporate hazards.

However, some hazards can hit both levels. Cyber hazards, for example, can hit both the corporate and operation levels.

Let us now look at an example of threat-from and threat-to concepts in natural resources, namely an operation with the following macro-elements:

- extractive area;
- plant;
- waste management facilities.

Tables 7.2, 7.3 and 7.4 show examples of threat-from and threats-to for each of the macro-elements.

Evidently the lists in the tables should be continued to reflect the operation's system.

We also want to go back to Fig. 6.3 (Anecdote 6.2) and display the sawmill example together with the identified hazards (Fig. 7.1).

Going further into details, hazard scenarios can be "elemental" or developed in long chains of "compounded/domino failures" (interdependencies) leading to

Table 7.2 Extractive area threat-from/threat-to hazards examples (to be completed on a case-by-case basis)

Threat-from	Threat-to
Protests, blockades, terrorism	Process (lack of raw material)
Geology (seismicity)	Population (town contamination)
Geology (insufficient knowledge)	Environment

Table 7.3 Plant threat-from/threat-to hazards examples (to be completed on a case-by-case basis)

Threat-from	Threat-to
Protests, blockades, terrorism	Workers (Health and safety)
Extractive area	Population (air, water and soil contamination)
Engineering (process and buildings)	Tailings (poor operational management)

Table 7.4 Waste management facility threat-from/threat-to hazards examples (to be completed on a case-by-case basis)

Threat-from	Threat-to
Protests, blockades, terrorism	Workers (Health and safety)
Environment/climate/seismicity	Population (air, water and soil contamination)
Project and design (lack of effort, excessive arrogance, errors and omissions)	Environment
Mill (poor operational management)	Production, workers (Health and safety)
Operations (insufficient inspections, monitoring, maintenance, repairs, etc.)	Production, environment, workers (Health and safety)

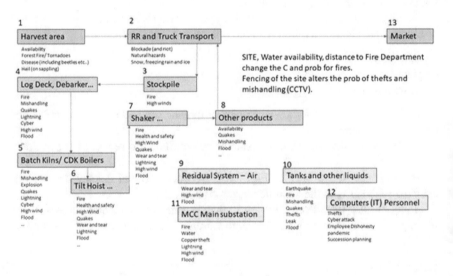

Fig. 7.1 Sawmill system (see Fig. 6.3, Anecdote 6.2) completed with identified hazards examples and hazard related notes

outcomes (consequences) of an impact larger than the initiating event. In order to enable proper consideration in a rational and transparent analysis, hazard scenarios must be:

- identified and classified in a proper hazard/risk register. The detail of the register will be compatible with the level of information available at the time of the study;
- characterized by at least (relative) likelihood and magnitude (intensity);
- evaluated in terms of potential targets, vulnerabilities and consequences;
- used to generate risk scenarios which have to be evaluated, ranked in the hazard/risk register and properly communicated to various audiences linked to the project to allow preliminary decision making.

Provided the steps above are developed in a consistent and scientific way, risks can then be compared to risk tolerance criteria (see Chap. 10) leading to rational and transparent risk prioritization, but even more importantly, to performing Risk-Based Decision Making (RBDM) or Risk-Informed Decision Making (RIDM) (see Sect. 12.2), which can be understood and shared with all the project's stakeholder. Although these terms will be discussed in Sec. 12.2, we note here that RBDM can be seen as a subset of RIDM, the two having a common point of departure which is a risk assessment. In this book we refer to RBDM when an analysis shores and supports decision making whereas we consider RIDM when we discuss tactical and strategic risks, extending into formal financial analyses of mitigative alternatives.

Table 7.5 displays the various "families" of elements with the potentially impinging hazards (not necessarily all apply to the specific case), scenarios, related effects and mechanisms, and finally their possible evolution to ultimate failure.

Hazard scenarios linked to information technology (IT) malfunctioning (for example sensors, SCADA systems or monitoring delivering erroneous information for technical/intrinsic reasons) are included in the line to which they pertain for each element as IT. Hazards linked to malevolent or criminal hacks on IT systems are included in the line to which they pertain for each element as "cyber". Of course the table can be more detailed, more granular, as pertinent with each study purpose, budget and schedule.

In this chapter we focus on the methods to be used to gather information about the potential hazards impinging on the system's elements and sub-elements. We will close the chapter by discussing how to ensure the hazard register is maintained current, how to check for future discrepancies, and how to decide if updates are necessary.

7.1.1 Leveraging Technology for Archival Discovery

As volumes of global data increase exponentially IDC (International Data Corporation) estimates that by 2025 worldwide data will grow 61% to 175 zettabytes!), risk analysts have to review larger volumes of information quickly and accurately to increase process efficiencies and avoid paralysis. Fortuitously, the capabilities of technology and machine learning to augment human review have been growing at a rate comparable to that of data. Today, we find ourselves with highly developed

Table 7.5 Examples of various "families" of general elements of a system with the potentially impinging hazards

General element of the system	Potential impinging hazard scenario	Effects and mechanism
Pipes, ponds, tanks	Traffic (includes contractors and snow removal collisions)	Spill from minor to full break and other collision effects
	Freezing (includes deformations at flanges)	Spill from minor to full break
	Sabotage/vandalism, cyber	
	Corrosion and intrinsic mechanical failures	
	Human (communication between mill and tailings operation), IT	
Buildings, confining structures (and their content)	Geotechnical/geological and climatological, hydrological	Settlement, deformation damages, seepage/piping, high pore pressure, cracks, static and\or dynamic liquefaction collapses
	Construction/building operations	
	Operation, maintenance and monitoring during service and record keeping, IT	
	Sabotage/vandalism, cyber	
Waste outflow structures	Geotechnical/geological and hydrological (capacity, durability)	Unsatisfactory performance
	Operations (service, emergency access), IT	
	Climate (includes extreme freezing)	Reduced capacity

technological approaches, but it can be difficult to know which technology to utilize and why.

In this section we highlight technologies and tools commonly implemented to drive proactive strategies and work in partnership with reactive tools, in order to aid in risk identification. We also touch on the importance of project management for effective technology implementation in this field.

Please note that the list provided is not exhaustive, as there are many other analytics and machine learning tools that can be deployed to help streamline the ability to identify risks. However, these are the most common tools we see used in the legal discipline and that we can tweak for risk assessment purposes.

Let's note for a start that there is no one information governance strategy, but professionals in this space commonly refer to it as "getting your data house in order". Common themes include migration from legacy systems (email and email archive, file share, SharePoint, etc.), compliance (data privacy, data retention, legal hold solutions, etc.), and data security (enablement, autoclassification, dark data, incident response

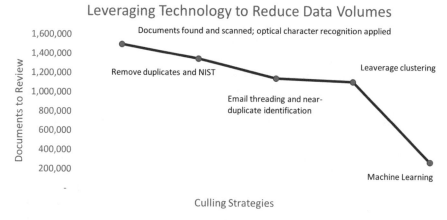

Fig. 7.2 Leveraging technology to reduce data volumes: stepped approach can reduce documents number five- to six-fold

and loss history). The process of information governance removes junk data (Fig. 7.2), reduces the risk of errors in the data and/or alleviates the downstream work of having to sift through stacks of electronic hay to find that proverbial needle.

Here is a list of analytic and machine-learning tools, what they do, and why we use them:

- **Optical Character Recognition (OCR)**: Processing a digital image in order to make it a searchable and editable text file. This is commonly used to recognize text that is scanned and works well with historical documents. This is often the first step in creating structured, labelled data out of unstructured documents.

- **Email threading**: Identifies and groups together emails that are part of the same conversation/thread. This allows duplicate content, such as the earliest emails in the thread, to be suppressed.

- **Near-duplication**: Groups documents that are highly similar to each other and identifies differences, similar to track-changes. It is useful for identifying documents that have undergone revisions, or for finding 100% text-similar documents of different formats such as a Word and PDF copy of the same document. Once documents have been grouped by text similarity, entire groups can be reviewed together, or documents above a certain level of similarity can be identified or suppressed.

- **Concept searching**: Instead of specific search terms, concept searching identifies documents that contain themes and language patterns similar to the example provided. Using key sentences or paragraphs, the database can be searched for additional records that contain similar concepts, facilitating identification of potentially relevant or privileged documents, or documents that address specific issues.

- **Categorization**: Allows many documents or paragraphs to be submitted as examples and returns documents that are conceptually similar to those examples. This is similar to concept searching, but on a larger scale.
- **Clustering**: Automatically groups documents based on language patterns and similarity. Clustering can be useful in unfamiliar datasets to provide a bird's-eye view of the content in the database and can be used to quickly promote or suppress large groups of documents from review.
- **Key expansion**: Identifies different language used to express the same or similar concepts. Keyword expansion can also be used on a word to identify other conceptually related terms and words in your index that you did not expect.
- **Supervised machine learning and continuous active learning**: Uses input from reviewers (relevant vs. not relevant for example) to categorize documents in the database and predict whether they are likely to be relevant. Data volumes can be substantially reduced using this technology, suppressing as much as 70% from review. Machine learning and AI technologies now have the ability to analyze sentiment (i.e., negative or positive patterns concerning employee behavior, trade secrets and other workplace matters), but we have not used this capability to date.

There are many reasons to incorporate tools into Quantitative Risk Assessments (QRA) workflows, including providing early access to key information, organizing information faster than ever before, decreasing the time needed for review, reducing costs associated with review, keeping companies competitive through fast-track risk assessments of new opportunities, and the ability to handle more work while keeping headcount the same.

Understanding the requirements before any deployment allows optimal application of these technologies and is one of the most important aspects of this process.

As stated in the *PMBOK Guide®* (https://www.pmi.org/pmbok-guide-standards/foundational/pmbok) by PMI (Project Management Institute), the five process groups defined for the life cycle of a project are: (1) Initiating; (2) Planning; (3) Executing; (4) Monitoring/Controlling; (5) Closing. Utilizing project management skills for technology implementation is crucial for the success of a project. Below, we briefly highlight the execution and monitoring/controlling aspects of the implementation of technology.

- "Garbage in, garbage out". The phrase skillfully articulates that, in the sense of training continuous active learning tools, the end result directly correlates to user input (human reviewer). The software continues to learn as more documents are coded by human reviewers and uses advanced statistics to determine when reviewers can stop based on the probability and predicted number of documents that directly correlate to human training may remain in the unreviewed set. Therefore, human input is the most essential piece of any machine learning workflow.
- As you embark on your technology implementation journey, it is important to remember that there is not a "magic button" that will completely remove the human element of review. There are tools to help augment human review so identifying

risk can be done faster and more accurately than ever before. The ultimate goal of incorporating technology into human review is to weed out irrelevant documents and focus on the pertinent issues, while minimizing time spent and concentrating efforts on high value tasks.

7.1.2 Workshops and Interviews

We have seen dozens of workshops where a few alpha individuals influenced everyone else and subordinates did not dare to talk. Furthermore, in general, these gatherings are meaningless because the glossary is not defined or adhered to, and the system is not defined beforehand. Below is our proposed remedy to alleviate these pains.

We consider disruption to be the key to unlocking better results. Indeed, if one watches Tim Harford's TED talk entitled "*How Frustration can Make us more Creative*" (https://www.youtube.com/watch?v=N7wF2AdVy2Q#_blank) a world opens up. Harford reports that psychological tests have shown that students who received handouts written with difficult-to-read fonts did better than students with handouts with easy-to-read fonts! Those tests show that a little difficulty leads to better results because it slows down the students and forces them to think more.

Harford then cites complex problem solving, which is generally considered to require a step-by-step procedure: prototype, tweak, test, improve. This widely applied procedure leads to incremental improvements but can also lead to painful dead-ends. If randomness is added (stupid moves, *mistakes*) (http://www.riskope.com/2013/03/28/no-failures-no-progress/#_blank) dead-end chances are reduced and quantum leaps often occur. He adds further that if a group of professionals is tasked with solving a problem, significantly better results are obtained by a group of strangers rather than a group of friends despite a generally stellar self-evaluation of the friends' group. In other words, friends' groups are complacent.

This last statement in particular leads us to avoid interviewing "buddies' teams" at our clients' operations. Furthermore, we always adopt a "Daddy what's that?" attitude. Indeed, we ask questions and never accept elliptical answers. We can tell by how the interviewee looks at us that after a while this drives them nuts. However, we do not care! Our goal is to disrupt workers' complacency. Our selection of interviewees and questions aims at introducing the element of randomness that psychological tests show to be so important.

It generally works this way:

- DAY 1: we ask questions and get elliptical answers: "I do not know"; "we have always done it that way"; "Joey knew, but is now retired", etc. Sometimes we actually run serious risks of getting a punch in the nose!
- DAY 2: people are better prepared: the disruption has provoked a tad of shame and lots of curiosity. They have discussed, possibly read old reports, called Joey… Sometimes we joke that the increase of knowledge on day 2 is likely one of the largest, immediate and most economical advantages of performing a risk assessment.

- DAY 3: interviews and meetings become more fruitful and *hazard iden-tification* (http://www.riskope.com/2016/02/18/intensive-risk-management-mod ule-at-the-university-of-turin-saa-business-school/#_blank) is easier as a creative environment is finally in place, and complacency has been mitigated.

Over the years the "Daddy what's that?" attitude has brought amazing benefits to our clients and great professional satisfaction to us.

Remember what you are aiming for during these sessions is not an audit. It is not a policing act. You must be earnest in your desire to understand how the system works. What you want to do is to gain as quickly as possible an unbiased and factual understanding on how the system (operation, process, team) works in real life. You want to see what lies beyond "official" flow charts and organizational schemes.

Here are four exercises that we use to engage the audience of workshops delegates.

1. **The deck of avatar cards**. This deck consists of eight cards, each representing an avatar typically encountered in organizations. For example, Mrs. Rozy Scenario is an avatar of the overconfident, hazard and risk-unaware character, while the Mr. Perryl Shield avatar represents the character who believes that technological, brute force mitigation can solve any present and future problem (Fig. 7.3).
   At the beginning of the interviews, whether they are one-on-one or group, we ask participants to select the avatar they believe most closely expresses *their attitude toward hazards (risks) and mitigations* (http://www.riskope.com/ 2015/05/28/why-when-approaching-strategic-tactical-operational-planning-one-needs-to-know-about-ostriches-denial-and-prayers/#_blank). During the course of the interview, if we detect a divergence between the explanations they give and the avatar character, we keep challenging the interviewee. It is fun, people enjoy it, and it helps to find weak spots in the stories without it being (too) personal.
2. **The Buddies Hazard Identification Role Play**. We ask the participants to split into three to four groups as they see fit. Generally this leads to "buddies groups" (possibly based on age, sex, cultural background, mind-set, etc.). We task each

Fig. 7.3 Left Mrs Rozy Scenario avatar; **right** Mr Perryl Schield

group to perform hazard identification on the system after it has been modelled and after delivering a detailed explanation of the hazard terminology.

As each participant in each group has previously selected his/her avatar, we engage the individuals and the groups based on their statements and avatars. We also ask the groups to discuss their findings. Currents of organizational thought emerge. Alpha representatives become visible independent of their hierarchical status. Concerns and hazards are sorted out (we are not really interested in concerns, unless they prove to have the potential to generate risk exposures).

3. **The Alternative Buddies Hazard Identification Role Play.** This is very similar to the previous exercise, but here delegates evaluate themselves in terms of their audacity and appetite for change using a scoring system before splitting in groups as they see fit. The rating system delivers a finer evaluation than the avatar cards. It opens the door to a specific test (see point 4 below) that is run during the second part of the hazard identification. In some cases we may ask delegates to use the deck of cards, then auto-evaluate and discuss blatant divergences.

4. **The Well Balanced Groups Hazard Identification Role Play.** In order to help us in deriving a personal objective audacity and appetite for change rating and related "talents", we present a questionnaire for participants to fill in. Delegates can now discuss the gap between their self-evaluation and the test rating, in view of understanding how this can impact their hazard awareness and mind-set. We then create working groups with teams that are balanced in terms of audacity and appetite for change.

As in exercises 2 and 3, we now task each group with performing hazard identification on the system. That is, of course, after we have modelled the system and provided a detailed explanation of what constitutes a hazard. As above, we engage the individuals and the groups, and ask the groups to launch into discussing on their findings. Organizational "currents of thought" emerge. Alpha representatives become visible independent of their hierarchical status. We sort out concerns and hazards (again, we are not really interested in concerns, unless they prove to have potential to generate risk exposures).

These four exercises summarize techniques we find useful during hazard identification interviews. It is not necessary to perform all four. The selection of which will be more effective in a given setting is made on the spot, on the basis of the preliminary interactions with the participants and their degree of willingness to participate.

7.1.3 Monitoring

Standard methods to acquire information on a given system include the whole set of design, production, supervision, distribution, geological, climatological, meteorological, geotechnical, social and environmental monitoring techniques. There are so many that even merely listing them is beyond our scope here, not to mention providing any details related to them.

When performing a comprehensive risk assessment, it is fairly common to have various forward (forecasts, predictive) and backward (historic, archival) information sources for data, including but not in any specific order:

- business information, business intelligence;
- media and third-party reviews;
- markets reviews;
- classic weather, ground-water, topographic, deformation instrumentation, manual, or automated to some level;
- drones and/or satellite imagery and data;
- reports and analyses;
- records of losses;
- documentation from engineering/technical parties;
- personnel who have worked for many years on site;
- traditional knowledge from the local population.

Each of these comes with their own level of credibility/uncertainty and is generally geared toward providing information to various stakeholders, but not the risk assessor. Thus, they rarely provide direct information about potential failure modes or failure causality, but give clues on areas of interest. Anecdote 7.1 offers an example.

Anecdote 7.1: Monitoring of Joey's corporation acquisition

Both forward and backward monitoring were used to evaluate the acquisition risk assessment for a specialty automotive company named Joey. The holistic monitoring program revealed that Joey's activities were afflicted by a number of day-to-day issues:
- poor H&S practices leading to 20 injuries in 2011;
- lack of contract discipline;
- uncapped liabilities have to be eliminated;
- poorly formulated force majeure clauses.

This last is especially important as have to be reviewed/reformulated, as they constitute, in our experience, one of the largest dormant risks in many industries (see Sect. 4.4).

Backward monitoring showed that Joey's operations had a documented history of problems (with a very wide range of monetary consequences, ranging from minor to several million dollars) including but not limited to:

- flooding;
- workplace accidents;
- poor workers noise protection;
- poor maintenance;
- human resources;
- permits and compliance, environmental (spills);
- poor contract protection and discipline;

- sensitivity to market's shifts;
- missing sales closings.

Closing point. Backward monitoring contributed to the hazard evaluation and forward monitoring ensured that corrective actions and mitigation would maintain the situation under control.

7.2 Methods for Hazard Identification

In this section we focus on satellite observation (generally called "space observation"), which is beneficial to Quantitative Risk Assessment (QRA), insofar as it makes it possible to feed enhanced data into an a priori risk assessment and deliver on a regular basis updated risk assessment with a significant economy of means while answering modern requirements. It also makes it possible to perform analyses of large portfolios such as those of large industrial companies with many systems scattered throughout vast territories and countries, or governments wishing to keep the industrial operations portfolio under their jurisdiction in check. Drone observation is more limited in scope and coverage, but has become so common that *the popular press has published articles on the subject* (https://thebossmagazine.com/mining-with-drones/) and some of our clients even have a drone division.

Nowadays publicly available tools such as Google Earth or Google Maps offer extremely valuable satellite-derived optical information. In many inhabited areas users can complement this with ground-based observations. These tools make it possible to gather information about land use, residential areas, and more. We use them routinely to inspect transportation corridors and infrastructures. However, commercially available solutions offer way more.

7.2.1 Satellites

Commercially available satellite imagery supplies necessary radar and optical data to enhance the evaluation of probability of failures and consequences. These solutions become attractive in cases of very large surfaces, or when publicly available information is very scarce. The best solutions are based on a combination of radar and optical imagery and proprietary algorithms. The observation should take advantage of historic imagery databases and look back as far as data is available.

Data and imagery derived from space observation can be used for interpreting a variety of civil systems observations (Oboni et al. 2018), including:

- where there are areas of potential land-use change or perimeter encroachment;

- deformations and topographic changes. Remarkably, the new *European Ground Motion Service* (https://land.copernicus.eu/pan-european/european-ground-motion-service) (EGMS) platform aims to provide reliable information regarding ground deformations over Europe, with great accuracy.
- quantitative differences in soil wetness or standing water;
- current vegetation health in comparison to that of similar vegetation of past years;
- pre- and post-accident site analyses.

Space observation generally involves both manual and automated methods of interpretation, which complement each other, and limit residual risks related to this element of monitoring. A manual search by an experienced interpreter of high-resolution imagery is aimed at identifying the nature of the change, to conduct a checklist search to identify any other issues of concern, and to ensure that items that may have been missed by the algorithms are caught. The automated methods are effective in drawing the interpreter's eyes to areas that might otherwise be overlooked.

One of the great advantages of space observation lies in being able to tap into historic data that can go as far as the beginning of *cold war spy satellites* (https://www.nytimes.com/2021/01/05/science/corona-satellites-environment.html?referringSource=article). We strongly recommend starting any space observation with a backward-looking analysis, covering as long a period as imagery is available, while recognizing that imagery quality was lower in the past than it is today.

Available space observation tools relating to the monitoring of civil systems includes Synthetic Aperture Radar (SAR), Interferometric SAR (InSAR) and specialized satellite sensors supplying high-resolution imagery in the visible and near-visible wavelengths. The use of SAR to study the earth's surface was made popular by the launch of the Earth Resources Satellite (ERS-1) and RADARSAT in 1990 and 1991, respectively. Products available in the marketplace today that relate these observations to applications include a number of publicly available free resources. The key is to select a resource compatible with the requirements of each project and not to fall for too good to be true promises.

InSAR solutions derive ground movement from a precise observation of the time that it takes for a radar pulse to be returned to the satellite sensor. Radar reflections (Ulaby et al. 1986) are the result of the type of surface and the geometry of the observation. For instance, flat calm water will redirect the radiation away from the radar and an image will show low backscatter values. A metal roof oriented to reflect radiation toward the radar is a very bright target, while a ceramic structure would be basically invisible to a radar. With the understanding of the reflective properties of the structures to be observed, the dependability of radar means that it is possible to make series of observations over long periods of time that allow the automated detection of year-to-year changes from, for instance, ongoing construction activities, soil wetness or standing water, and vegetation health issues. Reproducible measurements of ground movement have been demonstrated to within 2 mm/month in specific cases (Henschel and Lehrbass 2012; Henschel et al. 2015; Mäkitaavola et al. 2016). These are not averages or common results as vegetation, snow, ice and other disturbances can reduce the precision. Nevertheless, in various occasions, InSAR has made it possible to identify anomalies that would have gone undetected in traditional on-site observations.

There are a multitude of optical satellite systems available for monitoring both large and small areas at a very high resolution. Optical satellites provide images of the planet in visible and near-visible frequencies. Visible pictures of the surface are available with resolutions of a few centimeters and can provide instantaneous descriptions of a structure and its surroundings. Optical satellites can indeed be powerful. The immediate recognition of objects from the imagery helps the human brain to provide context and quickly exploit the image information. While change from one image to the next can be understood very quickly, understanding the difference caused by a persistent change or a particular type of anomaly can be more challenging. Some questions pertinent to risk assessments include:

- When and where did historic change first or most recently occur?
- Which change is definite versus less likely based on the strength of the spectral difference and other factors?
- What is the probable type of change?

All of this information can be directly used within a risk assessment procedure to update the probabilities and consequences values.

The techniques can be integrated with probabilistic analyses to automate or semi-automate the updating of systems' probabilities to selected extents (Oboni and Oboni 2020 Sect. 10.1.2). Of course, like all techniques of this kind, caution must be exerted during all phases of the deployment. This is not a universal panacea but, to use an automotive metaphor, it provides a good set of lights to drive through the night and a couple more instruments on the dashboard to alert the driver in case of an emerging problem, thus furnishing impartial, fact-driven, emotionless aids to driving. Thus, while we cannot ensure that these automated systems will detect divergences, we can certainly say that they will enhance business-as-usual planning and mitigation capabilities. It is possible to develop probabilistic updating (see Sect. 8.1.4) of various types of data which may include, among other things, deformation velocity (e.g., cm/year), and number of events of a certain magnitude (e.g., number of events exceeding a certain magnitude per year), etc. The updating then makes it possible to re-frame probabilities present in the quantitative risk register and to re-evaluate the risks.

The various examples above show the benefits found in linking multi-temporal objective space observation with a dynamic convergent quantitative risk assessment platform in projects and operations. The two-pronged approach enables us to measure and make sense of a complex problem. Among other things, it allows us to:

- transparently compare alternatives;
- discuss rationally and openly the survival conditions;
- evaluate the premature failure of a structure.

Connecting a dynamic quantitative risk assessment platform with a high-performance data gathering technique reduces costs, avoids blunders, and constitutes a healthy management practice, especially for long-term projects requiring short- or long-term monitoring, including, of course, site restorations.

7.2.2 Big Data, Thick Data and AI

Big data and the Internet of Things (IoT) are becoming common features in all sorts of business activities. They will help define better ranges for reliability and failure of a system's elements, and make it possible to search world-wide occurrences of near-misses, losses, news, etc. At the other end of the spectrum, thick data are useful to understand deep motivations and can foster SLO and CSR by fostering proper communication (see Chap. 5). Anecdote 7.2 provides an example of thick data that we experienced in the Swiss Alps a while ago.

Anecdote 7.2: An Alpine story

Distinguishing knowledge in risk assessment and risk-informed decision making is paramount. Once upon a time we were studying large landslides in the Alps. We were working on a Swiss Federal Research Project in the capacity of geotechnical engineers and risk (hazard) specialists, as members of a multidisciplinary team comprising geologists, hydrogeologists, and monitoring specialists. The research project focused on landslide-prone areas characterized by continuous (slow) movement, where failure is defined as a brutal and sudden acceleration of the sliding movement.

We deployed brand new methodologies for investigation, testing and probabilistic slope stability analysis to build specific knowledge. The aim was to establish new guidelines for tactical and strategic planning of land use of landslide-prone areas.

Although it was not yet fashionable to use terms such as risk and risk-informed, the idea was precisely to provide risk-informed guidelines based on continuous real-time monitoring (today we would call that a big data approach).

As a matter of fact, we looked at better ways to forecast potential accelerations, using monitoring and probabilistic approaches; areas that would accelerate within a complex landslide; complex "multi-dimensional" consequences and, finally, mitigation in terms of preserving structural integrity and minimizing discomfort of inhabitants. That is to say, all the ingredients of the risk equation were present, but rather than calling it "risk" we called it good engineering sense and practice.

We soon discovered that specific knowledge is not enough, even in the presence of large budgets and, accordingly, big data. General knowledge is required to foster understanding of complex situations. Indeed, big data is not enough to understand phenomena and requires parallel deep data understanding.

Acquiring deep data may be easy, pleasant and cheap. General knowledge also reveals truths that specific knowledge oftentimes fails to reveal. To acquire general knowledge cheaply there is one universal recipe: talk to locals and ask them questions they can understand and answer.

For instance, one day we were looking at a chalet sitting in a particularly distressed slope. The structure was unscathed whereas all the neighbors had worrisome cracks in the stone walls. Geologists and monitoring experts were scratching their heads. So, we spoke with the farmer who candidly said, "Well, my house is solid because my Grandfather built it on rock". Now, the whole team was shaking their heads. The landslide was likely hundred meters deep and there was certainly no bedrock in that location. The farmer then took us to the cellar and he was right! The whole cellar was in solid rock! We finally understood what specific knowledge did not reveal: Grandpa knew by observing his cattle grazing in the area that there was rock in that corner of his property. Thus he decided to build the house on it. What he did not know is that the rock was a huge rock slab, likely detached from bedrock after last glaciation, behaving like a raft in a river (the slide).

That was not an isolated case. We found numerous perfectly preserved churches high in the Alps, apparently sitting in actively sliding slopes. Each time general knowledge reasons explained that blatant anomaly in the damage level.

When we worked for the *GICHD* (https://www.gichd.org/) we supported the development of a *hybrid general-specific knowledge-based method* (https://web.archive.org/web/20120115024955/http:/www.gichd.org/fileadmin/pdf/publications/Lao-PDR-Risk-Management-March-2007.pdf) for predicting the presence of landmines in suspected locations. The methodology showed promising results after pilot testing.

Closing point. Whether one deals with large slope instabilities, infrastructural projects, dams or other projects, it is paramount to acquire general and specific knowledge. We would even say that general knowledge should guide specific knowledge acquisition. That is why in the ORE (see Fig. 11.2) methodology we call the starting block "rich data", and why we also tell our clients we want to study existing documentation, reports, media, and publicly available information before going to site. Finally, that is the reason why we generally go to site with a preliminary version of our a priori *ORE risk assessment* (https://www.riskope.com/company/riskopes-methodology/).

Big data and Thick data are actually two sides of a coin (Fig. 7.4). It is essential to understand their differences.

Big data is a term for large or complex data sets that traditional software has difficulties processing. Processing generally involves, for example, capture, storage, analysis, curation, searching, sharing, transferring, visualizing, querying, updating, etc. However, big data also often refers to the use of predictive analytics, user behavior

Fig. 7.4 Big data versus thick data

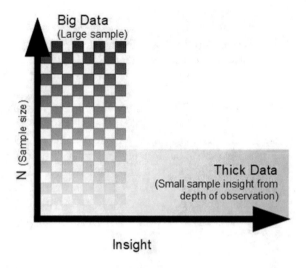

analytics or certain other advanced data analytic methods. Analysis of data sets can find new correlations to spot business trends, prevent diseases, combat crime and so on but focusing solely on Big data can reduce the ability to imagine how the world might be evolving. Big data only is not sufficient for risk assessment, and in particular *hazard identification.* (http://www.riskope.com/2017/04/05/hazard-ide ntification-science-art/#_blank) It can create a distorted view of the risk landscape surrounding an entity. Big data relies on machine learning, isolates variables to identify patterns, reveals insight. Big data gains insight from scale of data points, but loses resolution details. It does not tell you why those patterns exist and is unprepared to cope with new extremes, divergences.

Thick data is generated by ethnographers and anthropologists, adept at observing human behavior and its underlying motivations. Thick data is qualitative information that provides insights into the everyday emotional lives of a given population. Thick data relies on human learning, accepts irreducible complexity, reveals social context of connections between data. Thick data gains insight from anecdotal, small sample stories, but loses scales. It tells you why but misses identifying complex patterns or future behavior.

To date, big data and thick data have been used and supported by different groups. Organizations grounded in the social sciences tend to use thick data, while corporate IT functions tend to favor big data. This constitutes a perfect example of silo culture. Ideally, big data and thick data should "talk to each other", but most of the time do not because of siloed approaches.

If one is seeking a map of an unknown risk territory (risk landscape) and data are scarce, then thick data is the tool of choice. As data availability grows on its way to becoming big data, integrating both types of data becomes important. In the case of innovative companies, that combined insight can be highly inspirational.

When performing risk assessments, we always collect and analyze stories, anecdotes and loss reports to gain insights into pre-existing states of the system. The combined insight may tell us that a system that looks wonderful actually has a congenital defect that may raise the probability of failure. Big data would not be capable of showing that but could probably reveal a pattern between third-party observations and, say, meteorology. In fact, it could reveal patterns among any other groups of variables, which could sound an alarm on shorter-term emergent hazards.

Working successfully with integrated big and thick data certainly enhances any risk assessment. Over the years we have found ways to integrate data from multiple sources and of various natures in our risk assessments. We routinely use incomplete thick data sets in conjunction with expert opinions and literature to generate a first, a priori estimate of the probability of occurrence of hazards and failures (see Chap. 8). This immensely increases the value of the first-cut risk assessment, which can then be updated using big data and Bayesian techniques (see Sect. 8.1.4).

The combined approach also makes it possible to enhance the value of big data, avoid capital squandering, and reduce the running cost necessary to obtain big data. Recent studies have shown that without that approach data oftentimes remain virtually *unused* (https://www.inc.com/jeff-barrett/misusing-data-could-be-costing-your-business-heres-how.html).

Integrating big data and thick data brings value and should be fostered. Thus, it is crucial to explore how big data and thick data can supplement each other. This demands the integration of qualitative evaluation and expert-based judgments with hard quantitative data.

While we recognize that merging big and thick data isn't easy, we do it on a daily basis.

We could not close this section without relating a case where thick data showed all its power and avoided doubling the Fukushima disaster (Anecdote 7.3). One man did it, soon to be forgotten. In the business of risk management, it is sad but true that success is not as flamboyant as failure.

Anecdote 7.3: Onagawa nuclear power plant

There is a good reason why you probably never heard of the *Onagawa Nuclear Power Plant* (https://en.wikipedia.org/wiki/Onagawa_Nuclear_Power_Plant#2011_T%C5%8Dhoku_earthquake). Although it was the nuclear power plant closest to the epicenter of the 2011 Tōhoku earthquake, way nearer than the Fukushima I power plant, it did not fail, even though the town of Onagawa, to the northeast of the plant, was severely damaged by the tsunami.

This was because civil engineer and corporate executive in the electric power industry Yanosuke Hirai, against the advice of all his colleagues, pushed for a higher breakwater wall. Hirai also proposed changes to the plant's water intake cooling system pipes so they could still draw cooling water during the low-tidal movement preceding a tsunami. He prevailed, and Tōhoku Electric spent the extra money.

Hirai Yanosuke had been around the location proposed for the plant. He walked up into the mountains where he saw the stone markers indicating past events flooding levels. He read and understood a centuries' old message.

Closing point. When reporting on Fukushima and other accidents, many experts erroneously or falsely claimed it was a completely unpredictable black-swan event (Shrader-Frechette 2011, also see Sect. 1.1). Relying almost completely on industry-supplied data, government representatives also repeatedly made a black swan claim, as when the Japanese Atomic Energy Commission said the Fukushima Daiichi disaster was 'unpredictable' (Broad and Jolly 2011). While such reports show a blatant misuse of terms, Hirai's investigation is a brilliant example of thick data which debunks the black swan statement made by sensationalists as discussed in Sect. 1.1.

7.3 AI and Machine Learning

As said above, big data relies on machine learning. But *machine learning and AI can only work if big data exists* (https://www.forbes.com/sites/joemckendrick/2019/09/14/artificial-intelligence-only-goes-so-far-in-todays-economy-says-mit-study/?sh=473b1d821162). A recent MIT study has shown this brilliantly. AI uses neural network algorithms and machine-learning techniques to develop a sense for the interested problem, because it cannot a priori compute all possible available choices. It needs to carry out a large number of sample operations to conclusion and then select which ones encompass good solutions. Of course if every learning sample were to stop halfway, then the machine wouldn't know if what it did was good or no, that is, it could not learn what to do with anything beyond business-as-usual, thus divergence.

Let's use an example from the game of Go. The *AI program AlphaGo* (https://deepmind.com/research/case-studies/alphago-the-story-so-far) defeated professional player Lee Sedol in a five-game Go match. The program used a neural network algorithm and machine-learning techniques to develop a sense for the game, playing matches against itself, and learning from its mistakes.

Now, to go back to AI and monitoring. In many industries and infrastructures applications failures are relatively rare events despite what sensationalistic media and activists state without delving with detailed numeric analyses or by biasing results. Old reports on accidents were generally censored, possibly biased and based on expert assumptions. It is indeed difficult to perform flawless root cause analyses in the aftermath of a catastrophic failure when extant documentation is insufficient.

Those are precisely the conditions where machine learning struggles. AI learns from failures but, happily, these are rare events. A good illustration of the problem can be found in tailings dams. We know that most dams will be observed for the longest time with no occurrence of catastrophic failure. Even if we were to start monitoring every site to death, using IoT, etc., AI would still not be able to learn. Mind-you this occurs for us humans as well and we use to say: "no failure, no progress".

An old professor used to say that many sophisticated monitoring systems are only very expensive thermometers. Our experience over three decades has confirmed this, and not only as it relates to temperatures, so, please beware of AI over enthusiasm.

7.4 Hazards from Divergent Phenomena

Does climate change generate new hazards and hence risks? Because of the complexity of the issue, let us first say that we will discuss climate change effects, but not climate change causes. In addition, let us note that climate change may generate new large-scale, generalized hazards, such as for example:

- methane releases in permafrost areas (domes, pingos);
- ocean acidification and level increase;
- expanded territories of parasites due to frost-free seasons;
- large scale regional fires.

Finally, climate change will generally alter the probability of "known-knowns" (see below): localized hazardous events (see Sect. 2.2), such as:

- precipitation (rain, snow, hail) and linked flooding;
- electrical storms,
- drought;
- local fires;
- storm surges;
- wind (including dust storms).

Recent papers *highlight these changes* (https://www.nature.com/articles/s41 558-019-0666-7.epdf?author_access_token=4M8-EcJtFxH_jmyWCAoz39RgN 0jAjWel9jnR3ZoTv0OdMx1oJ3ZWa7BKzSg7sgojrZkS3XyaoGGEprx6mTbk-I7n zwcz-JiwcWUvc-q-6L4q6CtnA_imZNvKYWRoRWhHRJb6VkSFg-Fe06c24Ih fwQ%3D%3D) (Technical note 12.2). Correspondingly, the *Russian plan* (https:// www.dw.com/en/russia-unveils-plan-to-use-the-advantages-of-climate-change/a-51894830) to manage the risks and the effects on the *Mississipi "economy"* (https:// www.pbs.org/newshour/show/climate-change-is-jeopardizing-trade-along-the-mis sissippi-river) eloquently show the importance of climate change challenges and potential losses.

Any of the hazards in the list above is likely to be already present in your risk register, provided it is the result of a step-by-step approach, that is, after careful hazard identification within a well-defined system.

New large-scale, generalized hazards such as those listed earlier bring us into the realm of "known-knowns" versus "unknown-knowns. These identifiers have been used and misused by all sorts of people, including politicians. Technical note 7.2 shows their origin and discusses a few details.

Technical note 7.2: The Johari window

Back in 1955 the psychologists Joseph Luft and Harrington Ingham developed a tool called the Johari window (Luft and Ingham 1955). It splits the space of self-perception and public perception into four quadrants called Arena, Blind Spot, Façade and Unknown. The Johari window became world-renowned when the then US Secretary of Defense Donald Rumsfeld quoted parts of it. You probably remember the "known-unknowns" and the "unknowns you do not know you don't know" used by D. Rumsfeld in Department of Defense news briefing. There are interesting applications of the *Johari window in risk management* (https://www.riskope.com/2017/06/21/johari-window-application-risk-management/).

Let's first consider that risk assessments should start by recognizing that the absence of evidence is not evidence of absence. Indeed, if something has never occurred it does not mean it does not exist.

Arena corresponds to "known-knowns", facts that are likely the realm of service life/operational incidents; the things that we all know occurs time to time.

Blind spots are where we find "known-unknowns". These are unexpected evolutions of known events; they should be considered as visible emerging risks.

Façade is where we find corporate "unknown-knowns", i.e., issues management prefers not to see and know, in other words, to ignore. Companies that continue to project rosy scenarios and censor their risks have a strong façade.

Finally, Unknown is where we find the "unknown-unknowns". These are issues we truly "do not know we do not know". Only swift adaptive management working with a sustainable, highly resilient system can solve these.

Closing point. Appropriate risk assessment and management quite obviously expand the area of Arena while reducing those Blind spots and Unknowns.

Appendix

Links to more information about the Key terms from the Authors

A,B	*Act of God* (https://www.riskope.com/2020/12/09/act-of-god-in-probabilistic-risk-assessment/) *Black swan* (https://www.riskope.com/2011/06/14/black-swan-mania-using-buzzwords-can-be-a-dangerous-habit/) *Business-as-usual* (https://www.riskope.com/2021/01/13/business-as-usual-definition-in-risk-assessment/)
C,D	*Convergent* (https://www.riskope.com/2021/01/20/convergent-risk-assessments/) *Divergent* (https://www.riskope.com/2020/11/18/tactical-and-strategic-planning-to-mitigate-divergent-events/) *Drillable* (https://www.riskope.com/2020/01/15/probability-impact-graphs-do-not-fly/)
F	*Foreseeability/foreseeable* (https://www.riskope.com/2021/01/06/foreseeability-and-predictability-in-risk-assessments/) *Fragile/fragility* (https://www.riskope.com/2020/04/01/antifragile-resilient-solutions-for-tactical-and-strategic-planning/)
P,R	*Predictability/predictable* (https://www.riskope.com/2021/01/06/foreseeability-and-predictability-in-risk-assessments/) *Resilient, Resilience* (https://www.riskope.com/2016/11/23/resilience-cannot-based-instinctual-decision-making/)
S	*Scalable* (https://www.riskope.com/2015/04/16/how-system-definition-and-interdependencies-allow-transparent-and-scalable-risk-assessments/) *Societal risk acceptability* (https://www.riskope.com/2014/01/09/aspects-of-risk-tolerance-manageable-vs-unmanageable-risks-in-relation-to-critical-decisions-perpetuity-projects-public-opposition/) *Sustainability/sustainable* (https://www.riskope.com/2019/01/16/improving-sustainability-through-reasonable-risk-and-crisis-management/) *Survivability* (https://www.riskope.com/2011/03/17/ale-fmea-fmeca-qualitative-methods-is-it-really-what-we-need/) *System* (https://www.riskope.com/2017/07/26/three-ways-to-enhancing-your-risk-registers/)

(continued)

(continued)

Links to more information about the Key terms from the Authors	
T,U	*Tolerance* (https://www.riskope.com/2020/04/29/risk-tolerance-thresholds/) *Uncertainty/uncertainties* (https://www.riskope.com/2015/12/10/3-decision-making-truths-derived-from-uncertainty-taxonomy-scheme-of-classification-and-a-road-sign/) *Updatable* (https://www.riskope.com/2020/01/07/climate-adaptation-and-risk-assessment/)
Other linked information (https://www.riskope.com/blog-news/) search Riskope blog and use the search box	

Third Parties links in this section	
Threat-to threat-from taxonomy	https://croninprojects.org/Ethics-RFG2018/GeoEthics-Across-Property-Lines/Cronin-RFG2018-PropLines-LowRes.pdf
PMBOK by PMI	*PMBOK Guide	Project Management Institute (pmi.org)* (https://www.pmi.org/pmbok-guide-standards/foundational/pmbok)
Frustration makes us more creative	https://www.youtube.com/watch?v=N7wF2AdVy2Q#_blank
No failure no progress	http://www.riskope.com/2013/03/28/no-failures-no-progress/#_blank
Hazards identification	http://www.riskope.com/2016/02/18/intensive-risk-management-module-at-the-university-of-turin-saa-business-school/#_blank
Avatars to open conversation	http://www.riskope.com/2015/05/28/why-when-approaching-strategic-tactical-operational-planning-one-needs-to-know-about-ostriches-denial-and-prayers/#_blank
Drones observation	https://thebossmagazine.com/mining-with-drones/
European Ground Motion Service	https://land.copernicus.eu/pan-european/european-ground-motion-service
Cold war spy satellites	https://www.nytimes.com/2021/01/05/science/corona-satellites-environment.html?referringSource=article
GICHD	https://www.gichd.org/
Lao PDR UXO Risk Management	https://web.archive.org/web/20120115024955/http:/www.gichd.org/fileadmin/pdf/publications/Lao-PDR-Risk-Management-March-2007.pdf
ORE methodology	https://www.riskope.com/company/riskopes-methodology/
Hazard identification	http://www.riskope.com/2017/04/05/hazard-identification-science-art/#_blank

(continued)

(continued)

Third Parties links in this section	
Misusing data	https://www.inc.com/jeff-barrett/misusing-data-could-be-costing-your-business-heres-how.html
Onagawa nuclear power plant	https://en.wikipedia.org/wiki/Onagawa_Nucl ear_Power_Plant#2011_T%C5%8Dhoku_earthq uake
Bid data requires big data!	https://www.forbes.com/sites/joemckendrick/ 2019/09/14/artificial-intelligence-only-goes-so-far-in-todays-economy-says-mit-study/?sh=473 b1d821162
Alpha go	https://deepmind.com/research/case-studies/alp hago-the-story-so-far
Climate change highlights	https://www.nature.com/articles/s41558-019-0666-7.epdf?author_access_token=4M8-EcJ tFxH_jmyWCAoz39RgN0jAjWel9jnR3ZoTv0O dMx1oJ3ZWa7BKzSg7sgojrZkS3XyaoGGEprx 6mTbk-I7nzwcz-JiwcWUvc-q-6L4q6CtnA_imZ NvKYWRoRWhHRJb6VkSFg-Fe06c24Ih fwQ%3D%3D
Russian plan to benefit climate change	https://www.dw.com/en/russia-unveils-plan-to-use-the-advantages-of-climate-change/a-518 94830
Climate damages economy along Mississippi	https://www.pbs.org/newshour/show/climate-cha nge-is-jeopardizing-trade-along-the-mississippi-river
Johari window	https://www.riskope.com/2017/06/21/johari-win dow-application-risk-management/

References

Broad W, Jolly D (2011) UN's nuclear chief says Japan is 'far from the end of the accident.' The New York Times, March 27, A10

Henschel MD, Dudley J, Lehrbass B, Shinya S, Stockel B-M (2015) Monitoring slope movement from space with robust accuracy assessment. In: Proceedings of the 2015 international symposium on slope stability in open pit mining and civil engineering. The Southern African Institute of Mining and Metallurgy, Johannesburg, pp 151–160

Henschel MD, Lehrbass B (2012) Operational validation of the accuracy of InSAR measurements over an enhanced oil recovery field. In: Proceedings of 'Fringe 2011 workshop', Frascati, Italy, 19–23 September 2011 (ESA SP-697) https://earth.esa.int/documents/10174/1573054/Operational_validation_accuracy_InSAR_measurements_enhanced_oil_recovery_field.pdf

Luft J, Ingham H (1955) The Johari window, a graphic model of interpersonal awareness. In: Proceedings of the western training laboratory in group development, Los Angeles, University of California

Mäkitaavola K, Stöckel B-M, Sjöberg J, Hobbs S, Ekman J, Henschel MD, Wickramanayake A (2016), Application of InSAR for monitoring deformations at the Kiirunavaara mine. In: 3rd

international symposium on mine safety science and engineering, Montreal, August 2016 https://
isms2016.proceedings.mcgill.ca/article/view/73/25

Oboni F, Oboni C, Morin HR, Brunke S, Dacre C (2018) Space observation, quantitative risk
assessment synergy deliver value to mining operations & restoration. In: Symposium on mines
and the environment, Rouyn-Noranda, Québec, June 17–20, 2018 https://www.riskope.com/wp-
content/uploads/2018/06/Symposium-2018-article-MDA-riskope_2018-03-03.pdf

Oboni F, Oboni C (2020) Tailings dam management for the twenty-first century. Springer

Owen JR, Kemp D, Lèbre É, Svobodova K, Murillo GP (2020) Catastrophic tailings dam failures
and disaster risk disclosure. Int J Disaster Risk Reduction 42: 101361

Shrader-Frechette K (2011) Fukushima, flawed epistemology, and black-swan events. Ethics Policy
Environ 14(3):267–272

Ulaby FT, Moore RK, Fung AK (1986) Microwave remote sensing: active and passive: volume 2:
radar remote sensing and surface scattering and emission theory, Norwood MA, Artech House

Chapter 8
Defining Probabilities of Events

This chapter discusses how probabilities of single events can be evaluated and updated, and delivers some examples of portfolio analyses. A box with links to key terms is included in the references at the end of this chapter to facilitate the read.

Probabilities measure the chance an event will occur. In risk assessments and project evaluations we are not dealing with absolute probabilities, but with relative probabilities (within a portfolio) over a specific time period. Generally, the time period is one year and we therefore use annual probabilities.

Probabilities are perishable goods. They change as the considered system evolves, hence they require updates. Frequencies, which are linked to long-term averages of occurrences, also vary but are slower to change. In heavy industrial applications we may well consider a horizon of two to five years and initially assume probabilities will remain "fresh" that long, but we need to keep an eye on this and perform updates as soon as needed, especially if divergence is likely.

Operations oftentimes have service lives that span over multiple decades and their closure and post-closure lives may very well span over respectively another set of multiple decades and centuries. Nowadays, there are plenty of cases where the "P" word "Perpetuity" is used. Just remember, perpetuity is way longer than a long time! In one of our earlier books (Oboni and Oboni 2016) we used the following examples to provide a sense of the scale of time.

If our company had existed 500 years ago, we could have learned about Mr. Columbus's recent discovery. Total world population had increased to 500 M souls.

If our company had existed 200 years ago, we would have been concerned about the Battle of Trafalgar, Napoleon's retreat from Moscow, and the Battle of Waterloo. The world's population had reached 950 M souls.

If we had closed our operation in those earlier times, 500 or 200 years ago, would we expect it to still be right there where we had left it, unattended, unmaintained, unmonitored? Had we left a Standard Operating Procedure and Maintenance Manual for future generations, the manual would now be in a language difficult (if not impossible) to understand. The documents might have turned to dust or have been heavily damaged, if fires, floods, and wars had not destroyed them earlier. In addition, even

F. Oboni and C. H. Oboni, *Convergent Leadership—Divergent Exposures*,
https://doi.org/10.1007/978-3-030-74930-9_8

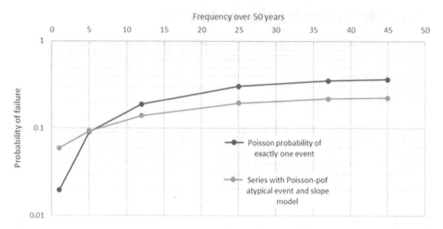

Fig. 8.1 Poisson probability of exactly one event (blue curve). Combined p_f of the slope under business-as-usual and a varying number of typhoon (grey curve), assuming the typhoons increase the probability of failure of the slope to $p_f = 0.5$

if someone at the dawn of the digital age had thought to transform our original document into a digital transcription, a solar flare (such as the Carrington event of 1859, see Sect. 3.2.2) would probably have erased them all.

Keep this information in mind as you go through the rest of this section.

The theme of the long-term survivability of man-made structures has long been studied and is the object of a significant body of literature. Some years ago we published a paper in which attention was focused on modeling the aging process of geo-structure as a series of discrete hits by hazardous conditions (these could be anything, from an earthquake to flooding, etc.) (Oboni et al. 2014).

We used a similar approach at a client's site to evaluate the impact of the combination of an atypical event on the overall probability of a slope accident evaluated to have a business-as-usual annual probability of failure $p_f = 0.05$ (5%).

We looked at a service life of 50 years and we combined the $p_f = 0.05$ business-as-usual slope probability of failure, with an extreme event, such as a typhoon increasing the slope failure to $p_f = 0.5$. We considered a possible number of occurrences of the typhoon, varying over the service life between 1 and 45 (typhoon frequency varies between $1/50 = 0.02$ and $45/50 = 0.9$). Figure 8.1 displays the results of the evaluation as follows:

- Blue curve: Poisson probability (see Sect. 8.1.3) of exactly one typhoon event occurring if the frequency (expressed as number of events over 50 years operation's life) varies between 1 and 45.
- Grey curve: combination (series, see Sect. 8.2.1) of the business-as-usual slope probabilities of failure with the Poisson probability of the typhoon assuming, as stated above, that typhoons generate $p_f = 0.5$ on the slope.

In Fig. 8.1 it can be noticed that in case of:

- one typhoon over 50 years, the impact of that one typhoon on the combined probability is insignificant, as p_f remains very near to the 0.05 business-as-usual.
- Five typhoons over 50 years, the combined probability of failure increases to approximately 0.09.
- A number of typhoons varying between 1 and 5 over 50 years, the combined probability of failure roughly doubles.

Thus, in this example it can be concluded that as typhoons were considered at that time and in that location, following available information as low frequency events (say 1 to 3 over 50 years), the calculated impact on the combined probabilities would be less than the uncertainties due to geotechnical and other driving parameters of a potential slope failure.

In (Oboni et al. 2014) an attempt was also made to draft a multidimensional estimate of future consequences. Increased population and changes of land use may indeed significantly alter the consequences of a mishap in the future.

8.1 Probabilities of One Event

The scope of this text does not include development of mathematical skills or a review of one of the numerous possible techniques to evaluate probabilities given a specific type and quality of available data. There is indeed a vast body of literature discussing this subject (Ang and Tang 1975,1984; Lipscomb et al. 1998; Clemen and Winkler 1999; Garthwaite et al. 2005).

However, there are a few mathematical tools whose concepts are important in the deployment of sensible risk management when statistics are poor or non-existent and projects evolve over many years, decades.

Probabilities allow us to consider the various sources of uncertainty and evaluate their impact on the big picture. Thus we are of the opinion that even rudimentary probabilistic analysis is better than working deterministically. In fact, the inclusion of uncertainties using stochastic variables is far superior to deterministic parametric studies (as, for example, varying one or two parameters at a time to see their influence on the overall results).

The key question of what constitutes the essential (understood as basic and indispensable) and ideal (understood as perfect) data set to use is frequently asked by users of risk assessments. There is no simple answer to that question, as we often deal with facilities that may not even have been commissioned yet and past performances may not reflect future behaviour because of system or climatic changes or divergent risks. Indeed, any internal or external change to the system has the potential to invalidate the assumption that past experiences are sufficient to understand and calibrate future implementations.

Let's also remark that no risk assessment ever has the ideal data set available. Indeed, available data are generally gathered for other purposes, may be censored and biased and, most importantly, they reflect the past, not the future. This is the case even in extremely regulated environments. Therefore, the analyst must rely on his/her skill and specific knowledge to use available data, either specifically from the site(s), or from specific technical literature, to define framing probabilities ranges.

Of course, any factual data (for example, climatic data, market data, changes and deformations) will help immensely in framing a reasonable range of probabilities. Records of near-misses can also be considered very important. Accident records can be considered essential, although in many studies done in the past, there were no such records simply because the system was not even in service. For future facilities, "essential" means "reported in the literature", and in some cases "collected expert opinion" together with an encoding methodology that makes it possible to transform knowledge into a probability.

8.1.1 Initial Estimates

When dealing with business-as-usual, framing probabilities is easy. Even without performing any modelling or probabilistic calculations, one can use the tables in Appendix A or probabilities statistics and databases, as illustrated by Technical note 8.1.

Technical Note 8.1: Probabilities Statistics and Rate of Failure Data

Probabilities statistics and rate of failure data exist in the literature both as vendor data and as "unbiased" statistical resources. All data in the literature demand caution even when using reputable databases. Classic unbiased sources examples are (IAEA 1988; Denson et al. 1991). A possible approach when using these references may develop as shown in the examples below.

Let us start with the case of a tank storage failure rate. The databases report the following:

JTFTH tank storage FWST

Component boundary: detail n/a. Operating mode: all. Operating environment: normal

Generic failure mode: rupture. Original failure mode: rupture during operation

FAILURE RATE OR PROBABILITY: mean $2.68*10^{-8}$/hr ...

Operating experience: $1.36*10^5$ h of operation, no failure

Based on that mean failure rate estimate the return period (years) is:
$1/(2.6 * 10^{-8}/h * 8760 \, h/year) = \sim 4400$ years.

The mean failure rate appears as $2.6*10^{-8}$/hr. (i.e., a very low value), and likely derives from the fact that operating experience (last line) showed 136,000 h (15 years) without failures.

This value shows how careful an analyst must be when using databases results. Indeed, we doubt there is a tank anywhere that has resisted longer than the Pyramid of Cheops to the aggression of time and wear. Good sense and a little more thinking are necessary to arrive at a reasonable value leading to a meaningful risk assessment. Indeed, the devil is in the detail.

We can now review a pump motor driven failure, using again values from extant reputable databases. Two cases appear in the database in a normal operating environment: the first is "failure to run", the second is "failure to run given start".

PMYRS pump motor driven

Component boundary: pump and motor. Operating mode: all. Operating environment: normal

Generic failure mode: fail to run. Original failure mode: failure to run

FAILURE RATE OR PROBABILITY: mean $7.90*10^{-6}$/hr ....

PMYRW pump motor driven

Component boundary: detail n/a, include motor. Operating mode: all. Operating environment: normal

Generic failure mode: fail to run. Original failure mode: failure to run given start

FAILURE RATE OR PROBABILITY: median $3.0*10^{-5}$ ....

Pushing the literature-based approach a bit further, we can see below the pump motor driven failure rate probability, but when operated in a heavy chemical environment. The rate of failure in such an environment is 100 to 1000 times greater than shown above:

PMUR pump motor driven

Component boundary: pump and motor, excludes control circuits. Operating mode: all Operating environment: extreme.

Generic failure mode: fail to run Original failure mode: failure to run given start.

FAILURE RATE OR PROBABILITY: mean $3.03*10^{-3}$/hr ....

PMURW pump motor driven

Component boundary: detail n/a, include motor Operating mode: all Operating environment: extreme, post-accident inside containment.

Generic failure mode: fail to run. Original failure mode: failure to run given start.

FAILURE RATE OR PROBABILITY: median $1.0*10^{-3}$ ....

Closing point. One can conclude that taking data at face value is a major blunder, as conditions and peculiarities significantly alter rates of failure.

In some cases, specific approaches can be developed. For instance, in a study we performed years ago we developed with IT experts a scale of vulnerability for IT systems shown in Table 8.1.

One can use the vulnerability factors in two manners:

- If the probability p_a of an attack on the specific system can be evaluated, then the probability of successful attack is the multiplication of $p_a * V_f$;

Table 8.1 Protection level and vulnerability factor for IT systems from a quantitative risk assessment

Protection level	Description	Vulnerability factor V_f
Non-existent	No anti-virus, inexperienced user	1
Minimal	Anti-virus, informed user	1/5
Fair	Anti-virus + firewall skilled user	1/10
Reasonable	Uses all the controls defined in (CIS 2009); updated skills training	1/100
Protected	As above plus daily threat information plus monthly analysis on indirect metrics such as new users and issues and a personnel management guideline. Suppliers also vetted	1/1,000
Highly protected	As above but daily analysis of indirect metrics such as new users, issues and a personnel security management guideline. Suppliers also vetted and facilities audited	1/10,000
Protected to credibility threshold	After extended security audits using various best available methodologies, fulfillment of the previous measures (see above) no known vulnerability found or suspected	1/1,000,000

- If the number of expected attacks N_a on the system is evaluated, then the number of successful attacks is the multiplication of $N_a * V_f$.

The question is now: what do we do with the numbers we have developed in the examples above in case of divergent hazards? For example, when geopolitical changes, climate change, etc., alter long term "normal" patterns? Oftentimes that occurs with seemingly repeated extreme events. The probabilities in the "New normal" may significantly alter the risk landscape around a project or a corporation. They may transform tolerable risks into intolerable ones, tactical risks into strategic ones (see Figs. 10.5, 12.1, Technical note 12.2). To ensure that decision-makers and management can keep optimizing tactical and strategic planning a rational, emotionless update of the probabilities is paramount.

Appendix A shows how the probability of various events scales with respect to many real-life examples. The tables in Appendix A can be used to generate first estimates of the probability of failure in relative terms, when no or very little statistical data and history are available.

By selecting a wide range of probabilities (and consequences) for each event a risk assessment will become amenable to a Bayesian update of probabilities when new data become available. Bayesian updates are developed after a first, a priori evaluation is mathematically corrected using the Bayes theorem, as new data become available from ongoing monitoring and observations on site. Bayesian updates only cover the initial range the analyst has selected and therefore they are not predictors of divergent events.

Suppose we have two sets of values:

- Historic data (including the last few years of extremes) lead to an estimate of probability of occurrence of $p_f = 3\text{--}7\%$. Probability of exceedance p_{exd} (see Technical Note 8.2) based on historic data is say 3%.
- Predictive data based on the last few years show a rise of the probability of occurrence to 40–60% and a probability of exceedance at 33%.

We know by now that the Bayes theorem does not work for divergence, and we feel rather cornered. Indeed, because the extreme events are stochastic and very fresh, and there is no way to accept a single magic number due to uncertainties, we need to work by framing a min–max probability range to find a way out. The key is to look into the effects of divergence based on the risk tolerance criteria, leading to RIDM (Chap. 12). Let us note that these probability increases may push presently tolerable risks into the intolerable field (see Chap. 10 and Technical note 12.2), which will alter the risk landscape and lead to different risk prioritizations and mitigative roadmaps.

How can we deal with the case where there is a sudden outlier? Sect. 8.1.2 and Technical note 8.2 describe a simple approximate procedure.

8.1.2 First Estimate of Probabilities After an Event Following a Long Uneventful Period

Let's suppose the price of a key material has varied monotonously over the last 30 years. In other words, the price as followed business-as-usual evolution with its "noise".

In the last couple years, however, the price has skyrocketed due to some geopolitical and technical reasons. It is too early to say if the variation is here to stay. Perhaps it is a fluke? Perhaps some tweets or political decisions that created the panic will be reversed tomorrow? Or perhaps a policy change may even permanently change the future price of the commodity?

The uncertainty has to be somehow resolved to allow a risk assessment to be performed. Indeed, managers and decision makers need to update their ERM using available information while avoiding knee-jerk over-reactions to support rational decisions.

Technical Note 8.2: Avoiding Knee Jerk Over Reactions When Evaluating Probabilities

After 30 years of business-like-usual and one extreme event, one can say the frequency of such an event is $1/31 = 0.032$ (3.2%). At the second occurrence frequencies can be corrected with some confidence to $2/32 = 0.063$ (6.3%). Another interesting approach is based on the probability of exceedance evaluation. When the number of existing observations is not sufficient to estimate the probability density function of the existing observations, the exceedance probability p_{exd} (i.e., probability that the observed largest value among n existing observations will be exceeded in N future observations) calculated as in Eq. (8.1) (Frangopol and Kim, 2014):

$$p_{exd} = N/(n + N) \qquad (8.1)$$

Considering the last two years and the occurrence of the extreme "last year" the Exceedance probability of the extreme is $(1/(2 + 1)) = 0.33$ or 33%. Considering the last three years the same probability is 25%. If the whole thirty years of records is considered, the probability of exceedance is $1/31 = 3.2\%$, the same result as above.

Closing point. At the time of analysis, after the first occurrence, the range can be estimated at min $= 3.2\%$ and max 25%. Then, year after year the range changes depending on whether events occur or not.

Let's go back to steady environment. We know the frequency of a certain event and we are interested in the probability of that event occurring one, two, ... n times next year, or perhaps one time in the next 25 years. Section 8.1.3 comes to our aid.

8.1.3 Linking Frequency and Probabilities

Annual probabilities should not be confused with frequencies, which measure average number of occurrences over a certain time interval. The confusion between the two comes from the fact that the numbers expressing probability and frequency are very similar, once below say 0.1 (1/10). However while a probability of 0.1 is expressed with the same number—0.1—as a frequency of 1/10, conceptually they are not at all the same.

In particular, the use of frequencies, expressed as one event occurrence over n years (e.g., 1 flood in a period of 200 years), misleads people into thinking that that event will only occur every $n = 200$ years (see Sect. 4.3). Nothing could be more wrong: there are plenty of 1/200-year events that have occurred several times in a decade. With climate change, we are seeing this happening increasingly often, indicating that the statistics need to be updated.

Using the Poisson distribution, it is possible to link the number of occurrences of an event over a selected time t to the mean occurrence rate (frequency). For example, 15 catastrophic system failure events over 10 years have been observed equates to a measured frequency of 1.5 events/year. Using Poisson's distribution it is easy to

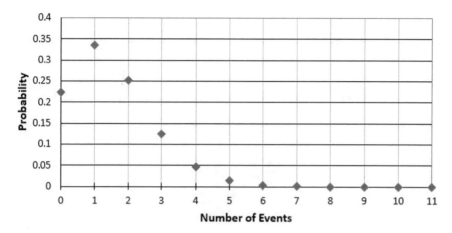

Fig. 8.2 Probability of occurrence (vertical axis) of 1, 2, 3 … *n* events in the next year (horizontal axis) if they have a measured frequency of 1.5 events/year evaluated using Poisson's distribution

compute and graph the probability of seeing any number of events (1, 2, … *n*) during, for example, a single year (Fig. 8.2).

In Fig. 8.2 the vertical axis shows the annual probability and the horizontal axis the number of events for f = 1.5 events/year. We can see that with that frequency one event per year has p = 0.33 to occur "next year", and three events have p = 0.12, to occur "next year", etc. As additional information delivers new occurrences of events, frequency and related probabilities can be updated in a new cycle, leading to updated risks.

As stated above for small frequencies (up to 1/10 (time units) one can assume annual probability approximately equal to frequency, as shown in Fig. 8.3. However, at frequency f = 1/5 time units (=0.2) the error of the approximation rises to 20%. After that, you need Poisson's distribution.

The probability of an event occurring one or more times within its return period is always 0.63. This is because the likelihood of seeing no event within its return period is 0.37, which is p = e^{-1}.

Two facts must be kept in mind when applying the Poisson process:

- the constant rate (frequency) of the events needs to be known, even though no one will ever evaluate it perfectly;
- events occur independently of the time since the last event.

As a result, the nature of the change dictates the applicability of the Poisson approach. If a fluke occurs (see Technical note 8.2) we can assume that our constant rate was initially incorrect. But when we apply the correction to 6.3% (second occurrence in Technical note 8.2) we can use Poisson. However, in case of a thread of panic-creating tweets, the events are obviously not independent as they obey the design of the author of the tweets. Finally, if we look to a two-year period, one with normal, the other with extreme event, then we have an estimate of 40%. If we look

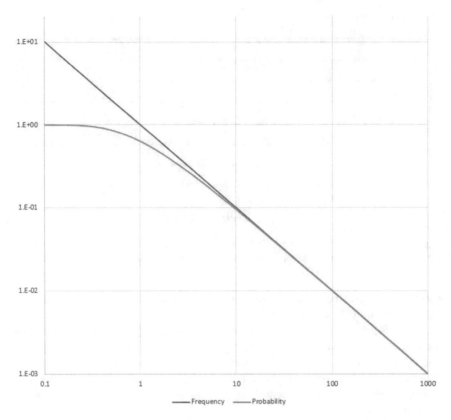

Fig. 8.3 Frequency versus probability plot. The vertical axis shows probabilities (by definition bounded to one), the horizontal axis represents frequencies

to a three-year period, one year normal and two years extreme, then the probability is roughly 50%.

So, there is a relatively simple way to convert frequencies into probabilities and vice versa: probability and frequency are not related linearly for frequencies higher than 1/20.

8.1.4   Updating Probabilities (Bayesian Approaches)

It is possible to develop probabilistic updating of various types of data, such as change rate (e.g., cm/year), number of events of a certain magnitude (e.g., number of events exceeding a certain magnitude per year), etc. The updating makes it possible to re-frame probabilities present in a quantitative risk register and re-evaluate the risks. In what follows we present a few techniques that can be used to update probabilities:

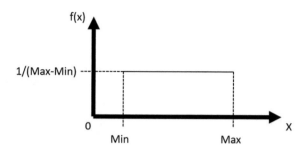

Fig. 8.4 Uniform distribution of f(x) between its estimated Min and Max values

- Exceedance probability (Technical note 8.2) updates. Forecasting the future exceedance of previously observed extremes is extremely important for risk assessments. Based on repeated observations it is possible to re-frame the probabilities of exceedance and thus to rationally update the risk register.
- Bayesian updates. Bayesian analyses make it possible to update frequencies and probabilities as new data are generated (Ang and Tang 1975; Straub and Grêt-Regamey 2006) by space observation, IoT, or other means. Consider, for example, the case where the available information is a set of observed n blackouts, which are described according to their duration and the time during which they occurred. Note that the Bayesian update will be valid only insofar as the observations are free of error (i.e., all blackouts are recorded); this is the fundamental reason why regular monitoring is a necessity. In order to allow later Bayesian updates a quantitative risk register should include the a priori estimate of frequencies or probabilities. If no data are available beyond a min–max range defined by models or expert opinions, the simplest, safest and oldest rule is to assume a uniform distribution (Fig. 8.4). Unfortunately, many users of Monte Carlo simulations or commercial risk software embark on analyses by arbitrarily selecting distribution types and shapes, without understanding the error they are introducing in their results. However, if and when sufficient data is available, the risk assessment could also be set up with a more refined "prior" distribution and then a Bayes approach used to obtain the first "posterior" distribution, the second posterior, etc. The application of a Bayes approach shows that one single event provokes a shift of the distribution, as shown in Fig. 8.5.

The conclusion is that it is possible to update first estimates probabilities under certain conditions but Bayes theorem does not work for divergent events.

8.1.5 Summary of Elemental Probabilities

The above leads to a well consolidated procedure. We generally start a quantitative risk assessment (called the a priori assessment) using uniform distributions for probabilities and consequences. The uniform distribution leads to the most conservative estimate of uncertainty, as it gives the largest standard deviation (NIST/SEMATECH

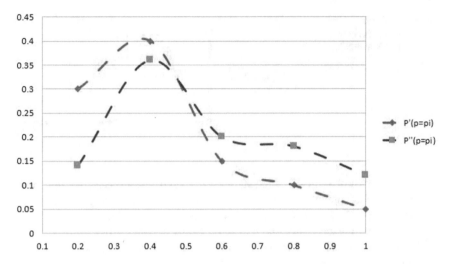

Fig. 8.5 A priori p' and a posteriori p" distribution of a parameter x between its estimated extreme values 0.2 and 1

2012). The uniform distribution makes it possible to evaluate first and second moments of functions of stochastic variables by hand (by means of direct formulae or using the point estimate method developed by Emilio Rosenblueth (1975), making it possible to bypass "black-box" solutions such as the Monte Carlo simulation, which, again, give a sense of false precision. Interested readers can go deeper into the theme of using uniform distributions as a priori distribution in a Bayesian approach by reading the literature on uninformative priors (Carlin and Louis 2008). The term "uninformative" is common but misleading, as the simple knowledge or estimate of min–max already constitutes a very valuable piece of information. In some cases— for example, when adding independent random variables or considering higher levels of information for a variable (e.g., min–max, first and second moment, that is, average and standard deviation) based on the central limit theorem—it is possible to assume that the result tends toward a normal distribution (informally, a bell curve) even if the original variables themselves are not normally distributed. The central limit theorem implies that probabilistic and statistical methods that work for normal distributions can be applied (with caution) to many problems involving other types of distributions.

The next step, as available information level increases, is to use an empiric distribution, such as a beta distribution (see Case Study 1, Chap. 14, Sect. 14.7.2).

Finally, when the level of knowledge increases further, the real distribution of a variable can be determined (exponential, Gumbel, etc.). However, each time an analyst feels the temptation to use a specific distribution, perhaps to perform a Monte Carlo simulation, the trade-offs and potential for a misleading result must be carefully considered.

In summary, for any risk assessment and any event we can use data from various local and external reputable sources available to us at the time of the analysis, as well as interviews with key personnel; we integrate these with any other factual data the client may deliver to us; we then cautiously adjust the values to take into account the specific location, habits, and possible future conditions; and finally, we obtain a framing range of probabilities and consequences for each type of event and each identified scenario.

Scenarios deemed to border on credibility should not be given probabilities of occurrence below the order of 10^{-6} to 10^{-5}, based on various industries' customary definition of credibility, as discussed earlier (see Sect. 1.1). We firmly discourage limiting risk assessments to credible scenarios, as this common practice leads to bias and censoring the results, because the credibility threshold may not be properly defined and people tend to instinctively place it orders of magnitude higher than the technical definition (Oboni and Oboni 2018). Events that are below credibility should end up falling out of the analysis as a result of the analysis itself, and not because they are the object of arbitrary decisions.

Any risk assessment should be updated if the frequencies, intensities, patterns of the hazards and/or vulnerabilities, or robustness of the systems change over time. In particular, sudden and significant changes should prompt a re-evaluation of the results of any risk assessment.

Bayesian updates can be developed, provided the a priori ranges of probabilities, consequences, and any parameter of interest are broad enough and the system is not facing divergence. The Bayesian inference model is indeed one of the cornerstones available to analysts among the tools for estimating probabilities and consequences in business-as-usual conditions. Allowing for rational updates when new data become available (from semi-static to real-time updates, depending on the applications and resources) will become the norm due to social and legal requirements, social and media pressure, etc. To allow Bayesian inference to be included (possibly at a later date of development of a risk management approach) a few conditions on top of business-as-usual are required:

(1)   Always start by identifying hazards using threats-to and threats-from (see Chap. 7).
(2)   All probabilities and consequences need to be expressed as ranges rather than single values, and uncertainty needs to be recorded. For example, for every parameter range there must be a justification recorded with the different opinions that led to its definition recorded as well.
(3)   Interdependencies need to be transparently implemented, i.e., the methodology has to provide results for single failures and domino effects.

Once these conditions are met, then on the basis of specific pertinent data reports, weather data, etc., a Bayesian inference model can be developed to determine posterior probabilities. In time, as data are gathered and changes occur, probabilities will be seamlessly updated.

At each new data entry, at discrete time intervals (as required, or in real time, depending on the application), it will be necessary to perform new probability-magnitude estimates for the hazards and their consequences. That will allow for Bayesian updates of the risks.

8.2   Probability of Failure in a Portfolio

8.2.1   Independent Elements

Let's use as an example a theoretical portfolio of mechanically independent machines constituting an assembly line. We will consider these machines as being fed by different power sources, so we consider them as truly independent. Each machine has its own probability p_f of incurring maximum foreseeable loss (MFL).

A series system is such that all its components are required for system success. Equivalently, the system fails if any component fails (Fig. 8.6). In the following case we define the top event as the failure, but in other circumstances the top event could be the success or just another event of interest.

If the failure of one machine blocks the entire line criteria the overall system's probability of failure X and success can be evaluated using the series probability formula shown in Eqs. 8.2 and 8.3.

Failure:

$$X = 1 - \prod_{1}^{N}\left(1 - X_j\right) \tag{8.2}$$

Success:

$$\overline{X} = \prod_{1}^{N}\overline{X}_j \tag{8.3}$$

A parallel system is a redundant system that is successful, if at least one of its elements is successful (although with less functionality/capacity) (Fig. 8.7).

The probability of failure X and success of a parallel system can be evaluated using Eqs. (8.4) and (8.5).

Failure:

$$X = \prod_{1}^{N} X_j \tag{8.4}$$

Success:

Fig. 8.6 Reliability diagram of a series system of elements, where the failure of one means the failures of the whole system. If Mode 1 OR Mode 2 OR … Mode n (not shown) fail, then the system has failed

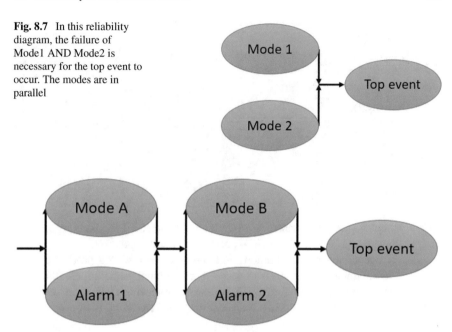

Fig. 8.7 In this reliability diagram, the failure of Mode1 AND Mode2 is necessary for the top event to occur. The modes are in parallel

Fig. 8.8 A system made of two parallel systems in series

$$\overline{X} = 1 - \prod_1^N (1 - \overline{X}_j) = \coprod_1^N \overline{X}_j \qquad (8.5)$$

Real-life system are generally hybrids encompassing parallel and series systems linked together. Technical note 8.3 shows such an example.

Technical Note 8.3: Using Series and Parallel Systems to Define Top Events

Let's look at the system of Fig. 8.8, where the occurrence of independent Modes A and B (NB: Mode could be an event or a series of events leading to one probability of failure for that Mode) is monitored by, respectively independent, Alarm 1, Alarm 2 which, in turn will stop the process in a safe manner. The Top Event (e.g. failure of the system) will not occur if both (Mode-Alarm) couples work. Let's note:

If a Mode doesn't fail then it is successful.

If a Mode fails and its Alarm functions one can (time permitting) implement corrective measures so the couple (Mode-Alarm) remains successful.

Only if a Mode fails and its alarm fails there will be a failure of the (Mode-Alarm) couple.

Mode A-Alarm 1 is a parallel sub system. Mode B-Alarm 2 is another parallel subsystem. They are linked in series. Below we will evaluate the probability of the top event to occur.

The following probabilities must be evaluated:

p_{ModeA}; p_{Alarm1}; p_{ModeB}; p_{Alarm2} meaning that Mode A,B happen with their respective probabilities and the respective associated mitigations (Alarm) could malfunction (with their own probability).

Closing point: the system failure (top event) probability:

$$p_{Top\ Event} = \left(1 - \left[1 - \left(p_{ModeA} * p_{Alarm1}\right)\right] \times \left[1 - \left(p_{ModeB} * p_{Alarm2}\right)\right]\right)$$

We cannot neglect to summarize two ubiquitous standard methods to evaluate probabilities: Event Tree analysis (ETA) and Fault tree analysis (FTA).

ETA is one of the most common methodologies to evaluate probabilities of complex cascading events in terms of series of conditional probabilities. In reality, ETA is a graphic-mathematical forward, bottom-up, logical modelling technique. It is a common mistake to develop excessively complex ETA giving their users a false sense of security and precision.

FTA is one of the most common methodologies to evaluate a system against single or multiple initiating faults. It a top-down, deductive failure analysis method in which an undesired state of a system is analyzed using Boolean logic (and/or gates) to combine a series of lower-level events. This method is mainly used in the fields of safety engineering and reliability engineering to understand how systems can fail, identify the best ways to reduce hazards, or determine (or get a feeling for) event rates of a safety accident or a particular system level (functional) failure. It has been shown on many occasions that the Boolean logic can be misleading, insofar as a correlation may exist between failures of similar elements implemented to increase redundancy (false redundancy in nuclear, aerospace applications, for example, also see CCF below). FTA is a failure analysis tool, and as such can be integrated in quantitative risk assessments. FTA is aimed at analyzing the effects of initiating faults and events on a complex system. This contrasts with failure mode and effects analysis (FMEA, see Appendix B), which is an inductive, bottom-up analysis.

8.2.2 Dependent Elements

The case of dependent elements is very complex. The dependencies can be physical, intrinsic or external, related to common cause failure (CCF), etc.

Dependent facilities also have the potential to generate MFLs which are larger than the single facility failure would generate through domino effects (interdependencies) and ripple effects (consequences, see Technical note 9.2). Thus, the reasoning must simultaneously cover the probability and the consequences, i.e., a full risk assessment has to be performed.

In a real-life portfolio, dependent failures must be studied with attention. Oftentimes risks from dependent failures may be higher or smaller than what intuition suggests. There is no one-size-fits-all intuitive solution to these estimations.

8.2.3 Summary of Conclusions on Portfolios of Elements

In a real-life system of independent elements, it is more efficient to define (and mitigate) the "bad apples" rather than to evaluate cumulative losses to justify more reserves. The elements are then to be evaluated jointly to determine the maximum portfolio potential loss and its probability.

Appendix

Links to more information about the Key terms from the Authors	
A, B	*Act of God* (https://www.riskope.com/2020/12/09/act-of-god-in-probabilistic-risk-assessment/) *Black swan* (https://www.riskope.com/2011/06/14/black-swan-mania-using-buzzwords-can-be-a-dangerous-habit/) *Business-as-usual* (https://www.riskope.com/2021/01/13/business-as-usual-definition-in-risk-assessment/)
C, D	*Convergent* (https://www.riskope.com/2021/01/20/convergent-risk-assessments/) *Divergent* (https://www.riskope.com/2020/11/18/tactical-and-strategic-planning-to-mitigate-divergent-events/) *Drillable* (https://www.riskope.com/2020/01/15/probability-impact-graphs-do-not-fly/)
F	*Foreseeability/foreseeable* (https://www.riskope.com/2021/01/06/foreseeability-and-predictability-in-risk-assessments/) *Fragile/fragility* (https://www.riskope.com/2020/04/01/antifragile-resilient-solutions-for-tactical-and-strategic-planning/)

(continued)

(continued)

P, R	*Predictability/predictable* (https://www.ris kope.com/2021/01/06/foreseeability-and-pre dictability-in-risk-assessments/) *Resilient* (https://www.riskope.com/2016/11/ 23/resilience-cannot-based-instinctual-dec ision-making/) *Resilience* (https://www.riskope.com/2016/11/ 23/resilience-cannot-based-instinctual-dec ision-making/)
S	*Scalable* (https://www.riskope.com/2015/04/ 16/how-system-definition-and-interdepende ncies-allow-transparent-and-scalable-risk-ass essments/) *Societal risk acceptability* (https://www.ris kope.com/2014/01/09/aspects-of-risk-tolera nce-manageable-vs-unmanageable-risks-in-rel ation-to-critical-decisions-perpetuity-projects- public-opposition/) *Sustainability/sustainable* (https://www.ris kope.com/2019/01/16/improving-sustainab ility-through-reasonable-risk-and-crisis-man agement/) *Survivability* (https://www.riskope.com/2011/ 03/17/ale-fmea-fmeca-qualitative-methods-is- it-really-what-we-need/) *System* (https://www.riskope.com/2017/07/26/ three-ways-to-enhancing-your-risk-registers/)
T, U	*Tolerance* (https://www.riskope.com/2020/04/ 29/risk-tolerance-thresholds/) *Uncertainty/uncertainties* (https://www.ris kope.com/2015/12/10/3-decision-making-tru ths-derived-from-uncertainty-taxonomy-sch eme-of-classification-and-a-road-sign/) *Updatable* (https://www.riskope.com/2020/01/ 07/climate-adaptation-and-risk-assessment/)

Other linked information (https://www.riskope.com/blog-news/) search Riskope blog and use the search box

References

Ang AH-S, Tang WH (1975) Probability concepts in engineering planning and design, Volume I, basic principles. Wiley

Ang AH-S, Tang WH (1984) Probability concepts in engineering planning and design, volume II, decision, risk and reliability. Wiley

Denson W, Chandler G, Crowell W, Wanner R (1991) Non-electronic parts reliability data, RAC report NPRD. Reliability Analysis Center, Griffiss AFB, NY https://www.mwftr.com/CS2/ NPRD-91_a242083.pdf

Carlin BP, Louis TA (2008) Bayesian Methods for Data Analysis, 3rd edn. CRC

Center for Internet Security (2009) Twenty critical controls for effective cyber defense: consensus audit guidelines, version 2.1: August 10, 2009. https://csis-website-prod.s3.amazonaws.com/s3fs-public/legacy_files/files/publication/Twenty_Critical_Controls_for_Effective_Cyber_Defense_CAG.pdf

Clemen R, Winkler L (1999) Combining probability distributions from experts in risk analysis. Risk Anal 19(2):187–203

Frangopol DM, Kim S (2014) Prognosis and life-cycle assessment based on SHM information. In: Sensor technologies for civil infrastructures. Woodhead Publishing, pp 145–171

Garthwaite P, Kadane J, O'Hagan A (2005) Statistical methods for eliciting probability distributions. J Am Stat Assoc 100(470):680–701

International Atomic Energy Agency (1988) Component reliability data for use in probabilistic safety assessment, IAEA-TECDOC-478. https://www-pub.iaea.org/MTCD/Publications/PDF/te_478_web.pdf

Lipscomb J, Parmigiani G, Hasselblad V (1998) Combining expert judgement by hierarchical modeling: an application to physician staffing. Manage Sci 44(2):149–161

National Institute of Standards and Technology (2012) NIST/SEMATECH e-Handbook of statistical methods. https://doi.org/10.18434/M32189

Oboni C, Oboni F (2018) Geoethical consensus building through independent risk assessments. In: Resources for future generations 2018 (RFG2018). Vancouver BC, June 16–21, 2018

Oboni F, Oboni C (2016) The long shadow of human-generated geohazards: risks and crises. In: Farid A (ed.) Geohazards caused by human activity. InTechOpen. ISBN: 978–953–51–2802–1, Print ISBN 978–953–51–2801–4

Oboni F, Oboni C, Caldwell J (2014) Risk assessment of the long-term performance of closed tailings. In: Tailings and Mine Waste 2014. Keystone CO, USA, October 5–8, 2014

Rosenblueth E (1975) Point estimates for probability moments. Proc National Acad Sci USA PNAS 72(10):3812–3814

Straub D, Grêt-Regamey A (2006) A Bayesian probabilistic framework for avalanche modelling based on observations. Cold Regions Sci Technol 46:192–203

Chapter 9
Evaluating Consequences

Some years ago we stated: "Especially for very large projects, risk assessments generally consider too simplistic consequences and ignore 'indirect/life-changing' effects on population and other social aspects" (Oboni et al. 2013) (see Technical note 9.1). We noted that simplistic consequences evaluations are misleading, unrealistic and can significantly affect ESG (see Sect. 6.1.2), SLO and CSR (Chap. 5). A box with links to key terms is included in the references at the end of this chapter to facilitate the read.

Papers discussing this ubiquitous phenomenon of oversimplification are rare. However, scholars have recently discussed environmental, social and governance risks for deep-sea mining (Kung et al. 2020). They point out that those risks are poorly defined. We would have liked to see a clearer distinction between risks and their consequences in their paper. The paper includes a holistic view of the consequences of deep-sea mining, including possible impacts on shipping lanes and fiber-optic cables, for example.

Technical note 9.1 shows how the consequence analysis can include indirect health effects on the public, like for example those linked to stress and violence (see Sect. 5.3.4).

Technical note 9.1: Human/personal impacts through life changing units (LCUs)

On various occasions it is advisable to consider the consequences on wellbeing of the public resulting from physical and moral chocks deriving, for example, from a terroristic attack, environmental damages, socio-political crisis (Norris et al. 2002). Widespread scares can generate drops in consumer demand. For instance, in the aftermath of the September 11 attacks there was a significant reduction of demand for air travel (FAA 2002), with numerous interdependencies (indirect losses and ripple effects as discuss later in this chapter) in addition to the direct fatalities and direct economic consequences.

© The Author(s), under exclusive license to Springer Nature Switzerland AG 2021
F. Oboni and C. H. Oboni, *Convergent Leadership—Divergent Exposures*,
https://doi.org/10.1007/978-3-030-74930-9_9

In 1967 two psychiatrists published a study evaluating what they called "life changing units" (LCUs) (Holmes and Rahe 1967). LCUs were attributed to a list of 43 events that could be either harmful or positive but were in any case stressful, based on a study of 5000 patients. Obviously, the correlation was found to be very weak, but nevertheless Holmes and Rahe are considered to be the fathers of the relationship between stress and stress-induced conditions. Their procedure consisted in assigning LCU values to life events, then adding the LCUs of events that had occurred during the previous year to reach a total score used to rank the chances of developing a stress-related medical condition, and potentially a disease.

Here are samples of stressful events and their LCUs:

- Death of a spouse = 100
- Death of a family member = 63
- Loss of job = 47
- Restructuring of business = 39
- Foreclosure = 30
- Change of standard of living = 25
- Change of residence = 20
- Change of work conditions = 20
- Change of social life = 18
- Change of food habits = 15.

Scores were correlated to chances of illness as follows:

- **300+** : strong chance;
- **150–299**: moderate chance (a 30% reduction from the line above);
- **<150**: slight chance.

A 2020 contribution by Melinda Wenner Moyer on Scientific American describes a *survey conducted by researchers at the Centers for Disease Control and Prevention* (https://www.scientificamerican.com/article/you-can-get-through-this-dark-pandemic-winter-using-tips-from-disaster-psycho logy1/). Over 25% of more than 5000 US adults had symptoms of anxiety while more than 24% had symptoms of depression—three to four times more than in the prior year. Other studies found similar results, showing an apparent correlation between poor mental health and income loss. Oftentimes physical symptoms were also present showing a correlation with emotional stress. These results were echoed by *studies in Israel and Italy* (https://www.fronti ersin.org/articles/10.3389/fpsyg.2020.586202/full).

Closing point. The validity of the LCU approach is confirmed by studies in various countries, across cultures and time. The approach is empirical and approximate, but sheds an interesting light on how to integrate stressful events into consequences analyses of any mishap, accident.

Using conceptually sound risk assessment methodologies makes it possible to consider the wide uncertainties surrounding failures driving parameters and related additive dimensions of consequences.

The usual general dimensions of losses we include in our day-to-day practice and deployments of our ORE methodology (see Sect. 11.3) are:

- human H&S (health and safety);
- fish, fauna and topsoil/vegetation;
- infrastructure;
- long-term economic and development;
- social impacts (including reputation, crisis potential, legal challenges, etc.).

These dimensions are additive, not stand-alone, and can possibly merge into one metric, while preserving the inevitable uncertainties.

9.1 Dimensions of Failures

As discussed earlier (Sect. 5.3.4 and Anecdote 2.1) each risk assessment requires the definition of consequences for the various scenarios considered for each specific element of the system under consideration. As stated earlier, losses and specific consequences are multidimensional, and the final result is the sum of the various dimensional components. It is a serious mistake to select the worst dimension as the value for the consequence, as is oftentimes suggested in risk assessments instructions to prepare FMEA, PIGs and risk matrix approaches.

Here is a list of four dimensions identified for various real-life studies with their sub-components (also additive):

Direct physical losses:

- fixed equipment;
- mobile equipment;
- infrastructures;
- business interruption.

Health and Safety: (none of these include chronic effects—see Environmental).

- harm to workers;
- harm to outsiders.

Environmental: (if dust and other by-products, such as contaminated water, are released at code limit values we generally assume there are no chronic effects).

- water releases;
- waste releases;
- hazardous materials releases;
- dust releases,
- noise

Reputational Damages: These will arise for example from:

- chronic releases—for example in high-density residential areas—of materials which are perceived as hazardous, even if they are code compliant. These can occur as Medusa, Pandora's box risk events following the German classification (see Sect. 4.2).
- Incidents and accidents that destabilize the public; These can occur as Medusa, Cassandra risk events following the German classification (see Sect. 4.2).
- Alterations of the perceived environment (for example, change of colours), even if they bear no measurable consequences. For example in Cassandra following the German classification (see Sect. 4.2)
- Lack of communication. Always a cause of poor reputation.
- Lack of empathy and care. Same as above.

Reputational damages may lead to crises of various types and depth (cost), legal proceedings and indictments (see Sect. 6.1.2), loss of SLO and CSR (see Chap. 5) and ultimately suspensions of licenses, business interruption, etc. The areas of the environment most prone to generating these damages are:

- highways,
- residential quarters,
- water courses receiving outflows (including accidental ones).

To provide some context for physical consequences, let's use a few examples based on cyberattacks, as shown in Anecdote 9.1.

Anecdote 9.1: Examples of physical consequences from cyberattacks

Successful hacking of industrial systems seemed to remain "isolated exploits" for a long time, but in recent years has become much more common. For example:

- in 2014 hackers struck a steel mill in Germany, manipulating and disrupting the industrial control systems (ICS) to such a degree that a blast furnace could not be properly shut down, resulting in *"massive damage"* (https://www.bbc.com/news/technology-30575104).
- In 2016 Goldcorp's payroll, trade secrets, and other intimate information leaked on a torrent free for anyone to download.
- In 2017 Triton and Trisis, malwares targeting the Triconex industrial control systems product line made by Schneider Electric came to light. The targeted systems provide emergency shutdown in critical industrial processes. Thus, they are Safety Instrumented Systems (SIS), comprising hardware and software components. Obviously, the impairment of such a safety system leads to unmitigated consequences of a malfunction. Power plants, gas refineries are among the users of the Triconex product line and SIS.
- In 2019 an *aluminum maker forced to shut down operations* (https://www.reuters.com/article/us-norsk-hydro-cyber/hackers-hit-aluminum-

maker-hydro-knock-some-plants-offline-idUSKCN1R00NJ) because of a ransomware attack.
Closing point
Predictable event: Do cyber attacks leave physical consequences? Yes.
Foreseeable event: Do cyber attacks leave physical consequences? Yes.

Table 9.1 depicts an example of an analysis of hazard scenarios and their potential consequences drawn from a real-life study, not a general rule. It defines the evolution of potential consequences from minor to extreme for each consequence dimension.

At this point it becomes paramount to discuss ripple effects. There is a fundamental difference between the interdependence we discussed in Sect. 8.2.2 (also see domino effects in Sects. 6.3.1, 7.1, 8.1.5) and ripple effects (Technical note 9.2), in that the first looks at how interdependencies can alter the probability of occurrence of more elements than the one under consideration. For example, the failure of a contract A (probability p_{fA}) causes other contracts B and C fail. Separately those would have probabilities of failure p_{fB}, p_{fC} but as domino effect (interdependency) of the failure of contract A their probability increases. The ripple effect looks at what happens to the consequences of contract A and how its cost propagates through the system (not necessarily causing other failures).

Technical notes 9.2: I-O models to study ripple-effects
The use of so-called "input–output" (Wassily Leontief's I-O) models to estimate far-reaching indirect effects of events has been adopted for environmental impact analyses and the fields of energy economy, terrorism attacks and disaster risk management (Cho et al. 2001; Rose 2004).

Various studies published at a later date showed that those models are useful for expressing far-reaching consequences (Santos 2008). Indeed, Leontief's I-O model can evaluate the ripple effects of an initiating event propagating through an interconnected, interdependent system of elements.

Input I is described by the expected performance change (say, for example, a drop in production) resulting from the occurrence of an event, and by a variation with respect to the business-as-usual value. Output O is the value of the expected change. It is generally expressed in monetary terms, but other metrics could be selected. Of course, the consequences model has to incorporate direct losses such as replacement costs, serviceability losses/business interruption, etc. This is paramount in order to avoid double counting.

Of course, the model can be as sophisticated as available data, time and budget allow.

In its simplest form the I-O model is written as:

$$q = A^* * q + c^* \tag{9.1}$$

Table 9.1 Example from a real-life study showing general elements of a system, hazard scenarios, vectors (threat-from codes) and potential consequences. Not a general rule

General element of the system	Hazard scenario	Potential consequences evolution from minor to ultimate			
		Direct physical losses	Health and safety	Environmental	Reputational damages
Pipelines ponds, tanks	Traffic (includes contractors and snow removal collisions)	From very limited (if caught in time) to major damages and finally dam break (interdependent failures)	From very limited to harming workers	From very limited (if caught in time) to major environmental damages and finally catastrophic break (inter-dependent failure)	Generally limited unless failure evolves into full break
	Freezing (includes deformations at flanges)				
	Sabotage/vandalism, cyber				
	Corrosion and intrinsic mechanical failures				
	Human (communication between mill and tailings operation), IT				
Buildings, confining structures (and their content)	geotechnical/geological and climatological, hydrological	As above with possible "sudden failures". Additionally inadequate storage capacity may lead to BI, costly system alteration requirements	From very limited to harming workers and possibly general public downstream	From very limited to major environmental consequences outside of the perimeter	We consider that any serious issue would be covered by local and provincial media and further, with great reputational damages
	Construction/building operations				
	Operation, maintenance and monitoring during service and record keeping, IT				
	Sabotage/vandalism, cyber				
Waste outflow structures	Geotechnical/geological and hydrological (capacity, durability)	From very limited (if caught in time) to major damages and finally catastrophic break (interdependent failure of dam)	From very limited to harming workers	From very limited (if caught in time) to major environmental damages and finally catastrophic break (inter-dependent failure)	Generally limited unless failure evolves into full break
	Operations (service, emergency access), IT				
	Water management				
	Climate (includes extreme freezing)				
	Sabotage/vandalism, cyber				

where:
- q is the "disfunction" vector expressed in normalized production losses (negative values). Its terms represent the ratio between loss of production and Status quo (thus Business-as-usual).

- A^* is the interdependence matrix within the system's elements. Its rows display the effect of the column header element on the elements of each row.
- c^* is the demand vector (valuated as q, but expressed in monetary terms).

Various countries have used the I-O model. For instance, such a study was completed in Switzerland in 2006 (Nathani and Wickhart 2006).

Closing point. Thanks to the I-O models it is possible, for example, to study, within the frame of a ERM of a large company with many operations and branches, a scenario where a disaster has caused a 30% reduction of throughput of the operations in a region and 15% global reduction of logistic capacity, and to evaluate the global impact both in terms of serviceability and economic losses.

9.2   Examples of Consequences Estimates

In this section we present a few examples of real-life consequences reports noting what was missing from them when they were delivered. Some information has been removed as indicated by [...] to ensure confidentiality.

Example 1, from a Chemical plant selling excess energy.

The operation manager gave us a list of items impacting runtime and performance efficiency:

- [...] failure in Q4 2018 resulted in claw back of all firm energy sales for that contract year (5.1M USD).
- [...] and multiple other challenges in Q1 2019 reduced firm sales revenue in that quarter (3.4M USD lost opportunity) and resulted in claw back of some firm energy sales for that contract year (0.4M USD).
- 19-year high pricing for electricity for two days in Mar 2019 resulted in 1.7M USD excess penalties for shortfall against firm energy sales targets.
- A reduction of 10% in performance efficiency would reduce combined energy revenue by approximately 4.3M USD.
- 144M USD 50% BI Boiler accident; 1–5 fatalities depending on "who is there".
- [...] accidents easily generate 44M USD losses.
- Lightning strikes and washer drum failure cost approximately 2M USD.
- Loss of lube oil in generators, 4M with an average of 9.5M USD.

We noticed this was an operation-centric list, with no consideration of the impact on health and safety, and no information about what may have happened outside of the battery lines of the operation (see Sect. 6.2.2). We therefore asked for historical incidents and accidents to the personnel, and after some discussion we learned that a number of times in the last 15 years the operation had been targeted by a group of rioters blocking the transit of goods.

Example 2, from a shipping terminal.

The manager provided us with replacement and value costs of what he saw as the exposed areas of the operation:

- Buildings 173M USD;
- Machinery 368M USD;
- Stock 4M USD;
- Business interruption (BI) 47M USD /month.

Again, this was an operation-centric view. We had to ask how many people were on site at a given time, and what the environmental damages may occur under various scenarios, along with their legal, reputational and crisis potential. We realized that in the event of a spill the operation would almost certainly lose its SLO, given the vicinity to residential areas and major transit roads.

Example 3, road transport.

We were asked to analyze the risk of a private access road (160 km long) in order to understand what the local population and environment would be exposed to. Thus, business interruption (BI), cost of replacement and other dimensions were irrelevant to our analysis. The consequences of accidents can be based on *geometric and climatological criteria along the road* (https://reviewboard.ca/upload/project_document/EA1415-01_Risk_Assessm ent_Technical_Report_submitted_by_Oboni_Riskope.PDF) such as:

- how far the vehicles involved in accidents may fall;
- the presence of water/sensitive areas;
- the difficulty of retrieving spilled material due to the cross slope;
- the nature of the material being transported.

Appendix

Links to more information about the key terms from the authors	
A, B	*Act of god* (https://www.riskope.com/2020/12/09/act-of-god-in-probabilistic-risk-assessment/) *black swan* (https://www.riskope.com/2011/06/14/black-swan-mania-using-buzzwords-can-be-a-dangerous-habit/) *business-as-usual* (https://www.riskope.com/2021/01/13/business-as-usual-definition-in-risk-assessment/)

(continued)

(continued)

Links to more information about the key terms from the authors	
C, D	*Convergent* (https://www.riskope.com/2021/01/20/convergent-risk-assessments/) *divergent* (https://www.riskope.com/2020/11/18/tactical-and-strategic-planning-to-mitigate-divergent-events/) *drillable* (https://www.riskope.com/2020/01/15/probability-impact-graphs-do-not-fly/)
F	*Foreseeability/foreseeable* (https://www.riskope.com/2021/01/06/foreseeability-and-predictability-in-risk-assessments/) *fragile/fragility* (https://www.riskope.com/2020/04/01/antifragile-resilient-solutions-for-tactical-and-strategic-planning/)
P, R	*Predictability/predictable* (https://www.riskope.com/2021/01/06/foreseeability-and-predictability-in-risk-assessments/) *resilient, resilience* (https://www.riskope.com/2016/11/23/resilience-cannot-based-instinctual-decision-making/)
S	*Scalable* (https://www.riskope.com/2015/04/16/how-system-definition-and-interdependencies-allow-transparent-and-scalable-risk-assessments/) *Societal risk acceptability* (https://www.riskope.com/2014/01/09/aspects-of-risk-tolerance-manageable-vs-unmanageable-risks-in-relation-to-critical-decisions-perpetuity-projects-public-opposition/) *Sustainability/sustainable* (https://www.riskope.com/2019/01/16/improving-sustainability-through-reasonable-risk-and-crisis-management/) *Survivability* (https://www.riskope.com/2011/03/17/ale-fmea-fmeca-qualitative-methods-is-it-really-what-we-need/) *System* (https://www.riskope.com/2017/07/26/three-ways-to-enhancing-your-risk-registers/)
T, U	*Tolerance* (https://www.riskope.com/2020/04/29/risk-tolerance-thresholds/) *Uncertainty/uncertainties* (https://www.riskope.com/2015/12/10/3-decision-making-truths-derived-from-uncertainty-taxonomy-scheme-of-classification-and-a-road-sign/) *Updatable* (https://www.riskope.com/2020/01/07/climate-adaptation-and-risk-assessment/)

Other linked information (https://www.riskope.com/blog-news/) search Riskope blog and use the search box

Third parties links in this section	
Survey on stress related to disasters	https://www.scientificamerican.com/article/you-can-get-through-this-dark-pandemic-winter-using-tips-from-dis aster-psychology1/
Italian and Israeli studies	https://www.frontiersin.org/articles/10.3389/fpsyg.2020. 586202/full
Hacking a blast furnace	https://www.bbc.com/news/technology-30575104
Aluminum producer hacked	https://www.reuters.com/article/us-norsk-hydro-cyber/hac kers-hit-aluminum-maker-hydro-knock-some-plants-off line-idUSKCN1R00NJ
Highway risk assessment	https://reviewboard.ca/upload/project_document/EA1415-01_Risk_Assessment_Technical_Report_submitted_by_ Oboni_Riskope.PDF

References

Cho S, Gordon P, Moore JE II, Richardson HW, Shinozuka M, Chang S (2001) Integrating transportation network and regional economic models to estimate the costs of a large urban earthquake. J Reg Sci 41:39–65

[FAA] Federal Aviation Administration, (2002) Aviation industry overview fiscal year 2001. FAA Office of Aviation Policy and Plans, Washington, DC

Holmes T, Rahe R (1967) Social readjustment rating scale. J Psychosom Res 11(2):213–218

Kung A, Svobodova K, Lebre E, Valenta R, Kemp D, Owen JR (2020) Governing deep sea mining in the face of uncertainty. J Environ Manage 279:111593. https://doi.org/https://doi.org/10.1016/j.jenvman.2020.111593

Oboni F, Oboni C, Zabolotniuk S (2013) Can we stop misrepresenting reality to the public. CIM

Nathani C, Wickhart M (2006) Estimation of a Swiss input-output table from incomplete and uncertain data sources, intermediate international input-output meeting on sustainability, trade and productivity July 26–28, 2006, Sendai, Japan

Norris FH, Byrne CM, Diaz E, Kaniasty K (2002) The range, magnitude, and duration of effects of natural and human caused disasters: a review of empirical literature, Fact Sheet, National Center for PTSD, White River Junction, VT

Rose A (2004) Economic principles, issues, and research priorities in hazard loss estimation. In: Okuyama Y, Chang S (eds) Modeling spatial and economic impacts of disasters. Springer , New York, pp 13–36

Santos JR (2008) Inoperability input-output model (IIM) with multiple probabilistic sector inputs. J Indus Manage Optim 4(3):489–510

Part IV
Tactical and Strategic Planning
for Convergent/Divergent Reality

In Part IV, we discuss a quintessential element of modern risk-informed decision making: the risk tolerance threshold both in historic and modern terms (Chap. 10).

We then review (Chap. 11) the "how to" procedure leading to the ability to prepare rational mitigative roadmaps as well as rules to enhance systems' resilience (Sect. 11.1.1).

Chapter 12 shows how to distinguish tactical from strategic families of risks. Finally, we touch on risk-adjusted project cost estimates (Sect. 12.2).

Appendix B complements this Part IV and many points delivered in Part III as it concentrates on the DON'Ts related to risk assessments. Readers should refer to it every time they feel like shaking their heads while reading the following sections.

Chapter 10
Tolerance and Acceptability

In this chapter we discuss risk tolerance and acceptability, how they used to be defined, how they can be defined. A box with links to key terms is included in the references at the end of this chapter to facilitate the read.

Tolerance and acceptability thresholds must be developed independently from risks to ensure unbiased results. The acceptability threshold denotes an unsuitable risk level quality and is always lower than the risk tolerance threshold, which is an upper boundary of permissible risk exposure (Kalinina et al. 2016). These thresholds can be set either in the form of constant values or curves.

Everyone has a different pain threshold; likewise, everyone has different risk tolerance thresholds. We use the plural because each one of us has various thresholds, for example, a perceived one and a financial one. Each of us decides every day to undertake some activities and consciously or unconsciously assumes risks we consider acceptable and/or tolerable or, sometimes, intolerable. Over the course of a lifetime, from cradle to grave, during any given activity some incidents will inevitably occur. Some will be benign; some might be more significant and evolve into accidents. Accidents involve consequences of greater significance.

Think about various activities:

- hunting;
- fishing;
- driving a heavy vehicle;
- cooking at home.

They all have possibly unpleasant consequences such as, respectively:

- encountering an aggressive bear or being shot by a fellow hunter;
- capsizing the boat;
- veering off road;
- starting a fire.

Each activity has some probability of ending in an accident and a range of consequences. For instance, the consequences of a car accident can range from bending a fender to totaling the vehicle, to damaging other vehicles; the preparation of a flambé might end in burning the stove, or the kitchen, or even the whole house.

We evaluate risk by combining the probabilities of an activity's ending in an accident with their consequences. We reach our tolerance threshold when we say, "I am unwilling to undertake that activity with that probability of an accident paired with those consequences". Corporations and operations, like individuals, have risk tolerance and acceptance thresholds. Obviously, accidents of various types will happen during the service life of the operation. Indeed, zero risk is unachievable, not even in highly controlled industries such as civil aviation, and certainly not in generic industrial settings.

Now, you may find a risk intolerable from a financial point of view: "I cannot afford to total my truck, ever, because I can't pay for a replacement". Or you may find a risk intolerable from the point of view of your own perception: "If I total my truck my spouse will kill me!"

The first step to gain a better understanding of risks and go from perception to rationality is to perform a convergent risk assessment. A non-convergent risk assessment and poorly defined tolerance thresholds can lead to many unwanted consequences, as Anecdote 10.1 shows.

Anecdote 10.1: Poorly defined tolerance, non-convergent risk assessment
An interesting story was reported (de Ruiter, 2020) from the Philippines. The Philippines see on average around twenty tropical storms per year, and they oftentimes cause significant flooding and landslides. Flooding risks were perceived as intolerable by local communities. Thus they tried to mitigate flooding by building houses in concrete and stone. However, the country is also exposed to damaging earthquakes. The July 2019 earthquakes in the northern Philippines significantly damaged these reinforced structures, which had brought perceived flooding risks to tolerable levels but proved to provide poor resistance to earthquakes.

Closing point. This example of non-convergent risk assessment shows that residual risks had not been understood. Mitigation was not the result of convergent thinking based on well-defined tolerance thresholds, and thus inappropriate.

As a manager of an operation or a government officer you should ask yourself:

- Are the risks tolerable and acceptable without mitigation?
- Are the risks tolerable and acceptable with mitigation?
- What are the residual risks if I mitigate risks?
- Is the proposed mitigation appropriate and sufficient?

To answer these questions we have to define the operation's risk tolerance thresholds. Note that the literature, legislations and guidelines do not help to complete the definitions of these thresholds as public thresholds address generic, large-scale societal tolerance concerns.

Risk tolerance thresholds are always project- and culture-specific. Therefore, we need to define specific thresholds by bringing to light the perception and the financial resilience of various stakeholders through appropriate questions. This will allow us to develop a reasonable, negotiated acceptability threshold model that can be used to define which risks are tolerable, intolerable, manageable, and strategic in view of tactical and strategic planning of the operations (see Chap. 12).

What constitutes acceptable/tolerable risk depends on the perspectives of the organizations or persons involved. A family or community living in the blast area of a chemical plant will have different perspectives than the company whose executives have been told their designs and risk management practices are based on best practices. The level of risk deemed acceptable for a corporation is not necessarily what may be considered acceptable for government or the public, even if it is based on a collaborative engagement process. The tolerance and acceptability thresholds are set by government, authorities, regulators, or others who are directly concerned (Darbre 1998). A determination of the acceptability of risk also must consider the implications of closure (end of service life) as well the broad benefits to society that flow from economic development. The level of protection (mitigation) to be applied to any system is therefore a function of the comparison between the various risk scenarios and the risk acceptability and tolerance thresholds.

Modern risk assessment methods help immensely to see through the complicated series of scenarios and tolerances and then make a balanced decision based on a meaningful public engagement process (see Sect. 5.2, 5.3, 5.4). From a corporate perspective, the financial consequences of a system failure are significant (see Chap. 9). It may be the case that possible alternatives to a given system have not yet been fully studied or understood. Thus, it is paramount that the risk assessment used to compare the alternatives be able to explicitly tackle the different levels of uncertainties (see Chap. 2, Sect. 4.2) and lead to a result that should be approximately right rather than precisely wrong.

If the aggregated risks of project or an alternative are too high and/or intolerable, it may be time to drop the project or alternative rather than build anyways and then try to mitigate.

From a public perspective there are two kinds of risk: voluntary and involuntary. Voluntary risk can be defined as the risk proceeding from one's own will, choice or consent, while involuntary risks are generated by situations that are beyond the public's will, choice or consent. In our litigation-prone societies, the distinction becomes fuzzier every day and is linked to the knowledge of the existence of the hazard. A perfectly ignorant person and/or company can only incur involuntary risks, as this person cannot formulate any choice based on potential hazard exposure. At the other end of the spectrum a perfectly knowledgeable person and/or company only incurs voluntary risks. Incurring voluntary risks and not mitigating them can be considered criminal negligence (see Chap. 4, Sect. 6.1.2), thus leading to the paradox of good companies and individuals performing good risk assessments but concealing them in order to avoid exposure to legal proceedings.

Selecting what constitutes an acceptable level of involuntary risk a project or activity exposes the public to cannot be the choice of the designer or the owner, operator. Rather, it has to be the choice of the government or regulatory authority, based on an acceptance of the identified risks of that project or activity. In the case of chemical processes, for example, the assurance statement required as part of the regulatory safety review report verifying that "the process is reasonably safe" means nothing more than the level of risk of the process is no worse than the level of risk previously approved by the government. Clearly the government is the final arbiter of acceptable risk on a case-by-case basis.

When discussing acceptability, a distinction must also be made between location-based risks and societal risks. Location-based risks derive from the annualized likelihood of a person being killed at a given location as a direct result of an accident associated with hazardous activities undertaken there. It is an expression of the risk that someone who lives or works in a place where a hazardous activity takes place is exposed to. In contrast, societal risks represent the likelihood of a group of people who are not directly engaged in an activity being killed in an accident arising out of that activity. It looks at the consequences of mishaps from the very broad point of view of an entire society, possibly affecting even those who are physically and emotionally removed from the mishap itself; as such, it is of interest mainly to public administrators (Geerts et al. 2016).

The determination of the acceptable risk level also generally differs depending on whether a phenomenon is natural or man-made, or a private or a public (societal) issue.

Acceptable risk and corporate commitment are inextricably linked. A company that is able to demonstrate its commitment to the public through the application of a strong governance framework will stand a better chance of having its proposed plans accepted by the public and maintain its SLO, CSR (Chap. 5) and ESG status (see Sect. 6.1.2). A company that can demonstrate its commitment to its own employees will more likely develop a proposal that involves less and/or lower risks. A company that is able to demonstrate its commitment to the regulatory authorities and the public will find the process of obtaining permits much easier to navigate. A company that has done all of the above will also stand an excellent chance of not experiencing a catastrophic system failure.

The ideal outcome is to have all parties agree, based on informed opinions, that the risks, and their mitigating measures for a proposed plan are acceptable. Informed opinions are only possible when all parties, including the public, have been provided with:

- the opportunity to participate in a meaningful communication and engagement process (see Sect. 5.4) which clearly states what the risks are, what mitigations will be implemented, by whom and when, with a clear roadmap, ethics (see Sec. 5.5) principles and accountability;
- the consequences evaluation (Chap. 9) of the proposed system, including the results of a catastrophic failure simulation and an unbiased risk assessment;
- information that:

- supports the selection of the processes implemented within the system and site location;
- demonstrates beyond a reasonable doubt that the owner is committed to the management of the system and its critical risks through the establishment of a comprehensive management framework and assurance program;
- information that demonstrates that the government has established and will be committed to an effective compliance and enforcement regime.

10.1   Historic tolerance thresholds

10.1.1   Examples of Constant-Value Acceptable and/or Tolerance Thresholds

In the past different criteria for simple societal constant-value risk acceptability have been developed (Wilson 1984; Comar 1987; Renshaw 1990). These are shown in Table 10.1. For landslides the Australian Geomechanics Society suggested tolerable loss of individual life risk (AGSLT 2007). These are shown in Table 10.2.

Table 10.1 Historic societal constant value risk acceptability criteria

Author	Units	Unacceptable risk/Upper bound	Negligible risk/Lower bound
Renshaw	Probabilities of fatality of one individual per year of exposure to the risk	$1 * 10^{-5}$	$1 * 10^{-7}$
Comar		$1 * 10^{-4}$	$1 * 10^{-5}$
Wilson		$1 * 10^{-3}$	$1 * 10^{-6}$
Renshaw	Probabilities of fatality of ten individuals per year	$1 * 10^{-5}$	$1 * 10^{-7}$

Table 10.2 AGS Suggested Tolerable loss of life individual risk

Situation	Suggested tolerable loss of life risk for the person most at risk per year
Existing slope / Existing development	$1 * 10^{-4}$
New constructed slope / New development / Existing landslide	$1 * 10^{-5}$

10.1.2   Examples
of Acceptable—and/or Tolerance-Threshold Curves

Countless curves have been developed for different industries/hazards. In log-log scale these curves appear as straight lines, while plotted in linear scale they assume a hyperbolic shape. Let's immediately remark that in a log-log graph a straight line of slope -1 (-45°, e.g., a curve passing through the point with coordinates $p=10^{-4}$, $C=10,000=10^{+4}$ will also cross the point $p=10^{-1}$, $C=10$) corresponds to a centered hyperbolic function of constant risk. That would be the tolerance curve of a "hyper"-rational being who does not feel any emotion toward small but extremely likely losses (the "pain in the neck" type of maintenance nuisance) or very large losses at very low likelihood (say the Fukushima type of phenomenon). As the mathematics have not changed over time some passages below are strongly inspired by an earlier book (Oboni and Oboni 2007).

The straight line in the log-log plot was called the boundary line in early literature (Farmer 1967; EDI 1989). Its slope, if less than 1/1, indicates risk averseness or, if greater than 1/1, risk appetite (Plattner 2005). Generally human beings have an appetite for voluntary risks, but an averseness to involuntary risks. Moreover, many Perception Affecting Factors (PAF) enter into the delicate equilibrium between risk averseness and appetite, as described in Table 10.3. Whereas Plattner (2005) considers perceived risks to be generally higher than real, objective risks, we take the stance of endemic possible misestimation (overestimation and/or underestimation) of risks, as factors as familiarity, knowledge, controllability etc. blatantly lead people to biased estimates of the risks they are exposed to.

Well-known examples of log-log straight thresholds are found in several sources (Christou et al. 2006; NSW 2011; HKEPD 1994; ANCOLD 2003; Whitman 1984). They are depicted in Fig. 10.1.

By simply examining the slopes of the graphs of Fig. 10.1 one can immediately see that:

- Whitman's (1984) tolerance thresholds slopes depict a risk-prone society;

Table 10.3 The delicate equilibrium between risk averseness and appetite

PAF	Represents
Voluntariness	Voluntariness
Reducibility	Reducibility, Predictability, Avoidability
Knowledge	Familiarity, Knowledge about risk, Manageability
Endangerment	Controllability, Number of people affected, Fatality of consequences, Distribution of victims (spatio-temporal), Scope of area affected, Immediacy of effects, Directness of impact
Subjective measure of extent of damage	Extent of damage
Subjective measure of frequency of event	Frequency of event

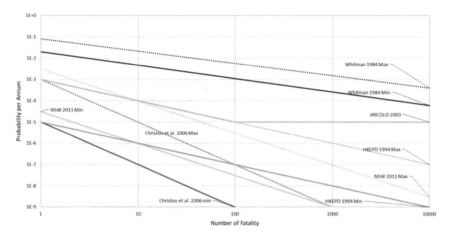

Fig. 10.1 Comparison between various risk tolerance thresholds, consequences expressed in fatalities. Notice that the vertical axis (probabilities per annum) extends well below the credibility threshold for a number of examples.

- HKEPD (1994) curves are ISO-risk;
- NSW (2011) curves are risk-averse;
- Christou et al. (2006) curves are strongly risk-averse;
- ANCOLD's curve first depicts an ISO-risk behavior and then, from 100 casualties onward, depicts a risk-indifferent behavior.

The Whitman tolerance is depicted in Fig. 10.2 together with data from accidents that had occurred at the time it was defined. In the graph in the original paper (Whitman 1984), reproduced here as is, the cost of a life was equated to 1M USD with no further emotion or discussion. That was considered just fine in those times. Today we do not attribute a cost to human life, a notion often considered repugnant and ethically unacceptable. Rather, we look at accepted capital expenditure in various countries to save the life of a citizen potentially exposed to natural hazards to find a "bounding value". In other words, in this book we use the mitigative investment a society is ready to make to save a life or its "willingness to pay" attitude (WTP, see Sect. 10.2.1) (Marin 1992; Mooney 1977; Jones-Lee 1989; Lee and Jones 2004; Pearce et al. 1996).

Societal risk tolerance is generally based on a fatalities generated by accidents.

Figure 10.3 shows *various accident types, from various industries, and their comparison to a published life/societal tolerances criteria*(https://www.riskope.com/2015/03/12/a-risk-perspective-on-the-concerns-raised-by-the-liability-of-hauling-dangerous-goods/) (Federal Highway Administration (FHWA) data, cited in Baecher 1987), namely: road transportation, US and Canada railway accidents and tailings dams failures in the decades around 1979 and 1999, Nuclear 5+ accidents over 14,000 years-reactors (Oboni and Oboni 2013).

Fig. 10.2 Whitman
tolerance data from accidents
that had occurred at the time.
NB: The cost of casualty is
assumed to be 1M USD

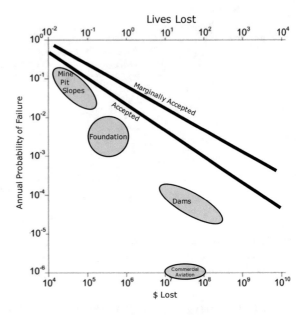

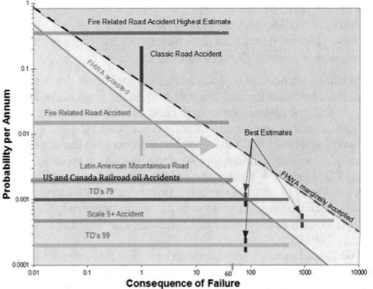

Fig. 10.3 FHWA life/societal tolerance criteria (Baecher 1987), and a sample of accidents ranges
in various industries.

10.1.3 Examples of Monetary Acceptable and/or Tolerance Thresholds Curves

For a monetary-only tolerance threshold the curves have a totally different appearance because in the event of a certain consequence being realized, the company will fail, or lose its freedom, no matter what. It is therefore modelled as an activation function. For a set of clients we developed the set of tolerance thresholds shown in Fig. 10.4. The curves are bundled to preserve confidentiality.

10.2 Modern Risk Tolerance

Risks can be sorted according to decreasing value in order to deliver a prioritized list. However, that prioritization is far from best practices as it lacks a vital piece of information: what risks actually do have the potential to do harm to the system under consideration, both internally and externally? It is also vital for the client to distinguish between high-probability low-consequence events and their opposite, i.e., low-probability, high-consequence events, something that common practice approaches like matrices, FMEAs, PIGs do not allow (see Appendix B) simply because their coloring schemes have been arbitrarily defined without proper understanding of their meanings. Indeed, in many common-practice approaches a Fukushima-type of event receives the same color as, say, the next migraine of the plant director, which is obviously a source of ill-conditioned decisions and conclusions. In order

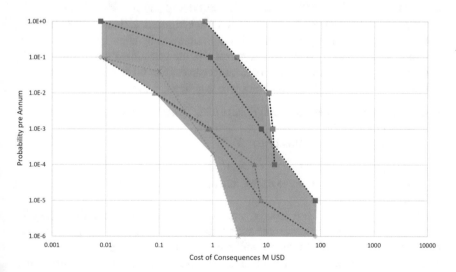

Fig. 10.4 Risk-tolerance thresholds developed empirically (by means of facilitated workshops) for a set of medium to large corporations. The envelope of acceptable tolerances is shaded in blue.

to perform sensible prioritization, risks must be compared to corporate and societal risk tolerances.

Why is risk tolerance so important? One example related to a classic trend is over-estimating outcome severity after one mishap. Let's consider a system that causes, on the average, one accident every 100 years with generally relatively small consequences, say one fatality each. However, once in a while there may be a catastrophic event generating ten fatalities. If the catastrophic event happens to occur, the public (or regulatory agencies) may believe that all accidents have catastrophic outcomes, thus they demand more safety measures than are justified by the actual damage expectation. Such a claim is not restricted to the particular facility that caused the accident; improvements are required for all other facilities of this type.

10.2.1   Corporate Risk Tolerance (CRT)

Corporate Risk Tolerance (CRT) defines the financial consequences a corporation views as tolerable at various levels of probability. Thus CRT should be seen primarily as a perceived tolerance risk threshold, the actual financial limit being generally higher.

CRTs are unique to each corporation, operation, project, and timing (within the life cycle) and can be determined through facilitated questions/answers using a methodology we have developed over the years, after plotting point-by-point curves in workshops with boards of directors around the world and looking at examples developed by others. Of course, when looking at risks involving human wellbeing, the concepts of ALARA, ALARP and BACT should be introduced and considered (Sect. 4.1).

Once the risks incurred by an operation are estimated, rational and sustainable decisions about risk mitigation are generally requested by clients wishing to adopt risk informed decision making and to maximize the investment they have made by performing a risk assessment. Such decisions can only be taken after an explicit risk tolerance function is defined. The tolerance function can be:

- derived formally (mathematically) from client's financial data;
- defined empirically;
- derived from public opinion tests;
- "negotiated".

Generally, the final tolerance curves selected by clients are the result of a mix of these various approaches.

Tolerable risk curves are always project- and owner-specific and indicate the level of risk which has been deemed acceptable by an owner for a specific project or operation (possibly taking public opinion into account (see Chap. 5)). This means, as an example, that within large companies' corporate risk tolerance may differ quite substantially from that of a branch operation. Case Study 3 (Chap. 16) shows an example of this.

Risks which plot to the left and below the tolerance curve are deemed bearable. Risks which plot to the right and above the curve are deemed unbearable and some measures of mitigation are considered necessary to reduce their likelihood. Reducing the likelihood of an impact may be, for example, as simple as imposing "no stop" zones on a road.

When working empirically, two curves should be developed, one representing the optimistic, the other the pessimistic view of tolerance. The area between the curves represents a range of uncertainty on tolerance defined by an organization. When data are available theoretical curves can be developed and then discussed with key personnel.

The development of empirically estimated tolerance curves requires caution and continuous calibration as the extent of correlation between an individual's estimate or ranking of probabilities and the true value/ranking is usually quite weak, sometimes even in the order of zero (Gordon 1924; Peterson and Beach 1976). However, it has been demonstrated that pooled judgments correspond better with the truth as the number of individuals increases. For instance, the average correlation between individual judgments and the correct rank order may increase twofold when pooled across seven individuals, and twenty individuals may have an excellent correlation. No wonder juries are made out of twelve people! This is one of the reasons why risk assessments, and in particular risk tolerance curves, should always be defined by a group and not by an individual (Hofstätter 1986; Wilde 2001).

In qualitative risk assessments (FMEA, PIGs, matrices, etc.) tolerance and acceptability implicitly emerge through the coloring of the probability-impact matrix cells, although their interpretation may be difficult and even misleading (Cox 2008; Ball and Watt 2013; Porter et al. 2019; Terzi et al. 2019).

However, explicit tolerance and acceptability thresholds are required to prioritize and thus efficiently mitigate risks. In this case, ethics is also fostered because cognitive biases and irrational fears are filtered through this type of prioritization.

Figure 10.5 shows the kind of quantitative quadrant depicting pairs of probabilities and consequences of various scenarios that we prepare BEFORE plotting the various risks. This particular example is for a major mining operation, but the concept applies to any industry, system, project. It is paramount to prepare the tolerance thresholds beforehand in order to avoid biases, and for the same reason it is even better to avoid showing any risk result until we plot them on the graph.

The following lines/curves are depicted on the graph:

- Yellow depicts the operations perceived risk tolerance threshold obtained through a facilitated process of questions and answers, using client's responses and data.
- Orange is the so-called "iso-risk" threshold, a line representing identical risk tolerance for any probability also derived from client's information. It would correspond to an emotionless acceptance where there is no shying away from even the highest consequences, provided their probabilities are very low, i.e. leading to uniform risks.

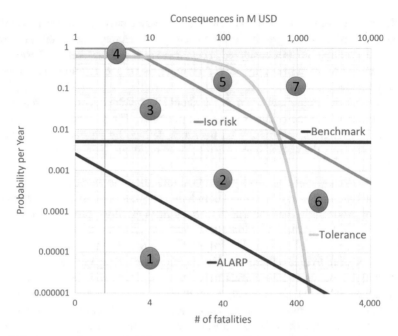

Fig. 10.5 An example of a probability-consequences quadrant. Vertical axis: annual probability of potential scenarios; upper horizontal axis: multidimensional consequences values expressed in monetary terms (other metrics are may be selected); lower horizontal axis: potential consequences expressed in number of fatalities. The seven areas allow risk categorization.

- Purple is an example benchmark obtained by reviewing historical performances in the system's industry. Indeed, anything above it is riskier than the world-wide portfolio of similar systems. Anything below is less risky than the portfolio.
- Burgundy defines the ALARP (Sect. 4.1) concept expressed in terms of Willingness To Pay (WTP=2.5M USD), based on specific literature (Marin 1992; Mooney 1977; Jones-Lee 1989; Lee and Jones 2004; Pearce et al. 1996). For higher values of consequences ALARP generally displays a limit value (vertical line), as shown in *F-N*(https://www.ngi.no/download/file/6259) curves.

Both corporate and iso-risk tolerances correspond to the top horizontal axis (consequences in USD). The ALARP tolerance corresponds to the bottom axis, the number of fatalities, but can also be read on the top axis, consequences in USD, with WTP=2.5M USD adopted in this example. Note the vertical line placed in the graph that corresponds to one fatality.

The different lines and curves define specific areas numbered from 1 to 7 denoting specific nature of the risks they may include. Here are descriptions of their properties.

1.   Below ALARP (fatalities expressed in WTP) and tolerable.
2.   Tolerable for the considered operation and below world-benchmark.
3.   Tolerable for the considered operation, but above world-benchmark.

4.   Corporately intolerable: high-probability low-consequence scenarios which create "irritation", such as having a piece of equipment which breaks down often, but does not really have any significant consequence. Although the corresponding risks are lower than the constant risk curve, clients will often decide to mitigate, e.g. replace the equipment, either because the impact is overestimated and/or simply because "that piece of equipment was a pain in the neck, so we changed it!". Definitely a manageable risk.
5.   "False comfort", tendency to "ignore, dismiss" because overestimating tolerance.
6.   Iso-risk tolerable, corporately intolerable, ALARP intolerable.
7.   Entirely intolerable.

NB: Risk in areas 6 and 7 can be either manageable or unmanageable. The distinction lies in the effect of possible mitigation and is an important one because it helps defining strategic risks, i.e. intolerable and unmanageable risks (see Chap. 12). If it is sustainable to mitigate a risk (reduce the probability of the scenario, i.e. pushing the risk downward, below the tolerance threshold), then the risk is tactical. If the risk has to be pushed below the tolerance threshold by reducing the consequences (shifting the scenario toward the left), this generally means that the system has to be altered, and therefore the risk is strategic. For instance, in Fig. 10.5 risks with consequences above 1B USD cannot be brought under tolerance (pushed downward) before reaching the credibility threshold ($p=1*10^{-6}$), hence they would be strategic.

Let's look at examples of risks in different areas of the risk tolerance thresholds shown in Fig. 10.5.

• A single heavy vehicle accident would likely plot left of the vertical "one fatality" line. Thus, it would land in Areas 1, 2 or 3 depending on the level of training, road maintenance, truck maintenance, etc. for those influence the probability. Truck damages, if they are high frequency, low consequences, would land in Area 4.
• Accidents that have even very significant financial consequences but remain ALARP compatible would land in Area 1. For instance, this could be the case of multiple tank failures where monitoring and alert systems allow personnel to evacuate the danger area in time.
• Area 2 would receive all those accidents that are fine with respect to world-benchmark and are corporately tolerable. Their consequences in terms of expected fatalities make them H&S intolerable.
• Area 3 receives similar risks to those of Area 2, but with probabilities higher than the benchmark.
• A failure could plot in Area 2, 3, 5 or 6 depending on the number of people and equipment in the area (including access control to these areas), the business interruption it would generate; and extant mitigation and monitoring and defenses.
• Note that Area 6 also includes events with extremely high consequences but low probabilities.
• A catastrophic explosion in the system would certainly fall in Area 7 if the probability is above world benchmark, and the cost of its consequences extremely high.

To close this discussion we will note that corporate risk tolerance is not constant in time. It is obviously a function of an organization's wealth, economic environment, etc. In the case of some industries—extractive industries in particular—this translates into a function of time. For example, as reserves are depleted, tolerance decreases because the company has less future wealth to buffer a hit (the operational safety margin decreases with time). Corporate wealth also has to do with the attitude a corporation may have in defining risk tolerance.

10.2.2 Societal Risk Tolerance

In modern times it has become customary use statistical approaches to loss of life based on industries results. Below is a review. To ensure good understanding, we have to define a few parameters and terms (Rausand 2011):

- Individual Risk Per Annum (IRPA) = probability an individual is killed during one year's exposure. Table 10.4 defines IRPA for various activities in the UK up to the year 2000 (see Table 3 *HSE* 2001(https://www.hse.gov.uk/risk/theory/r2p2.pdf)).
- IRPA* = observed no. of fatalities / Total no. of employee-years exposed
- Potential Loss of Life (PLL) = n * IRPA, where n is the number of people potentially exposed and IRPA is, as stated above, the individual risk per annum.
- Fatal Accident Rate (FAR) = is the expected number of fatalities per $1*10^8$ hours of exposure, a common indicator for H&S practitioners. If 1,000 persons work 2,000 hours per year for 50 years, their cumulative exposure time will be $1*10^8$ hours. FAR is then the estimated number of these 1,000 persons that will die in a fatal accident during their working life. Table 10.5 shows fatal injuries in the US in 2018 in relation to worker characteristics, occupations, and industries and civilian workers. In this case, the *statistics* (https://www.bls.gov/iif/oshcfoi1.htm#rates)

Table 10.4 IRPA for various activities in the UK up to the year 2000. (Table3 in HSE 2001) with provisional figures used for 2000/2001

Activity	IRPA=1 in ….	IRPA
Employees	125 000	$8 * 10^{-6}$
Self-employed	50 000	$2 * 10^{-5}$
Mining and quarrying of energy producing materials	9 200	$1 * 10^{-4}$
Construction	17 000	$6 * 10^{-5}$
Extractive and utility supply industries	20 000	$5 * 10^{-5}$
Agriculture, hunting, forestry and fishing (not sea fishing)	17 200	$6 * 10^{-5}$
Manufacture of basic metals and fabricated metal products	34 000	$3 * 10^{-5}$
Manufacturing industry	77 000	$1 * 10^{-5}$
Manufacture of electrical and optical equipment	500 000	$2 * 10^{-6}$

Table 10.5 US worker characteristics, occupations, and industries, civilian workers, in 2018

Characteristic	Total fatal injuries	Total hours worked (millions)	Fatal injury rate
Total	5,250	292,528	3.5
Natural resources, construction, and maintenance occupations	1,685	28,854	11.6
Construction and extraction occupations	1,003	16,397	12.2
Mining machine operators	9	163	11.0
Other extraction workers	5	118	8.5
Production, transportation, and material moving occupations	1,668	36,448	9.1
Transportation and material moving occupations	1,443	19,099	15.0
Industrial truck and tractor operators	44	1,289	6.8
Industry			
Private industry	4,779	253,458	3.7
Natural resources and mining	704	6,645	20.9
Oil and gas extraction	13	177	14.7
Mining (except oil and gas)	34	454	15.0
Coal mining	11	147	15.0
Nonmetallic mineral mining and quarrying	17	224	15.2
Support activities for mining	83	1,209	13.7

on which the analyses are based are solid but they have to be considered with care due to the assumptions made (e.g. age exclusion, time worked and industry characteristics).

- $FAR^* =$ (Expected no. of fatalities / No. of hours exposed to risk) $* 1*10^8$. Table 10.6 shows selected FAR^* values for the Nordic countries for the period 1980–1989 (from slide 14/24 in Haugen and Rausand 2011).

There are many published governmental and non-governmental H&S (fatalities) tolerance models, among which Hong Kong and Australia.

If we consider a model (Fig. 10.6) which defines a maximum risk tolerance threshold at 1 fatality at $p_f = 1*10^{-3}$/yr., and 10 fatalities at $p_f = 1*10^{-4}$/yr. (top green threshold), we are stating that an IRPA of $1*10^{-3}$ is at the upper limit of tolerance. Note that following the lower bound limit (bottom green threshold) the ALARP domain is two orders of magnitude lower (Amoushahi et al. 2018; Read and Stacey 2009).

Table 10.6 Selected FAR* values for the Nordic Countries for the period 1980–1989 (from slide 14/24 in Haugen and Rausand 2011).

Industry	FAR* (Fatalities per 10^8 working hours)
Agriculture, forestry, fishing and hunting	6.1
Raw material extraction	10.5
Industry, manufacturing	2.0
Utilities	5.0
Building and construction	5.0
Transport, post and telecommunication	3.5

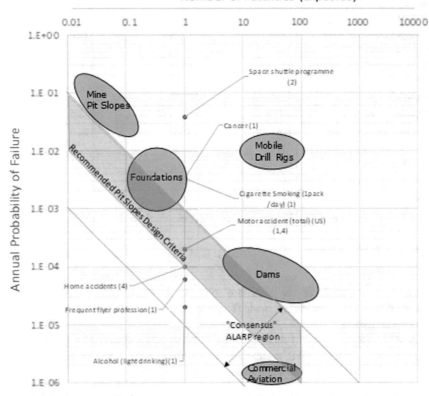

Fig. 10.6 Tolerance thresholds expressed as number of fatalities vs. annual probability of failure for various examples. The examples aligned on the vertical line (1 expected fatality) correspond to those of Fig. 10.7 and related numbered references.

Figure 10.7 shows tolerance thresholds expressed as number of fatalities vs. annual probability of failure. Risks for involuntary Orange, voluntary Blue according to: Wilson and Crouch 1987 (1); Philley 1992 (2); Hambly and Hambly 1994 (3); Baecher and Christian 2003 (4).

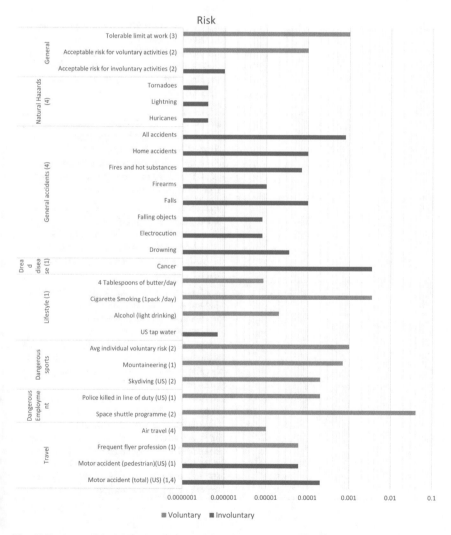

Fig. 10.7 Acceptable risk for involuntary, voluntary and work activities (top three lines)

Appendix

Links to more information about the Key terms from the Authors	
A,B	*Act of God* (https://www.riskope.com/2020/12/09/act-of-god-in-probabilistic-risk-assessment/) *Black swan* (https://www.riskope.com/2011/06/14/black-swan-mania-using-buzzwords-can-be-a-dangerous-habit/) *Business-as-usual* (https://www.riskope.com/2021/01/13/business-as-usual-definition-in-risk-assessment/)
C,D	*Convergent* (https://www.riskope.com/2021/01/20/convergent-risk-assessments/) *Divergent* (https://www.riskope.com/2020/11/18/tactical-and-strategic-planning-to-mitigate-divergent-events/) *Drillable* (https://www.riskope.com/2020/01/15/probability-impact-graphs-do-not-fly/)
F	*Foreseeability/foreseeable* (https://www.riskope.com/2021/01/06/foreseeability-and-predictability-in-risk-assessments/) *Fragile/fragility* (https://www.riskope.com/2020/04/01/antifragile-resilient-solutions-for-tactical-and-strategic-planning/)
P,R	*Predictability/predictable* (https://www.riskope.com/2021/01/06/foreseeability-and-predictability-in-risk-assessments/) *Resilient, Resilience* (https://www.riskope.com/2016/11/23/resilience-cannot-based-instinctual-decision-making/)
S	*Scalable* (https://www.riskope.com/2015/04/16/how-system-definition-and-interdependencies-allow-transparent-and-scalable-risk-assessments/) *Societal risk acceptability* (https://www.riskope.com/2014/01/09/aspects-of-risk-tolerance-manageable-vs-unmanageable-risks-in-relation-to-critical-decisions-perpetuity-projects-public-opposition/) *Sustainability/sustainable* (https://www.riskope.com/2019/01/16/improving-sustainability-through-reasonable-risk-and-crisis-management/) *Survivability* (https://www.riskope.com/2011/03/17/ale-fmea-fmeca-qualitative-methods-is-it-really-what-we-need/) *System* (https://www.riskope.com/2017/07/26/three-ways-to-enhancing-your-risk-registers/)

(continued)

(continued)

Links to more information about the Key terms from the Authors	
T,U	*Tolerance* (https://www.riskope.com/2020/04/29/risk-tolerance-thresholds/) *Uncertainty/uncertainties* (https://www.riskope.com/2015/12/10/3-decision-making-truths-derived-from-uncertainty-taxonomy-scheme-of-classification-and-a-road-sign/) *Updatable* (https://www.riskope.com/2020/01/07/climate-adaptation-and-risk-assessment/)
Other linked information (https://www.riskope.com/blog-news/) search Riskope blog and use the search box	

Third Parties links in this section:	
Comparing risks from various hauling system	https://www.riskope.com/2015/03/12/a-risk-perspective-on-the-concerns-raised-by-the-liability-of-hauling-dangerous-goods/
F-N curves	https://www.ngi.no/download/file/6259
HSE 2001	https://www.hse.gov.uk/risk/theory/r2p2.pdf
Accidents statistics	https://www.bls.gov/iif/oshcfoi1.htm#rates

References

[AGSLT 2007] Australian Geomechanics Society Landslide Taskforce [Walker B, Davies W, Wilson G (2007) Practice note guidelines for landslide risk management, landslide practice note working group. J News Australian Geomech Soc 42(1)

Amoushahi S, Grenon M, Locat J, Turmel D (2018) Deterministic and probabilistic stability analysis of a mining rock slope in the vicinity of a major public road—case study of the LAB Chrysotile mine in Canada. Can Geotech J 55(10):1391–1404

Australian National Committee on Large Dams (2003) Guidelines on risk assessment. ANCOLD, Sydney

Baecher G B (1987) Geotechical Risk Analysis User's Guide, Report no. FHWA-RD-87-011

Baecher GB, Christian JT (2003) Reliability and statistics in geotechnical engineering. John Wiley and Sons

Ball DJ, Watt J (2013) Further thoughts on the utility of risk matrices. Risk Anal 33(11):2068–2078

Christou M, Struckl M, Biermann T (2006) Land use planning guidelines in the context of article 12 of the Seveso II Directive 96/82/EC as amended by Directive105/2003/EC. European Commission Joint Research Centre, Institute for the Protection and Security of the Citizen. EUR 22634 EN FR DE

Comar C (1987) Risk: a pragmatic de minimis approach. In: De Minimis Risk, C. Whipple (ed), xiii-xiv, New York, Plenum Press

Cox LA Jr (2008) What's wrong with risk matrices? Risk Anal 28(2):497–512

Darbre G (1998) Dam risk analysis. Report, Federal Office for Water and Geology. Dam Safety, Bienne de Ruiter MC (2020) Dynamics of vulnerability: from single to multi-hazard risk across spatial scales, Ph.D. thesis, Vrije Universiteit Amsterdam

[EDI 1989] Eidgenössisches Departement des Innern (1989) Verordnung über den Schutz vor Störfallen (Störfallverordnung, SFV), Entwurf, Bern

Farmer F (1967) Siting criteria—a new approach. Containment and siting of nuclear power plants. International Atomic Energy Agency (IAEA), Vienna, pp 303–329

Geerts R, Heitinka J, Gooijerb L, van Vlietb A, Scheresa R, de Boerc D (2016) Societal risk and urban land use planning: creating useful pro-active risk information. Chem Eng Trans 48:955–960

Gordon K (1924) Group judgments in the field of lifted weights. J Exp Psychol 7:398–400

Hambly EC, Hambly EA (1994) Risk evaluation and realism. Proc Inst Civ Eng Civ Eng 102(2):64–71

Haugen S, Rausand M (2011) Risk assessment Chapter 4 how to measure risk, RAMS Group, Department of Production and Quality Engineering, NTNU, https://www.ntnu.edu/documents/624876/1277591044/ch4-risk-metric.pdf/1cd0e979-d0b4-49fc-b594-0b94e9f53b08

[HSE] Health and Safety Executive (2001) Reducing risks, protecting people, HSE's decision making process, ISBN 0 7176 2151 0

Hofstätter P (1986) Gruppendynamik. Rowohlt, Hamburg

[HKEPD] Hong Kong Environmental Protection Department (1994) Practice note for professional persons. ProPECC PN 2/94 https://www.epd.gov.hk/epd/sites/default/files/epd/english/resources_pub/publications/files/pn94_2.pdf

Jones-Lee MW (1989) The economics of safety and physical risk. Blackwell, Oxford

Kalinina A, Spada M, Marelli S, Burgherr P, Sudret B (2016) Uncertainties in the risk assessment of hydropower dams state-of-the-art and outlook, Zurich, ETHZ https://www.ethz.ch/content/dam/ethz/special-interest/baug/ibk/risk-safety-and-uncertainty-dam/publications/reports/RSUQ-2016-008.pdf

Lee EM, Jones DKC (2004) Landslide risk assessment, Thomas Telford

Marin A (1992) Costs and benefits of risk reduction. Appendix in risk: analysis, perception and management, Report of a Royal Society Study Group, London

Mooney GM (1977) The valuation of human life. Macmillan, New York

[NSW 2011] NSW Government: Hazardous Industry Planning Advisory Paper No 4 Risk Criteria for Land Use Safety Planning, January 2011. https://www.planning.nsw.gov.au/-/media/Files/DPE/Other/hazardous-industry-planning-advisory-paper-no-4-risk-criteria-for-land-use-safety-planning-2011-01.pdf?la=en

Oboni C, Oboni F (2013) Factual and foreseeable reliability of tailings dams and nuclear reactors—a societal acceptability perspective, Tailings and Mine Waste 2013, Banff, AB, November 6 to 9, 2013

Oboni F, Oboni C (2007) Improving sustainability through reasonable risk and crisis management. JSO, Froideville, Switzerland. Book—Riskope ISBN 978-0-9784462-0-8

Pearce DW, Cline WR, Achanta AN, Fankhauser S, Pachauri RK, Tol RSJ, Vellinga P (1996) The social costs of climate change: greenhouse damage and the benefits of control. In: Climate change 1995: economic and social dimensions of climate change. Contribution of Working Group III to the Second Assessment Report of the IPCC, Cambridge, Cambridge University Press

Peterson CR, Beach LR (1976) Man as an intuitive statistician. Psychol Bull 68:29–46

Philley JO (1992) Acceptable risk—an overview. Plant Oper Prog 11(4):218–223

Plattner T (2005) Modeling public risk evaluation of natural hazards: a conceptual Approach. Nat Hazards Earth Syst Sci 5:357–366

Porter M, Lato M, Quinn P, Whittall J (2019) Challenges with use of risk matrices for geohazard risk management for resource development projects. In: Proceedings of the First International Conference on Mining Geomechanical Risk. Australian Centre for Geomechanics, pp 71-84

Rausand M (2011) Risk assessment: theory, methods, and applications. Wiley

Read J, Stacey P (2009) Guidelines for open pit slope design. CSIRO Publishing. ISBN: 9780643101104

Renshaw FM (1990) A Major Accident Prevention Program. Plant Oper Prog 9(3):194–197

Terzi S, Torresan S, Schneiderbauer S, Critto A, Zebisch M, Marcomini A (2019) Multi-risk assessment in mountain regions: A review of modelling approaches for climate change adaptation. J Environ Manage 232:759–771

Whitman RV (1984) Evaluating calculated risk in geotechnical engineering. J Geot Eng 110(2):145–188

Wilde GJS (2001) Target risk 2: A New Psychology of Safety and Health. Toronto, PDE Publications (First edition available at https://psyc.queensu.ca/target)

Wilson R (1984) Commentary: risks and their acceptability. Sci Technol Hum Values 9(2):11–22

Wilson R, Crouch EAC (1987) Risk assessment and comparisons: an introduction. Science 236(4799):267–70. https://doi.org/10.1126/science.3563505

Chapter 11
Convergent Risk Assessment for Divergent Exposures

In this chapter we will discuss the expectations for a convergent risk assessment of business-as-usual as well as divergent exposures. We will see for whom and how such a risk assessment should be prepared. Appendix B could usefully be read in conjunction with this chapter as it concentrates on the DON'Ts related to risk assessments. A box with links to key terms is included in the references at the end of this chapter to facilitate the read.

11.1 Expectations

In response to government and public expectations (see Part I) there is an increasing demand for the assessment of mitigation alternatives for the purpose of reducing site-specific risks and impacts and enhance resilience. For example, the Australian Government Tailings Management publication (AG TMH 2016) shows that nowadays design submissions are expected to demonstrate beyond a reasonable doubt that sustainable outcomes will be achieved during operations and after closure through the application of leading practice risk-based design that:

- fully assesses the risks associated at a particular site;
- compares the suitability of all available methods;
- demonstrates that the selected design will manage all risks to within acceptable levels in accordance to ALARP (ICOLD 2013, also see Sect. 4.1).

We might add that leading practice risk-based design also:

- considers closure requirements and its associated risks;
- enables the public to participate in a collaborative manner in the examination of alternatives and their related risk management strategies;
- aims at ensuring a reasonable resilience of the system.

Regarding permit conditions, leading practice requires that permits and permit amendments be granted on the basis of strict conditions related to critical operating parameters and risk mitigating strategies, and that these be measurable and enforceable.

As we have stated earlier, there is a blatant gap between the above requirements, similar to those discussed in Chap. 5, and the capabilities of common practices (see Chap. 6 and Appendix B). PIGS do not fly when systems need to be evaluated over any time horizon.

Thus, in this section we examine how to build an ISO 31000-compliant modern quantitative risk approach which is scalable, drillable, convergent, allows Bayesian updates as new data become available and can evaluate divergent exposures resulting, for example, from climate change. In other words, a risk assessment which covers the specs indicated in Chaps. 5 and 6. Furthermore, risk assessment should allow RIDM (see Sect. 12.2) as we will see in Part V.

We also discuss the use of risk tolerance thresholds (discussed in Chap. 10), which should be developed independently of the probability and consequence register in order to avoid biases and conflicts of interest. Incidentally, we will show that families of risk (i.e., tolerable risks; intolerable but manageable risks; intolerable and unmanageable risks) will become self-evident once tolerance thresholds are used. We will show that this latter family covers the strategic risks (see Chap. 12) and that these are the result of a risk assessment and not an instinctual, a priori, impulse characterization based on the most frightening scenario.

11.1.1 Enhanced Resilience

Any mitigation selection should be based on convergent RIDM and should aim at developing resilient solutions (Indirli et al. 2014).

The sustainability, value creation, financial soundness of any mitigation should be tested using risk-encompassing methods and not using, for example, net present value (NPV, see Sects. 12.2.1 and 12.2.2). As already stated, RIDM based on convergent, updatable, drillable, quantitative risk assessment helps foster proper communication and ethics (Chap. 5).

Resilience cannot be based on instinctual decision making either. Instead, it should rely on forward-looking risk assessments to avoid squandering private and public money in useless capital expenditure (Kaluarachchi et al. 2014). The question here is: Do we really need resilience enhancements?

The reply is a resounding YES, as code compliance and good design do not mean fostering resilience. As a matter of fact, in the *2018–2019 budget* (https://www.infras tructure.gc.ca/pd-dp/eval/2018-19-hgr-grh-overview-resume-eng.html) the Canadian government pledged 40 M CAD over five years to integrate climate resilience into building design, guidelines and codes, as they recognize that existing codes do not integrate the concept. The reason lies in the fact that codes customarily rely on statistics, business-as-usual, and experience, none of which being nimble enough to cope with climate changes and their unusual, and perhaps unknown, set of new parameters.

Think about the classic 200-year (or less, see Sect. 2.2.3) rain event engineers use to design all sort of hydraulic urban and industrial infrastructures or think about snow loading on roofs, service temperature extremes and many other classic "engineering" thresholds. Codes are not helping in coping with divergent hazard occurrences, as they offer too little or no guidance for divergence.

Just to prove the point, we have lately seen a chocolate factory being critically damaged by floods, many S&P 500 corporations facing *large financial impacts* (http://www.riskope.com/2014/06/19/extreme-weather-is-already-draining-the-big gest-u-s-companies/), and a *very promising startup go bankrupt* (http://www.ris kope.com/2016/04/06/ore-predicts-business-startups-3-financing-rounds-success/) because of temperature anomalies resulting from climate change. All of these systems were code-compliant, yet they failed one way or another.

As we discussed earlier (Sect. 4.4), contractual force majeure and the notion of negligence will "shift" because of climate change and other risk divergences. We have personally witnessed public officials, CEOs, CxOs in various countries challenged on whether they should have foreseen catastrophic events, or even simply hazardous events. We have acted as expert witnesses in civil and criminal courts in such cases. Insurance companies seem to react to divergence by denying renewal to some of their long-term insurees, based on arguments of excessive risk (see Anecdote 5.1). As for *cyber attacks* (http://www.riskope.com/2016/01/21/should-you-listen-to-your-insurer-for-your-business-cyber-risk-management/), actuaries are in trouble, for their science is based, like most codes and guidelines, on a rear-view mirror approach and not a forward-looking one.

Thus, clearly, managers and promoters must:

(1) be supported by forward-looking RIDM;
(2) accept that codes are insufficient;
(3) understand that business-as-usual thinking will almost certainly lead to significant negative outcomes.

Thanks to modern risk assessment methodologies it is possible to guide attention in a rational way toward areas that require resilience enhancements, as resilience must be shored up by carefully crafted forward looking risk assessments.

Resilience enhancements do not need to be capital-expenditure-intensive but, as stated above, instinctual decisions may be costly. Twenty years ago we convinced a world-renown producer of hazelnut products to move their electrical rooms away from the cellars and put them in the attic instead. The idea was to increase resilience in case of a flood. Our risk assessment had proven that was the best "bang for the buck". Although they had previously raised the whole unit, at great cost, by 1.5 m, building a sort of artificial island in case of flooding, the cellar remained potentially below the flood level. This was a total waste of capital expenditure in the name of instinctual resilience-enhancing decision-making.

Another industrialist whose operation was also exposed to flooding, had built, again at great expense, a "Great Wall of China" around his operation. The structure was equipped with automatic flood water-resistant gates. By doing so, he had actually reduced by 30% the hydraulic section of the valley in which his operation was located. We explained that while he may have saved the factory from waters, he was likely to see social turmoil and perhaps even a class-action suit because the reduced hydraulic section of the valley meant higher flooding outside his battery limit. Again, resilience enhancement does not pair well with instinctual decision-making. Instead we should shore it up by carefully crafted forward-looking risk assessments like those produced using ORE (Optimum Risk Estimates, ©Oboni Riskope Associates Inc., 2014-*, see Sect. 11.3.2).

The 10 commandments for resilient design we propose below are based on risk management concepts. They are a complement to the resilient design principles delivered by the Resilient Design Institute, and are based on our world-wide experience of risk-based decision making in risk mitigation and sustainability enhancement.

1. **Never rely on the properties of a single material/ component.**

   - **Why:** By relying on one single material or component you expose your project to sudden failures due to quality variability, supplier disruptions, etc.
   - **How to**: Design should include various lines of defense, based on different materials/components. Redundancy based on the same materials or components may be an illusion due to Common Cause Failure (CCF, see Sects. 6.3 and 8.2.2).

2. Ensure that redundancies are true.

   - **Why:** Perceived, but *untrue redundancies* (https://www.riskope.com/2015/08/27/integrating-mitigations-with-redundancies-in-an-industrial-process/) give a false sense of safety and control over the system.
   - **How to:** Reduce common cause failure of materials/components by selecting different suppliers, models and in some cases even different batches of similar components.

3. Avoid fragile systems at all costs.

   - **Why:** Fragile failures are difficult to monitor, prepare for and can catch anyone by surprise.
   - **How to**: Include in your design extra links (hyper-static structures, for example), sacrificial elements that will stress before the vital ones and work as early indicators of distress, elements that will delay the main failure.

4. Promote ductile (gradual damage increase) failures.

   - **Why:** Ductile failures absorb massive amounts of energy and slow propagation before final failure occurs, so they are safer than brittle (fragile) or sudden ones.
   - **How to**: Include ductile materials in your design, ductile links, and energy-absorbing devices and shapes.

5. Ensure that failure does not propagate.

   - **Why:** Propagation of failure (domino effects, interdependencies (see Sect. 6.3.1, 7.1, 8.2.2)) magnifies consequences (ripple effects (see Sect. 8.2.2, Chap. 9)) and makes it very difficult to resume operations.
   - **How to:** Insert controls that segment your system, avoid propagation, so as to limit domino effects and ripple propagation of consequences as far as possible.

6. Limit and control interdependencies.

   - **Why:** Propagation may occur without failure, because of proximity (geographic inter-dependency); for example, a chemical spill from a neighboring process might generate a stoppage.
   - **How to:** Study the environment very carefully with a holistic, 360° approach. Moreover, develop scenarios using the concept of threats-to/threats-from (see Chaps. 7 and 8) in your risk assessment.

7. Understand your system and keep an eye on its evolution.

   - **Why:** Systems can fail due to Threats-to/Treats-from impacts. It is important to understand the functional relationship of the system and its evolution.
   - **How to:** Wet-ware, software, hardware and energy, resources flows need to be understood and taken into account in holistic, convergent risk assessment.

8. **Avoid** normalization of deviance.

   - **Why:** Normalization of deviance can lead to over-looking a warning sign of a more serious event.
   - **How to:** Near misses and prior losses should be recorded and carefully analyzed to understand what and how they happened, so as to benefit from lessons learned.

9. Understand the limits of known-knowns and unknown-unknowns.

   - **Why:** As future may significantly diverge from the past, identifying new threats and unplanned scenarios is paramount (see Technical Note 7.2).
   - **How to:** List your assumptions and hypotheses while considering threats-to and theats-from scenarios (see Sect. 6.2.2, 7.1) and resulting failures in the risk assessment. This will help understanding if mitigation is possible or if *strategic* (https://www.riskope.com/2015/05/28/why-when-approa ching-strategic-tactical-operational-planning-one-needs-to-know-about-ostriches-denial-and-prayers/) shifts are necessary.

10. Adapt your design and maintain your systems to stay current.

    - **Why:** In the future conditions will likely diverge, and today's systems will become obsolete unless they are updated.

- **How to**: Information has to flow through evolution, so that lessons learned pass to future designers. Moreover, make it possible for anyone in the chain of command to propose a solution, which should then be openly discussed with key stakeholders.

11.1.2 How Often Should a Residual Risk Assessment Be Performed?

The classic question clients ask is: How often should a risk assessment be updated? There are no fixed rules on this. It all depends on the environment in which the system operates. For cyber, for example, one could foresee real-time updates based on the rate and quality of attacks. For an infrastructural system, the time may be way larger, as a function of how hazards, consequences, and the system itself evolve. For instance, for a mining operation a response time of one year may be valid.

However, beware of divergence! One should be able to test hypotheses on divergence and if significant divergent risks show up on the radar, updates should be carried out.

11.2 Who Should Perform a Risk Assessment?

In their 2016 guidelines, the International Council on Mining & Metals made it clear that enhanced efforts are required to ensure that:

> suitably qualified and experienced experts are involved in risk identification and analysis, as well as in the development and review of effectiveness of the associated controls (ICMM 2016).

Regarding qualifications, the Rio Tinto management system standards stated for example:

> Qualitative and quantitative risk analysis must be facilitated by competent personnel and include personnel with adequate knowledge and experience for the risk being evaluated (Rio Tinto 2015).

We believe the specs above should be reworded: risk assessment is now considered a discipline in its own right, one which requires particular skills and knowledge. It is not a hobby, or something that can be "facilitated" by someone who has good knowledge of the hazards.

Furthermore, one of the most important conditions is that the lead assessor be fully qualified to conduct such assessments. Expert knowledge in the design and operation of the specific system is not a requirement. Indeed the role of the lead assessor is primarily to lead the process, not to have a hand in influencing the outcomes as this could lead to conflicts of interest that must be avoided at all costs. With regard to the assessment team, it is important that a range of perspectives and experience be represented (Sect. 7.1.2). The lead assessor must not allow any member of the system's team (designers, owners, contractors, other stakeholders) to pull rank or dominate the discussions, or the debate to be marred by conflicts of interest.

Incidentally, we are not keen on conducting risks assessments workshops in the way this is commonly done: unprepared people, no system definition (Chap. 6), no success/failure criteria definition (Sect. 1.3, Anecdote 1.2), etc. We always propose performing a preliminary definition and knowledge gathering and then enhance efficiency by leading small groups or even one-on-one meetings and interviews (see Sect. 7.1.2).

In their *Tailings Governance Framework* (https://www.icmm.com/position-statements/tailings-governance), the International Council on Mining & Metals also remarked that performance criteria should be "…established for risk controls and their associated monitoring, internal reporting and verification activities". Regarding tailings management, they further suggest that "Critical control management has been identified as an approach to managing low probability, high impact events…" The identification of those issues that will require the highest level of attention is a necessary outcome of any risk assessment, and it cannot be the result of gut feelings.

Regarding risk management framework, leading practice would require that a company, working within the framework of ISO 31000, establish a corporate risk management standard that would include statements regarding the qualifications of audit assessment teams and require the identification of critical risks and their controls.

11.3 How to Perform a Risk Assessment?

We open this section by returning to typhoons in Japan (see Technical Note 2.1). Here Technical note 11.1 shows how a risk assessment could be built at national scale.

> **Technical note 11.1: Risk Assessment at the national scale, the example of Japan**
> Armed with historic probabilities and cost of consequences (see Technical note 2.1) defined for this example in terms of casualties it is possible to evaluate the evolution of risks generated by typhoons in Japan over the three considered periods (see Technical note 2.1). For each period we can evaluate the median risk for "next year" by using the probability of "1 or more typhoons next year" and the period median casualty count. It is also possible to evaluate for example the risk linked to "next year's" first typhoon, using the probability of exactly one typhoon and the "per event" casualty count. Because typhoons in Japan have such a high probability of occurrence it is the first alternative which is more interesting; in other cases the choice of the time horizon used in the

risk evaluation is of paramount importance. Indeed, if one considers very long time horizons, even small probabilities per annum will lead to high overall probabilities possibly dramatically changing the long term risk profile of an operation, organization or country.

The risk results (probability * consequence) (Eq. 1.1, 2.1, 4.1) are condensed in Table 11.1 and Fig. 11.1.

Closing point. If other hazards were considered (for example, earthquakes, volcanoes, climate change, etc.) in a more general study, then the risk figures would allow a comparison, ranking, and rational investment allocation among these hazards.

11.3.1 Synergistic Methodologies

In order to achieve the results required by modern corporations and outlined throughout this book, convergent Quantitative Risk Assessments (QRA, see Sect. 3.3, 5.1, 7.2)) are needed. They should be updatable, scalable, drillable.

Preliminary QRA deployments using multiple data sources deliver initial estimates of the probability of occurrence of various events, their consequences and preliminary alert thresholds. They also provide results that assist in the setup of emergency procedures. Thanks to new incoming data it is then possible to confirm

Table 11.1 Period and risk

Period	Risk (in term of Casualties)
I	120
II	33
III	22

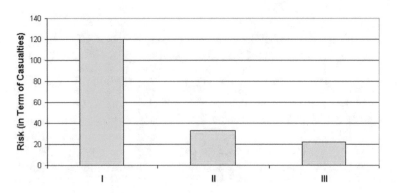

Fig. 11.1 Risk in term of casualties versus period

and gradually calibrate the QRA, as well as validate old reports and their assumptions. In the next two sections we briefly describe the methodologies and tools we deploy for single systems or portfolio studies.

After defining the system (Chap. 6) and identifying hazards (Chap. 7), probabilities (Chap. 8) and consequences (Chap. 9) can be evaluated following a number of methodologies and simulations, from the simplest to the most sophisticated, in function of the context of the study. In many countries there are specific codes defining how to approximately evaluate consequences for specific structures, in particular civil engineering projects such as dams.

The development of a portfolio-specific risk tolerance threshold (Chap. 10) allows users to determine which risks actually really matter in a portfolio.

As a multidimensional consequence function is foreseen (consequences are generally multifaceted, Chap. 9) it is possible to perform an integrated comparison of project execution, community, legal, environmental, financial, technical and H&S risks (or whatever dimension the user may want to define). This convergent risk vision fosters healthy discussions and helps organizations build consensus on decisions at the operational, tactical and strategic levels.

11.3.2   Using the ORE Platform

For our QRAs, we use our universal platform called ORE (Optimum Risk Estimates, ©Oboni Riskope Associates Inc., 2014-*). Although ORE is a proprietary platform, its principles are public and we regularly teach them. They are summarized below. You are welcome to apply them "by hand" to simple cases, but beware: pretty soon any system goes beyond the capabilities of an Excel worksheet and properly structuring the risk register will become a challenge.

ORE follows a continuous loop, as shown in Fig. 11.2.

ORE studies are generally used to support decisions (go/no-go, system changes, mitigation, M&A, supply chain, etc.), waste systems prioritizations, and insurance coverage decisions (business interruption, third-party liabilities, etc.) in industrial activities. ORE is the result of over twenty years of continuous development and applications, and makes it possible to include uncertainties, interdependencies, and societal and corporate risk tolerance, and is ISO 31000-compatible. To date ORE has been successfully deployed for processes, dam portfolios, linear facilities (roads, railroads, pipes), mining, etc. For tailings dam portfolios there is a specific ORE subset application named *ORE2-Tailings* (https://www.riskope.com/2018/05/02/ore2-tailings-dam-analysis/) that we routinely use to various degrees of sophistication. Technical note 11.2 is a continuation of Anecdote 7.1 and shows an example of intolerable risk summary delivered by ORE for Joey's acquisition.

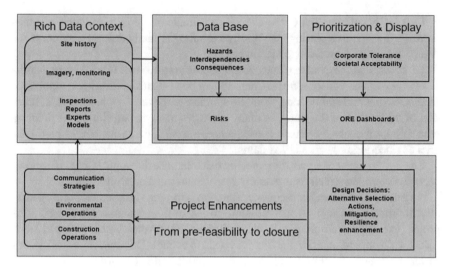

Fig. 11.2 The ORE (© Oboni Riskope Associates Inc.) continuous risk assessment/Risk-Based Decision Making (RBDM) (see Sect. 12.2) workflow

---

Technical note 11.2: Intolerable risks summary at Joey's Corporate Level (ORE)

The results shown in Table 11.2 are a continuation of Joey's acquisition risk assessment introduced in Anecdote 7.1.

It can be concluded that Joey's intolerable risks are split roughly half and half between business (first row, 56%) and technical/operational risks (all the other rows).

Intolerable risks are defined for 95 records out of a total of 406 records, i.e., 23.4%.

We also note that 89% (56 + 7 + 10 + 16 = 89) of the intolerable risks come from four families of scenarios:

- Administrative mismatch, miscommunication, hardware/software problems, missing, non compliant, invalid procedures, hardware, software

  - Closure Failure
  - Lifelines, suppliers (even if remote), power, gas, fuels
  - Natural
  - Vehicles, machines, tools, hanging objects, scaffolds, ladders, tire blowouts, drill rig accidents
  - Impact (collisions and shocks)
  - Wharves, Logistical Platforms, Storage

- Natural

Table 11.2 Joey's acquisition risk assessment cont'd from Anecdote 7.1

Causes of failure	Failure mechanism	Joey's total intolerable risk (%)
Administrative mismatch, miscommunication, Hardware/software problems, missing, non-compliant, invalid procedures, hardware, software	Contract closure failure	56
Lifelines, suppliers (even if remote), power, gas, fuels	Natural	7
Vehicles, machines, tools, hanging objects, scaffolds, ladders, tire blow-outs, drill rig accidents	Impact (collisions and shocks)	10
Wharves, logistical platforms, storage	Natural	16
Flooding, extreme events	Tailings	1
Quality assurance/Quality control (QA/QC), integrated products, a product from recently acquired company has problems	Liabilities	1
City, industrial, garbage dumps Availability, pricing, transport, storage Strikes, naval accident, airplane crash, etc Industrial chemicals, reactants Bankruptcy, currency, banking failure… Rain, winds Equipment, tools Aerosols	Fire Fuels Halt/Delay Hazardous substances Insolvency High wall and open pit Noise and vibrations Hazardous substances	9
	TOTAL	100

Following the ORE rational approach, mitigative funds would initially be allotted proportionally among these families, respectively $56/89 = 62.9\%$, $7/89 = 7.8\%$, $10/89 = 11.3\%$, $16/89 = 18\%$. If any of the families of risk could be fully mitigated with less than the allowable budget, then the positive balance would be transferred to the other areas as necessary.

Closing point. It is possible to accurately describe the risk landscape of a business and to design a mitigative roadmap based on risk informed decisions.

Appendix

Links to more information about the Key terms from the Authors	
A, B	*Act of God* (https://www.riskope.com/2020/ 12/09/act-of-god-in-probabilistic-risk-assess ment/) *Black swan* (https://www.riskope.com/2011/ 06/14/black-swan-mania-using-buzzwords- can-be-a-dangerous-habit/) *Business-as-usual* (https://www.riskope.com/ 2021/01/13/business-as-usual-definition-in- risk-assessment/)
C, D	*Convergent* (https://www.riskope.com/2021/ 01/20/convergent-risk-assessments/) *Divergent* (https://www.riskope.com/2020/11/ 18/tactical-and-strategic-planning-to-mitigate- divergent-events/) *Drillable* (https://www.riskope.com/2020/01/ 15/probability-impact-graphs-do-not-fly/)
F	*Foreseeability/foreseeable* (https://www.ris kope.com/2021/01/06/foreseeability-and-pre dictability-in-risk-assessments/) *Fragile/fragility* (https://www.riskope.com/ 2020/04/01/antifragile-resilient-solutions-for- tactical-and-strategic-planning/)
P, R	*Predictability/predictable* (https://www.ris kope.com/2021/01/06/foreseeability-and-pre dictability-in-risk-assessments/) *Resilient* (https://www.riskope.com/2016/11/ 23/resilience-cannot-based-instinctual-dec ision-making/) *Resilience* (https://www.riskope.com/2016/11/ 23/resilience-cannot-based-instinctual-dec ision-making/)

(continued)

(continued)

S	*Scalable* (https://www.riskope.com/2015/04/16/how-system-definition-and-interdependencies-allow-transparent-and-scalable-risk-assessments/) *Societal risk acceptability* (https://www.riskope.com/2014/01/09/aspects-of-risk-tolerance-manageable-vs-unmanageable-risks-in-relation-to-critical-decisions-perpetuity-projects-public-opposition/) *Sustainability/sustainable* (https://www.riskope.com/2019/01/16/improving-sustainability-through-reasonable-risk-and-crisis-management/) *Survivability* (https://www.riskope.com/2011/03/17/ale-fmea-fmeca-qualitative-methods-is-it-really-what-we-need/) *System* (https://www.riskope.com/2017/07/26/three-ways-to-enhancing-your-risk-registers/)
T, U	*Tolerance* (https://www.riskope.com/2020/04/29/risk-tolerance-thresholds/) *Uncertainty/uncertainties* (https://www.riskope.com/2015/12/10/3-decision-making-truths-derived-from-uncertainty-taxonomy-scheme-of-classification-and-a-road-sign/) *Updatable* (https://www.riskope.com/2020/01/07/climate-adaptation-and-risk-assessment/)

Other linked information (https://www.riskope.com/blog-news/) search Riskope blog and use the search box

Third Parties links in this section:	
2018–2019 Canadian budget	https://www.infrastructure.gc.ca/pd-dp/eval/2018-19-hgr-grh-overview-resume-eng.html
Financial impacts due to extreme weather	https://www.riskope.com/2014/06/19/extreme-weather-is-already-draining-the-biggest-u-s-companies/
Startup failures	https://www.riskope.com/2016/04/06/ore-predicts-business-startups-3-financing-rounds-success/
Insurance and cyber attacks	https://www.riskope.com/2016/01/21/should-you-listen-to-your-insurer-for-your-business-cyber-risk-management/
Untrue redundancies	https://www.riskope.com/2015/08/27/integrating-mitigations-with-redundancies-in-an-industrial-process/
ICMM tailings governance	https://www.icmm.com/position-statements/tailings-governance
ORE2_Tailings	https://www.riskope.com/2018/05/02/ore2-tailings-dam-analysis/

References

Australian Government (2016) Tailings management: leading practice sustainable development program for the mining industry. https://www.industry.gov.au/sites/g/files/net3906/f/July%202 018/document/pdf/tailings-management.pdf

International Commission on Large Dams (2013) Sustainable design and post-closure performance of tailings dams, Bulletin 153. ICOLD, Paris

International Council on Mining & Metals (2016) Position statement on preventing catastrophic failure of tailings storage facilities. https://www.icmm.com/website/publications/pdfs/commit ments/2016_icmm-ps_tailings-governance.pdf

Indirli M, Knezić S, Borg RP, Kaluarachchi Y, Romagnoli F (2014) Venice and its territory: multi-hazard scenarios, vulnerability assessment, disaster resilience, and mitigation. ANDROID Academic Network for Disaster Resilience to Optimise Educational Development

Kaluarachchi Y, Indirli M, Ranguelov B, Romagnoli F (2014) The ANDROID case study; Venice and its territory: existing mitigation options and challenges for the future. In: 4th international conference on building resilience, building resilience 2014. Salford Quays, United Kingdom, 8–10 September 2014

Rio Tinto Management System (2015) Standard. https://www.riotinto.com/documents/RT_Manage ment_System_Standard_2015.pdf

Chapter 12
Defining Manageable-Unmanageable and Strategic Risk

When a risk assessment is completed, among other things management must quickly evaluate which risks are:

- tolerable;
- intolerable but manageable;
- intolerable and unmanageable.

A box with links to key terms is included in the references at the end of this chapter to facilitate the read of this chapter.

Let us look at these terms closely with the aid of Figure 12.1, which is a graph drawn on the basis of the calculated risks (see Sect. 1.1, Eq. (1.1)) for an entire hazard portfolio.

The orange curve (Fig. 12.1) is the risk tolerance threshold (can be societal or corporate as discussed in Chap. 10 and in particular in Sect. 10.2). We see three different classes of risk in Fig. 12.1, namely, the blue, yellow, and red bubbles. The blue area represents the risks that are tolerable; therefore, by definition they can be put aside until there are no more yellow and red risks impinging on the system. The yellow bubble comprises the risks that are intolerable but manageable because their probabilities can be reduced by mitigation. Those risks are tactical. The third class, indicated in red, are the those that are intolerable and unmanageable because they cannot be mitigated and require changing the system. Hence those risks are of strategic nature. Going into more details: by using any of the explicit tolerance thresholds discussed earlier (Chap. 10) it is possible to provide a transparent definition of what constitutes a manageable risk. If a risk (probability, consequence) above the selected tolerance threshold can be lowered to below it, before hitting the credibility limit of, say, $p=10^{-6}$, then that risk is manageable (yellow bubble in Fig. 12.1). In order to ensure manageability, the mitigative investments and risk transfer must still preserve the economic livelihood of the company.

The key element here is the choice of what level of mitigative investment is effective yet preserves the economic livelihood of an entity. That is why improving the project cost evaluation, including risk adjustments, is paramount (see Sect. 12.2.2).

© The Author(s), under exclusive license to Springer Nature Switzerland AG 2021   247
F. Oboni and C. H. Oboni, *Convergent Leadership—Divergent Exposures*,
https://doi.org/10.1007/978-3-030-74930-9_12

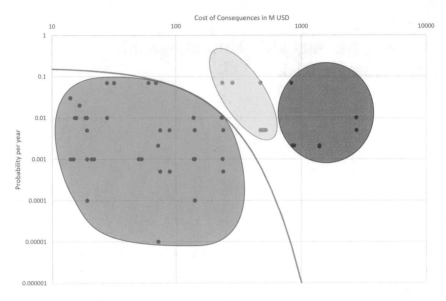

Fig. 12.1 Risk scenarios and tolerance thresholds. Blue: tolerable; yellow: intolerable but manageable; red: intolerable and unmanageable

If the risk cannot be brought under the tolerance threshold as described, then it must be considered unmanageable. Unmanageable risks cannot be mitigated; they require strategic shifts on the part of in the corporation/government (red bubble in Fig. 12.1). Insurers and lenders may use this notion to select projects and clients they agree to insure or create bundles of insured risks that solve the unmanageability of one or more specific project in a portfolio.

12.1   What to do with those risks families?

A risk assessment per se does not really help to make any decisions on risk reduction/accident prevention and other mitigative plans. It becomes rationally operational only when its results are compared with a threshold generally called "risk tolerance" (Fischhoff et al. 1982 and Chap. 10). Risks that can be mitigated by reducing their probability to values lower than the risk tolerance in a sustainable and economic way are tactical risks. In contrast, risks which require the system to be altered (mitigations to reduce consequences and get the risk under tolerance by shifting toward the left) are strategic risks. Tactical risks fall under the responsibility of operations' or projects' management, whereas strategic risks might require corporate upper management/owner to shift their objectives. To clarify this distinction with an example, we might say that reinforcing a bridge to reduce its probability of collapse

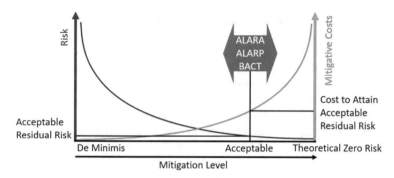

Fig. 12.2 Decrease of risk against increasing mitigative costs. The vertical line is set in the ALARP/ALARP/BACT zone. The position tends to slide toward the right, away from the optimum, due to the pressure of public opinion.

is a tactical mitigation, whereas changing the route of the roadway to avoid major geo-hazards in a certain area constitutes a strategic mitigation.

Figure 12.2 (identical to Fig. 4.1, Sect. 4.1, and reproduced here for the reader's ease) demonstrates the decrease of risk against increasing mitigative costs. When risks are very high, a relatively small investment generally makes it possible to swiftly reduce risks whereas investments increase asymptotically when risks are reduced beyond a certain level. The graph shows the point at which an acceptable threshold of risk mitigation might be settled—stating explicitly that the mitigative costs will realistically be too high to achieve a theoretical total abatement of risk. The vertical line is drawn in the zone where risks are as low as reasonably achievable (ALARA), as low as reasonably practical (ALARP), or in line with the best available control technology (BACT) concept (Sect. 4.1).

Needless to say, not many industries besides utilities, nuclear, chemical define these risk abatement levels. In summary, the vertical line of Fig. 12.2 is depicted "in the commonly-accepted reasonable risk abatement zone", and its intersection with the risk abatement and mitigative investment functions defines the residual risk and the investment necessary to attain it.

12.2 Beyond risk assessments

The risk assessment is followed by the phase of risk-based decision making (RBDM) and risk-informed decision making (RIDM), mitigative roadmaps and risk-adjusted cost estimates.

Following the Federal Energy Regulatory Commission (FERC) *Risk Guidelines*(https://www.ferc.gov/industries/hydropower/safety/guidelines/ridm.asp) definition, Risk Informed Decision-Making (RIDM) applied to dams uses the likelihood of loading, dam fragility, and consequences of failure to estimate risk. Risk estimates are then used to decide if dam safety investments are justified or warranted.

For Risk Based Decision Making (RBDM) we follow the definition given by the Committee on Improving Risk Analysis Approaches (NRC US 2009). RBDM organizes and orders information about potential harmful events to shore up decision makers' management choices. RBDM asks questions like: What can go wrong? How likely are potential problems to occur? How severe might the consequences be? Is the risk tolerable? What can/should be done to mitigate the risk?

When comparing the two definitions it results that RBDM can be seen as a subset of RIDM, the two having a common trunk. In this book we refer to RBDM when an analysis of a system shores up and supports decision making, whereas we consider RIDM when we discuss tactical and strategic risks, extending into formal financial analyses of mitigative alternatives (see Sec. 12.2.2).

RBDM uses the results of a convergent risk assessment (see Sec. 11.3, Fig. 11.2) to inform about possible mitigations, scenarios, including interdependencies and complex direct and indirect consequences. This is done by comparing the risks with the societal and corporate tolerance thresholds, not only at the level of corporate ERM, but also at the level of operations (see Part V, and in particular Chap. 15).

Again, using the sample of Japan, Technical Note 12.1 shows how to develop a rational efficiency evaluation.

Technical note 12.1: Mitigative Measures efficiency evaluation at national scale, the example of Japan
The immense damage caused by the typhoon Ise-wan in 1959 was a turning point for disaster management in Japan, giving rise to a movement to plan and prepare a comprehensive disaster management system. Thus, the Disaster

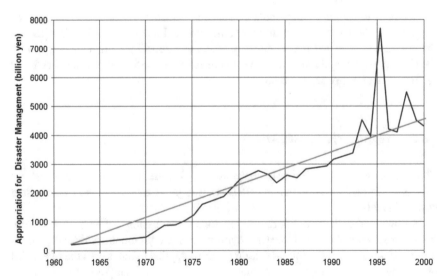

Fig. 12.3 Amount allocated to the disaster management program between 1961 and 2000 in Japan

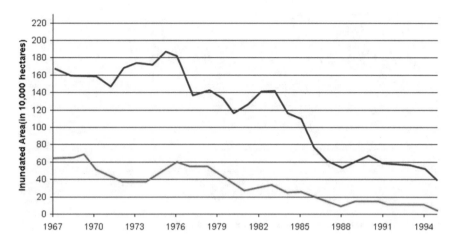

Fig. 12.4 Per-year comparisons of total flooded area (blue line) versus flooded residential and other areas

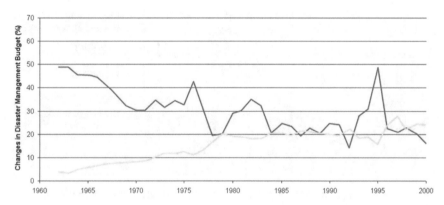

Fig. 12.5 Comparison of per-year reductions relative to proactive disaster prevention (yellow line) versus reactive disaster recovery construction (green line)

Countermeasures Basic Act was enacted in 1961. This led to a three-pronged approach consisting in:

1. an increase in the overall spending for disasters;
2. the implementation of wide-spread structural flood controls;
3. a passage from a reactive to a proactive policy.

The blue line in Figure 12.3 shows the amounts the Japanese government allocated for disaster management not adjusted for inflation. This assumption can be made because the inflation rate was historically very low in Japan during the 1970s and 1980s and even presented negative periods from 1993 to 2002, (Chiodo and Owyang 2003). As the exchange rate of the yen against the US

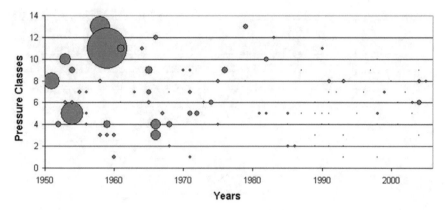

Fig. 12.6 Eye pressure classes versus Casualties versus year

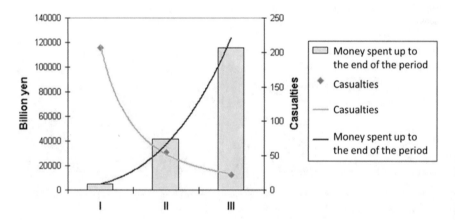

Fig. 12.7 Risk mitigation versus investments in the three periods (this graph shows a practical example of the concepts of Fig. 12.2)

Table 12.1 Differences in the damages caused by the 1964 and the 1990 typhoons	Damages	1964	1990	Reduction (%)
	Dead	47	40	-14.89
	Injured	530	131	75.28
	Houses destroyed	71269	16541	76.79
	Houses flooded	44751	18183	59.37

dollar presented large fluctuations (from 360 JPY per USD in the era of Bretton-Woods system to 80 JPY per USD in 1995 (Imai 2002), our study is performed in yen for the sake of simplicity.

Table 12.2 Differences in the damages caused in periods I and III

Damages	I	III	Absolute difference (I–III)	Reduction (%)
Dead	12583	828	11755	93.42
Injured	60844	8642	52202	85.80
Houses destroyed	2552115	402249	2149866	84.24
Houses flooded	3720251	384576	3156233	84.84

The orange line in Fig. 12.3 is a simplified model which takes into account the fact that disaster relief is also included in the blue line. Moreover, since the blue line also includes countermeasures against other disasters (tsunamis, earthquakes, etc.) these were filtered out. For example, the peak in 1995 is the Hanshin-Awaji earthquake and is therefore excluded from the orange line. The orange line equation was obtained by interpolation. Therefore, the amount of money spent for the period from 1961 to 2005 was evaluated at 115,714B JPY.

This amount corresponds to an "all disasters" preventative program, i.e. purged of disaster relief. If we make the assumption that the percentage allotted to the typhoons compared to the others disasters remained unchanged over the years, we can calculate the amount spent to mitigate typhoons.

Figure 12.4 depicts the effects of structural flood controls implemented in Japan during the period going from 1967 to 1995 in terms of flooded area. The blue plot is the total flooded area during each year. The green plot is the total flooded area of residential and other properties. As it can be easily seen, both plots display a downwards trend with time.

The change from a reactive to a proactive approach is studied in Fig. 12.5 where the green line is the budget percentage devoted to disaster recovery and reconstruction (reactive) while the yellow is the budget percentage allotted to disaster prevention (proactive). As it can be seen, the percentage of the budget allotted to recovery and reconstruction curves tends to diminish whereas the one allotted to prevention is increasing.

The increase in funding, changes in the policy from reactive to proactive, and the downward trend in observed damages lead us to ask two questions. The first one seems relatively straightforward and easy to answer: Did all the mitigative measures really bring the expected results, i.e. reduce the damage? A straightforward answer to the first question would be "Yes, really" (see the orange bubbles Fig. 12.6), as the size of the bubbles representing eye pressure classes versus casualties versus year after the 1970s significantly decreases, showing a tremendous decrease in casualties per event (please refer to Technical note 2.1, Fig. 2.4 for an explanation of the term eye pressure class).

The second one is a thorny one: Was the staggering amount invested in mitigations money well spent? To be able to answer this one, considering the simultaneous investment in mitigation, increase in population, and different type of losses (human lives, houses, etc.) a more complex metric has to be developed. The loss data are quite abundant, including the categories of human damage (dead and injured persons) and the housing damage (destroyed or flooded).

To perform the comparison we selected two very similar events with significantly same paths and same minimum pressure at the eyes (based on available data and criteria defined above): one in *1964* (https://en.wikipedia.org/wiki/1964_Pacific_typhoon_season), the other in *1990* (https://en.wikipedia.org/wiki/Typhoon_Flo_(1990)#Impact). Table 12.1 summarizes the damages reduction data when comparing these two specific phenomena.

Of course we cannot compare these results with the amount of money spent by the government between the years, because even if we have the precise path we cannot say which mitigative measure had an impact and which did not for these specific typhoons. This is the reason why we have to perform a large-scale, country-wide study, which gives us the results described in Table 12.2.

Returning to our first question, Table 12.2 with the last period complete, allows us to answer, "Yes, the results are quite good, all damages were reduced by more than 80% on the different targets".

Now to be able to answer the second question explicitly we must compare these numbers to the considerable 115,714B JPY spent in the all-disaster management program cited earlier. Unfortunately, we do not have information related to the portion of the investments specifically allotted to typhoon mitigation. Thus the whole amount will be used for the calculations, and the final results discussed bearing in mind that only a portion of the investment has been used for the purpose.

Now that we have shown that the risk reduction was effective, we are going to focus our attention on the investment afforded by Japan to save a life. Note that here we do not attribute a cost to human life (Mooney 1977; Jones-Lee 1989; Marin 1992; Pearce et al. 1996), a notion often considered repugnant and ethically unacceptable, but instead we will look at how much money the government spent to save the life of a citizen potentially exposed to flooding. In other words, we will measure the mitigative investment a society is ready to make to save a life, that is, its Willingness to Pay (WTP) attitude.

Obviously the WTP is strongly influenced by cultural, religious, philosophical, and economic considerations as it simultaneously considers a society's view on risk and response to risk. Marin (1992) indicated that in the UK, at 1990 prices, WTP was in the order of 2M to 3M GBP, while Lee and Jones (2004) defined ranges between 1.9M to 9M USD for developed countries, with an average at approximately 3.5M USD.

Due to the lack of detailed data, this study is based on rough estimates of costs of consequences for physical damages as follows:

- price for an average destroyed house: 30M JPY;
- a flooded house and an injured person: 3M JPY.

These assumptions lead us to calculate that the Japanese government spent an unknown (to us) portion of 41, 593B JPY (net of damages to houses and injured people) to save a total of 11,755 lives (that is, the difference of the number of victims between Period I and Period III). Now, if a population increase of 53% is taken into account, then the number of lives saved must be increased proportionally, leading to an adjusted theoretical number of lives saved 18,425.

The resulting cost is 2.26B JPY per life, being understood that only one part of this was allotted to typhoon mitigation. If we now assume that 10% of this cost was actually devoted to typhoons, the value would be in line with ranges cited above formulated by Marin (1992) and Lee and Jones (2004). Additionally, as the percentage devoted to typhoons was most certainly higher, one can see that the investment per life saved (adjusted to demography) falls well within the order of magnitude (higher end) of the WTP considered by other developed countries.

Figure 12.7 shows the amount spent up to the end of period III versus the theoretically (adjusted to demography) number of lives saved. The curves clearly indicate a well-known phenomenon in risk mitigation shown in Figure 12.2: investments during the early program stages bring dramatic reductions of the risk level and thus present very high efficiency, whereas as risk levels decrease, increasingly larger investments bring increasingly smaller results.

Closing point. This kind of approach, completed by the cost-benefit analysis, constitutes an excellent tool to support rational decision making about mitigative and humanitarian programs. As the world evolves and climatic changes bring new challenges to humanity it will be necessary to bring transparency and accountability in the mitigative decision making, both at the strategic level (which hazards to tackle) and the tactical level (how to mitigate the selected hazards).

12.2.1 Net present value versus risk as a key decision parameter

While net present value (NPV) is still often used as a key decision parameter, it is less reliable than risk. In this chapter we will use a case study from our day-to-day practice in RIDM support in order to show how to use risk as a key decision parameter in RIDM and how the commonly used NPV can create distortions and biases when analyzing reclamation (or other) alternatives. To do this we will present a case study,

Table 12.3 Financial parameters of the status quo (in millions, denoted M CAD)

Cause/Hazard for status quo	Probability	Cost M CAD
Capital investment necessary at start on the treatment plant	90%	5
Energy cost (diesel for power plant)	30%	2*5
Climate change	15%	3*5

then show the differences between a traditional NPV analysis and a risk adjusted approach.

The case study concerns a very large, leaching, underground storage of a toxic water soluble compound with a potential to leach into the water table. We consider two alternatives: long-term pumping plus treatment plant (the status quo) and encapsulation (rehabilitation).

For the status quo, to prevent leaching a pumping system is installed. The permanent pumping system keeps the underground water level below the bottom of the storage. Water percolates from the surface, leading to the need to invest in a treatment plant to treat the leachate. Earlier risk assessments judged risk to the ecosystem and human health to be negligible. Table 12.3 summarizes the financial parameters and risks linked to maintaining the status quo.

The alternative to the status quo would be a rehabilitation of the site by means of encapsulation of the underground storage. The encapsulation would require a large capital investment (120M CAD), but afterwards would considerably reduce the permanent pumping and treatment (Table 12.4). As this encapsulation constitutes a first in the world, a risk assessment has been performed, showing that there is a significant likelihood (10%) that the encapsulation may cost twice as foreseen. Table 12.4 summarizes the "financial parameters" and risks linked to building and maintaining the Rehabilitation.

Finally, because of uncertainties (construction, long term climate change, etc.) there is also a probability that after developing the encapsulation as above (i.e. with the 10% likelihood it may cost twice the initially foreseen amount), it may be necessary to maintain pumping as in the Status Quo. This means that despite investing in the encapsulation the project would still not work properly. This is a failed rehabilitation, or worst-case scenario.

Table 12.4 "Financial parameters" of the Encapsulation alternative (in millions, denoted M CAD)

Cause/Hazard for Encapsulation alternative	Probability	Cost M CAD
Capital investment chance to double (additional 120M)	10%	120
Energy cost (diesel for power plant)	30%	2*5
Climate changes can force to require pumping just like the status quo, despite the work of encapsulation work	5%	3.6

Traditional NPV Analysis

Let's use a rate of return of 9% for this analysis and consider a life duration of 40 years. The NPVs are negative because the project generates only expenses and no profits:

- Status quo: 3.6M CAD/yr, 40-year life span, NPV = –42.33M CAD;
- Rehabilitation: 120M CAD construction, then 0.3M CAD/yr, 40-year life span, NPV = –123.23M CAD.

This case study is particularly strong in building an argument against using NPV. That is because most of the rehabilitation expenditure is upfront, and the yearly costs (as traditionally calculated, without the risks) are relatively small, while the duration is very long. In this case the NPV almost nullifies any expense coming after approximately 20 years.

Thus NPV mislead us suggesting that the status quo has by far a better NPV value than the rehabilitation. This a wrong conclusion because of the long life of the project. Additionally, risks need to be included. Indeed, there are two ways we could alter such an analysis to include risks. One would be to include yearly risks as additional costs. The other would be to increase the rate of discount to include uncertainties. However, both these attempts would fail to yield pertinent results. The NPV would strongly indicate the status quo as the most viable between the two alternatives.

Using Risk as a Key Decision Parameter

We can use innovative approaches which eliminate the pitfalls of NPV at preliminary design level (Oboni and Oboni 2007, 2008; Oboni 2005). Indeed, our Comparative Decision Analysis & Economic Safety Margin (CDA/ESM) compares alternatives by evaluating: (a) lifecycle balance encompassing internal and external risks over a selected duration and (b) project implementation and demobilization costs and risks.

Figure 12.8 displays CDA/ESM (average and spread) cumulative cost results at the 40 year time horizon for three cases: the status quo (–295M CAD), the rehabilitation (–140M CAD), and the failed rehabilitation (–405M CAD). In fact, the status quo

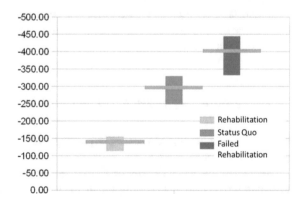

Fig. 12.8 For each analysis: min, max, average of the cumulative cost at forty years

will cost cumulatively twice as much as the rehabilitation. That is because of the risks afflicting each alternative, such as the possible increase in energy costs, which were included in this analysis.

If, for the sake of comparison, risks are now introduced in the NPV evaluations as described in the prior section it appears that the rehabilitation CDA (average) result is roughly equal to the NPV with risks and not far from the traditional "riskless" NPV.

This happens because the initial amount spent is very large compared to the yearly spending which indeed seems negligible. However, the NPVs of the status quo with risks (–90.2M CAD) and without risks (–42.3M CAD) are lower than that of the rehabilitation (–139M CAD). This blatant contradiction between CDA/ESM and NPV results confirms that NPV evaluations are plainly inadequate when integrating alternatives' specific risks in the comparison process because their discounted nature annihilates the effects of long-term expenditures and makes it essentially impossible to consider risks in a proper way.

Conclusions

This case story shows that risks should be used as a discriminant parameter from the beginning of any project for successful long-term planning and managing rational decisions.

We have seen that the use of innovative approaches at the preliminary design level eliminates the pitfalls of NPV, an obsolete financial concept still used by many. The evaluation of a project should of course include the annual risks potentially afflicting the project, construction risks, and risks of malfunctioning, and possibly also the costs of demolition/reclamation. Thus NPV can lead to erroneous conclusions in terms of the overall cost of a project, particularly for very long-term projects. Because of this, NPV is especially misleading when dealing with long-term environmental rehabilitations/reclamations.

We use CDA/ESM to avoid the NPV pitfalls. As a matter of fact, it compares alternatives in financial terms, including the lifecycle balance encompassing internal and external risks over a selected duration, and project implementation and demobilization costs and risks.

To date we have applied CDA/ESM to ropeways versus roadways transportation, surface versus underground solutions, water treatments alternatives, transportation networks, go/no-go decisions.

12.2.2   Improving project cost evaluations

Comparing alternatives for a project requires careful quantitative and convergent risk analysis. Without it neither short-term (construction, implementation) nor long-term (all the way to closure and wastes of waste) life cycle costs can be rationally evaluated. The long term is particularly important for water treatment plants, environmental cleanups and restoration/rehabilitation that are oftentimes foreseen for long service lives.

Technical note 12.2 illustrates how apparently small changes impact probabilities of overcosts, higher temperatures or any other considered parameter.

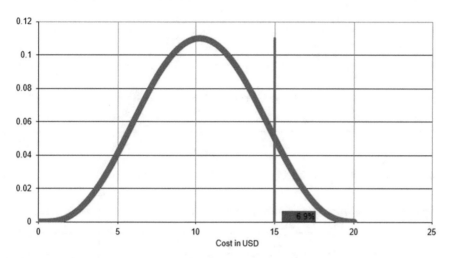

Fig. 12.9 Distribution of a risk scenario consequence (loss) and probability of the consequence exceeding 15M USD

Fig. 12.10 Frequency of summer temperature anomalies showing how often they deviated from the historical normal (1951–1980) over the summer months in the northern hemisphere. Transcribed from NASA (Hansen et al. 2012)

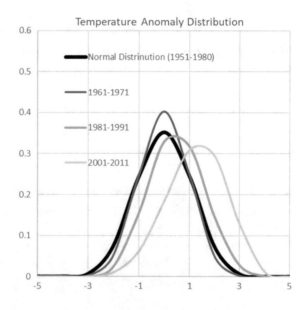

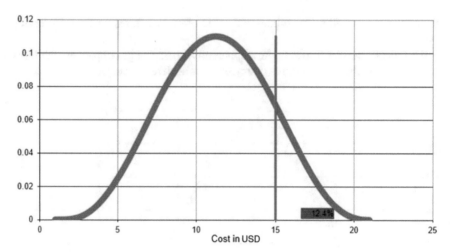

Fig. 12.11 Same distribution than Fig. 8, shifted to the right by 1M USD

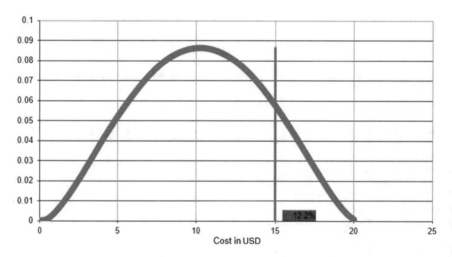

Fig. 12.12 Same average as Fig. 12.11, but standard deviation at 3.5M USD

Technical note 12.2: Apparently small changes impact probabilities

Have you ever been involved in a project whose final budget exploded as a result of apparently small, negligible schedule or material risks? Similarly, even though we record small change in weather averages, big impacts can be noticed. Although unrelated these observations have a lot in common as we explain below.

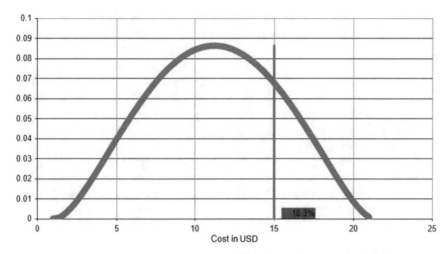

Fig. 12.13 Increase of average and wider standard deviation lead to even larger increase of the probability of overcost.

Let's start with cost exceedance. Suppose the consequences of a risk scenario vary between 0 and 20M USD with an average of 10M USD and a standard deviation of 3M USD. Management was clear that any cost over 15M USD was unacceptable, based on their tolerance threshold at 0.05 likelihood.

Based on the distribution of the consequence described above, we can compute the probability of the consequence (loss) being higher than the threshold (15M USD). It is the area right of the blue line under the orange curve in Fig. 12.9.

As can be seen in Fig. 12.9, the probability of the consequence being exactly 15M USD is 0.05% as readable on the vertical axis. The probability of an event costing more than 15M USD (the overcost probability) is 6.9%, given the assumed distribution of consequences.

Let's now look at probability of extreme events due to climate change.

Kodra and Ganguly (2014) illustrated how a shift and/or widening of a probability distribution of temperatures affects the probability of temperature extremes. In real life we can see many records featuring "small" increases in variance and mean, as exemplified in Fig. 12.10.

Now, if we look at the risk scenario we introduced above, we can calculate how small changes in average consequences can lead to dramatic changes in costs.

Let's follow the case a of the IPCC (2012), involving a shift the distribution to the right by 1M USD (Fig. 12.11).

This leads the event costing 15M USD to have a likelihood of 7% and an over-cost probability of 12.4% (Fig. 12.11). That is roughly twice the initial value. The event would also be intolerable, if the tolerance threshold was 15M USD at 0.05.

We can look at case b of the IPCC (2012), which maintains the average at 10M USD but increases the standard deviation to 3.5M USD. The overcost probability also increases to almost the double the initial state (Fig. 12.12).

Now if we look at case c of IPCC (2012) we combine the two prior cases, i.e. shifting and increasing volatility. The result is staggering (Fig. 12.13). Almost 20% probability of overcost.

Closing point. As you can see, small changes in average and volatility lead to significant changes in the probabilities of events. Divergence is lurking all the time. Its effects can be measured, predicted and there is no need to invoke "unprecedented" effects.

ORE risk-adjusted lifecycle cost analysis

We started using probabilistic-specific methodology for improving project cost evaluations at the beginning of the 2000s. Lately, we have used the "pelican beak" analogy (Fig. 12.14) to show how important it is to invest more in preliminary analyses and detailed a priori risk evaluations.

That is in contrast with common practice approaches, whose users satisfy themselves with "*risk-myopia*(http://reflectd.co/2013/07/12/psychological-myopia/)" and NPV.

Cost evaluations can be greatly enhanced by using ORE and risk-adjusted life cycle cost analysis, due to the following factors:

- NPV cannot differentiate between alternatives because it does not properly evaluate the long term, and it does not explicitly include risks. ORE does both.
- The selections of alternatives (short-term and long-term) and true lifecycle costs (cradle to grave) have to include risks. Those are, for example, construction and

Fig. 12.14 Graph explaining the pelican beak analogy

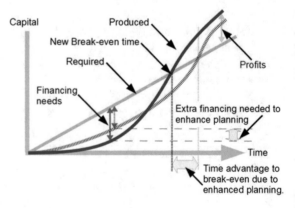

Table 12.5 Typical client conundrums and what should be expected from a convergent non-quantitative risk assessment allowing RIDM

Conundrums	What you should expect (risk assessment deliverables)
How many lines of defense? For example, buffer pond or not, buffer stock or not, how big, frequency of sampling, etc	Probability of occurrence of mishaps and failures, with or without mitigations and lines of defense, including common cause failures and interdependencies
Which mitigations and at what costs?	Different risk profiles of an operation depending on the implemented mitigation
Usual Design and contracting/operating path versus Design Built & Operate (DBO)?	Quantified life cost for each alternative
Risk transfers, long-term risks, exclusions and force majeure clauses?	A clear definition (quantified) of each in order to bring value and limit legal liabilities
How are true construction and lifecycle costs defined, including wastes of wastes, and related risks	Rational estimates including risks, explicit uncertainties, and possibility to update as new data become available

lifecycle risks including life-end operations and closure. Waste of waste handling costs have to be included as well. ORE allows that.

- The risk register needs to bring clarity to management and decision-makers and be easily updatable (because of the changing environment). We built ORE around that.
- ORE can study the effect of mitigations on the overall risk of an operation.
- Thanks to ORE, clear definition (quantified) of risks will limit time in court battles, especially in case of force majeure events (climate change).

Examples of typical conundrums and what ORE will deliver

Table 12.5 gives a series of sample questions from clients and an explanation of the kind of replies ORE can give. The example focuses on water treatment processes and plants.

Appendix

Links to more information about the Key terms from the Authors	
A,B	*Act of God* (https://www.riskope.com/2020/12/09/act-of-god-in-probabilistic-risk-assessment/) *Black swan* (https://www.riskope.com/2011/06/14/black-swan-mania-using-buzzwords-can-be-a-dangerous-habit/) *Business-as-usual* (https://www.riskope.com/2021/01/13/business-as-usual-definition-in-risk-assessment/)
C,D	*Convergent* (https://www.riskope.com/2021/01/20/convergent-risk-assessments/) *Divergent* (https://www.riskope.com/2020/11/18/tactical-and-strategic-planning-to-mitigate-divergent-events/) *Drillable* (https://www.riskope.com/2020/01/15/probability-impact-graphs-do-not-fly/)
F	*Foreseeability/foreseeable* (https://www.riskope.com/2021/01/06/foreseeability-and-predictability-in-risk-assessments/) *Fragile/fragility* (https://www.riskope.com/2020/04/01/antifragile-resilient-solutions-for-tactical-and-strategic-planning/)
P,R	*Predictability/predictable* (https://www.riskope.com/2021/01/06/foreseeability-and-predictability-in-risk-assessments/) *Resilient, Resilience* (https://www.riskope.com/2016/11/23/resilience-cannot-based-instinctual-decision-making/)
S	*Scalable* (https://www.riskope.com/2015/04/16/how-system-definition-and-interdependencies-allow-transparent-and-scalable-risk-assessments/) *Societal risk acceptability* (https://www.riskope.com/2014/01/09/aspects-of-risk-tolerance-manageable-vs-unmanageable-risks-in-relation-to-critical-decisions-perpetuity-projects-public-opposition/) *Sustainability/sustainable* (https://www.riskope.com/2019/01/16/improving-sustainability-through-reasonable-risk-and-crisis-management/) *Survivability* (https://www.riskope.com/2011/03/17/ale-fmea-fmeca-qualitative-methods-is-it-really-what-we-need/) *System* (https://www.riskope.com/2017/07/26/three-ways-to-enhancing-your-risk-registers/)

(continued)

(continued)

Links to more information about the Key terms from the Authors	
T,U	*Tolerance* (https://www.riskope.com/2020/04/29/risk-tolerance-thresholds/) *Uncertainty/uncertainties* (https://www.riskope.com/2015/12/10/3-decision-making-truths-derived-from-uncertainty-taxonomy-scheme-of-classification-and-a-road-sign/) *Updatable* (https://www.riskope.com/2020/01/07/climate-adaptation-and-risk-assessment/)
Other linked information (https://www.riskope.com/blog-news/) search Riskope blog and use the search box	

Third Parties links in this section:	
FERC risk guidelines	https://www.ferc.gov/industries/hydropower/safety/guidelines/ridm.asp
Pacific typhoon season	https://en.wikipedia.org/wiki/1964_Pacific_typhoon_season
Risk myopia	https://reflectd.co/2013/07/12/psychological-myopia/

References

Chiodo AJ, Owyang MT (2003) Symmetric inflation risk, monetary trends, The Federal Reserve Bank, St. Louis

Fischhoff B, Lichenstein S, Slovic P, Derby SC, Keeney R (1982) Acceptable risk. Cambridge University Press, Cambridge MA

Hansen J, Sato M, Ruedy R (2012) Perception of climate change. Proc Natl Acad Sci 109(14726–14727):E2415–E2423. https://doi.org/10.1073/pnas.1205276109

Imai R (2002) The inflation targeting. Kyushu University, Lecture on Economics

[IPCC] Intergovernmental Panel on Climate Change (2012) Managing the risks of extreme events and disasters to advance climate change adaptation: special report of the intergovernmental panel on climate change, Field CB et al. (eds) Cambridge University Press, New York. International Panel on Climate Change's Special Report on Extremes (IPCC SREX)

Jones-Lee MW (1989) The economics of safety and physical risk. Blackwell, Oxford

Kodra E, Ganguly AR (2014) Asymmetry of projected increases in extreme temperature distributions. Sci Rep 4:5884. https://www.nature.com/articles/srep05884

Lee EM, Jones DKC (2004) Landslide risk assessment. Thomas Telford

Marin A (1992) Costs and benefits of risk reduction. Appendix in risk: analysis, perception and management, Report of a Royal Society Study Group, London

Mooney GM (1977) The valuation of human life. Macmillan, New York

Oboni F (2005) Do risk assessments really add value to projects? CIM, Ottawa

Oboni F, Oboni C (2007) Improving sustainability through reasonable risk and crisis management, ISBN 978-0-9784462-0-8

Oboni F, Oboni C (2008) Oboni, https://learn.edumine.com/store?utf8=%E2%9C%93&st=franco&commit=R

Pearce DW, Cline WR, Achanta AN, Fankhauser S, Pachauri RK, Tol RSJ, Vellinga P (1996) The social costs of climate change: greenhouse damage and the benefits of control. In: Climate Change 1995: economic and social dimensions of climate change. Contribution of Working Group III to the Second Assessment Report of the IPCC. Cambridge, Cambridge University Press

Part V
Convergent Assessment for Divergent Exposures: Case Studies

Part V uses real-life case studies to showcase rational convergent assessment procedure under business-as-usual and divergent conditions aiming at risk-informed decision making (RIDM).

The case studies' details and sensitive data have been altered to avoid possible recognition of clients and their railroads, wharves, chemical plants in isolation and as a system.

Chapter 13
Objectives of the Case Studies

The objectives of the case studies that compose this and the following chapters in Part V is to demystify quantitative risk assessment fears and show how even with a limited amount of information one can provide solid answers to questions oftentimes asked by owners and their management.

The development of each of the three cases follows the same logic, as the risk methodology we encourage you to follow is linear and universal. Explanation of key terms are delivered in the references section. The stepped approach covers the following nine steps:

1. **The client's request.** This is a first description of what the client's, owner and/or their management, aims and needs are. Oftentimes the request may be as simple as "we would like a risk assessment done on our operation X" or even bear on some specific aspects of a single project. Sometimes the request may cover the entire ERM deployment. There are infinite possibilities (see Table 12.5 for some examples) and therefore selecting the themes for the following chapter was not easy. We finally opted for: a specific case where we were asked to evaluate a transportation business interruption; an entire commercial wharves operation and finally an entire system, including three production plants, the transportation network, and the shipping facility (wharves). When we transcribe the client's request in the scope of work description we always join the *glossary of technical terms* (https://www.riskope.com/knowledge-centre/tool-box/glossary/) in order to ensure that all the parties talk the same language.

2. **Success metric (failure) and consequences dimensions.** The definition of the success/failure criteria is a necessary preliminary step to any risk assessment (see Sect. 1.3, Anecdote 1.2). Without clearly defining success/failure, the risk assessment is likely to slide off theme and lose focus. Numerous risk assessments initiatives fail because failure is not properly defined. In addition, the analysis must include the consequences dimensions as discussed in Chap. 9.

3. **System definition.** Without a clear definition of the system (Chap. 6), the risk assessment will fail. (NB: the ISO 31000 definition of the context is a synthesis between system definition and success/failure definition. By defining the system

F. Oboni and C. H. Oboni, *Convergent Leadership—Divergent Exposures*,
https://doi.org/10.1007/978-3-030-74930-9_13

we understand where our job starts, where it ends, and what the battery lines (See Sect. 6.2.2) are. It is necessary that system limits be developed together with the client.

4. **Gathering existing information.** The client (management and personnel) has significant knowledge of most aspects of his operation, both in documentation and personal knowledge. We almost always interview employees individually (see Sect. 7.1.2) to get the most honest answers possible at all different levels to be sure to grasp the truest picture. Although a risk assessment looks forward (in contrast to an audit, which looks only at the past up to the present), it is very important to gain as much knowledge as possible about the history of the system, mishaps, near-misses, crises and all other available data. This information is used as a basis to formulate evaluations of probabilities and consequences projected toward the future.

5. **Requesting further necessary information.** This is the complement to the previous step 4. The client usually doesn't know everything about his operation: there are shadowy areas due to habit, normalization of deviance (see Sect. 11.1.1), common cause failure, etc. For example, neighboring third-party toxic material storages could provoke a business interruption if they incur an accident. Many environmental and man-made hazard are oftentimes neglected as their interdependencies with the considered operation may have simply gone unexamined.

6. **Hazard identification.** Without performing a proper hazard identification (Chap. 7) the risk assessment is almost meaningless. Brainstorming, energy-based analyses, and resource-flow analyses are all valid ways to complete the hazard list. This is not the time to censor the list, all ideas should be considered, as "crazy" scenarios will disappear by virtue of the analysis later on.

7. **Risk model design.** The risk model design is always dictated by the general ORE (see Chap. 11) flowchart. A semantically robust convergent, scalable and drillable hazard and risk register (see Sect. 1.1, 5.3.3, Chap. 6, 7, 8) has to be built. It must include: all the elements and sub elements of the system (Chap. 7), the impinging hazards based on threat-to, threat-from analyses characterized by their probabilities of occurrence and magnitude (Chap. 8) and the related consequences (Chap. 9) in all their dimensions. Interdependent scenarios have to be included in the register as well. Independently from the hazard and risk register the owner risk tolerance threshold (see Sect. 10.1.1) has to be developed and the societal risk tolerance (see Sect. 10.1.2) has to be adopted based on the jurisdiction of the operation/ project/ corporation. In some cases (see Chap14) with a limited scope and specific aspects of a request, the tolerance may be implicitly defined, e.g. a percentage of a maximum loss or an exceedance probability.

8. **Results and communications.** Plotting all the identified events on a same graph, with the tolerance thresholds, if applicable) is necessary to enable their first visual comparison. The plot also allow to carry out a last check the coherence of the success/failure criteria and consequences metric: if that is the case, corrections have to be made. There is an infinity of custom-tailored graphs that can be used to present the results by scenario/element or aggregated, such as: risks per

elements, risks per hazard, tolerable risks, intolerable risks, tactical and strategic risks, etc. (see Part IV). These graphs can be reunited in an ORE dashboard that delivers an immediate nuanced answer to the questions formulated by the client. The dashboard can be used to foster healthy communication with all partied involved. The results can be organized to reveal a roadmap to resilience, robustness and overall risk mitigation through Risk Based Decision Making (RBDM) and Risk Informed Decision Making (RIDM) which will foster ethical choices and therefore enhance the ESG, SLO and CSR (see Sect. 6.1.2) stance of the client.

9. **Recommendations and conditions of validity.** Adding conditions of validity is a step unfortunately oftentimes forgotten in risk assessment. Conditions of validity help the client to understand the assumptions that shore up the study. They also work as a reminder that the risk assessment is simply a model that by design cannot perfectly replicate reality.

Each of the three chapters illustrating the case studies that follow discusses each of these nine steps and reprises the reasons "why we do it" for ease of reference and as a reminder why each step must be performed.

Appendix

Links to more information about the Key terms from the Authors	
A,B	*Act of God* (https://www.riskope.com/2020/12/09/act-of-god-in-probabilistic-risk-assessment/) *Black swan* (https://www.riskope.com/2011/06/14/black-swan-mania-using-buzzwords-can-be-a-dangerous-habit/) *Business-as-usual* (https://www.riskope.com/2021/01/13/business-as-usual-definition-in-risk-assessment/)
C,D	*Convergent* (https://www.riskope.com/2021/01/20/convergent-risk-assessments/) *Divergent* (https://www.riskope.com/2020/11/18/tactical-and-strategic-planning-to-mitigate-divergent-events/) *Drillable* (https://www.riskope.com/2020/01/15/probability-impact-graphs-do-not-fly/)

(continued)

(continued)

Links to more information about the Key terms from the Authors	
F	*Foreseeability/foreseeable* (https://www.ris kope.com/2021/01/06/foreseeability-and-pre dictability-in-risk-assessments/) *Fragile/fragility* (https://www.riskope.com/ 2020/04/01/antifragile-resilient-solutions-for-tactical-and-strategic-planning/) *Foreseeability/foreseeable* (https://www.ris kope.com/2021/01/06/foreseeability-and-pre dictability-in-risk-assessments/) *Fragile/fragility* (https://www.riskope.com/ 2020/04/01/antifragile-resilient-solutions-for-tactical-and-strategic-planning/)
P,R	*Predictability/predictable* (https://www.ris kope.com/2021/01/06/foreseeability-and-pre dictability-in-risk-assessments/) *Resilient, Resilience* (https://www.riskope.com/2016/11/ 23/resilience-cannot-based-instinctual-dec ision-making/)
S	*Scalable* (https://www.riskope.com/2015/04/ 16/how-system-definition-and-interdepende ncies-allow-transparent-and-scalable-risk-ass essments/) *Societal risk acceptability* (https:// www.riskope.com/2014/01/09/aspects-of-risk-tolerance-manageable-vs-unmanageable-risks-in-relation-to-critical-decisions-perpetuity-pro jects-public-opposition/) *Sustainability/sustainable* (https://www.ris kope.com/2019/01/16/improving-sustainab ility-through-reasonable-risk-and-crisis-man agement/) *Survivability* (https://www.riskope. com/2011/03/17/ale-fmea-fmeca-qualitative-methods-is-it-really-what-we-need/) *System* (https://www.riskope.com/2017/07/26/three-ways-to-enhancing-your-risk-registers/)
T,U	*Tolerance* (https://www.riskope.com/2020/04/ 29/risk-tolerance-thresholds/) *Uncertainty/uncertainties* (https://www.ris kope.com/2015/12/10/3-decision-making-tru ths-derived-from-uncertainty-taxonomy-sch eme-of-classification-and-a-road-sign/) *Updatable* (https://www.riskope.com/2020/01/ 07/climate-adaptation-and-risk-assessment/)

Other linked information (https://www.riskope.com/blog-news/) search Riskope blog and use the search box

Chapter 14
Case Study 1: Railroad RR

Our client, Kryptonite Corp, uses a third-party railroad RR to ship chemical products from its chemical plants K1, K2, and K3 to its shipping port, Terminal.

The rail network has redundancies for the first part of the itinerary, meaning that RR has cooperation agreements with other rail companies, using different tracks, between the chemical plants and a junction located inland of the Terminal. The various rail companies constitute a parallel system (see Sect. 8.2) up to the junction. Thanks to these redundancies the chance of business interruptions (BIs) between the plants and the junction is considered at the border of credibility. However, the last stretch between the junction and the Terminal has no redundancies (circled in Fig. 14.1) and is therefore a service loss in that stretch I the sole significant driver for potential BI.

14.1 The client's request

The client needs to evaluate their insurance coverage for two cases:

Case (A) BIs related to rail transportation longer than a certain duration due to any hazard effecting RR including the effects of climate change, hence a convergent study.

Case (B) BIs generated by a major earthquake, hence a single-hazard evaluation.

Like in all our studies we ensured the client request was correctly understood and presented our *glossary* (https://www.riskope.com/knowledge-centre/tool-box/glossary/) and specified key terms (see links in the references).

Note that this is not a risk assessment of RR geared toward protecting the operation from its own hazards (landslides, rockfalls, river erosion, etc.). It is rather an exercise geared toward protecting Kryptonite from critical service outages and insuring against BI. This shows how understanding the point of view is paramount to focus the assessment effort.

© The Author(s), under exclusive license to Springer Nature Switzerland AG 2021
F. Oboni and C. H. Oboni, *Convergent Leadership—Divergent Exposures*,
https://doi.org/10.1007/978-3-030-74930-9_14

Fig. 14.1 Rail network system under consideration. Only the last RR stretch is analysed as various alternatives exist between the plants and the junction inland of the Terminal

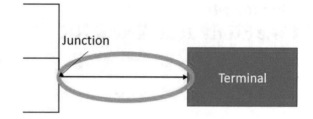

14.2   Success Metric (Failure) and Consequences Dimensions

This example covers a limited scope, specific aspects of client's request. Case A is multi-hazard, case B is mono-hazard and both bear on the single BI dimension of a potential RR failure. Failure is therefore defined in this study as any malfunction (due to multi- (case A) or single- (case B) hazard occurrence generating a BI measured in months and related M USD consequence.

Why we do it. The definition of the success/failure criteria (Sect. 1.3, Anecdote 1.2) is a necessary preliminary step to any risk assessment. Without clear definition of success/failure, the risk assessment is likely to consider the wrong hazards, lose focus and deliver misleading results. Indeed, numerous risk assessments initiatives fail because failure is not properly defined.

14.3   System Definition

The system is defined as the last segment of the railroad line (circled in Fig. 14.1) from the junction inland to the unloading station at Terminal, including all special segments such as but not limited to railroad crossings, bridges, tunnels, earth-retaining structures above and below grade, natural and man-made slopes around the railways, and watercourses and their banks.

The option to reroute (built-in extant mitigations) would normally be part of the system to be studied. However, in this particular segment, there are no rerouting options.

Why we do it. Without a clear definition of the system the risk assessment will fail. NB: the ISO 31000 definition of the context is a synthesis between system definition and success/failure definition. By defining the system (see Chap. 6), we also understand the battery lines (see Sect. 6.2.2) which must be developed together with the client to avoid misunderstandings that might prove costly.

14.4 Gathering Existing Information

Information was mostly gathered by interviewing RR key personnel and reviewing topographic maps, aero-photos and satellite optical imagery. Insurers and loss control department were interviewed and history of losses and, when possible, near-misses was compiled.

Why we do it. Clients (management and personnel) have significant knowledge of most aspects of their operation, both in terms of documentation (in some instances with big data) and personal knowledge (thick data) (Sect. 7.2.2). We almost always interview employees individually to get the most honest answers possible (Sect. 7.1.2) at all different level to be sure we grasp the truest picture.

Although a risk assessment looks forward (in contrast to an audit, which looks only at the past, up to the present), it is very important to gain as much knowledge as possible about the history of the system, mishaps, near-misses, crises and all other available data. This information will be used as a basis to guide hazard identification, formulate probabilities and consequences evaluations projected toward the future, including possible effects of climate change and other divergences.

14.5 Requesting Further Necessary Information

After reviewing the information gathered we requested additional information and clarifications from the client about selected issues and past accidents/issues. Sites visits to specific spots were conducted. The client usually doesn't know everything about his operation: there are shadow areas due to habits, normalization of deviance (see Sect. 11.1.1), Common Cause Failure (CCF), etc., which we try to explore with the request for additional information.

14.6 Hazard Identification (HI)

Hazards are events that can generate failures as defined in Sect. 14.2 on the system defined in Sect. 14.3. Hazardous events were identified based on experience, interview, and archival data (gathered information) and using the threat-from and threat-to concept (see Sect. 6.2.2). For example, in this system one possible hazard is a flood capable (threat-from) of damaging a bridge which will then generate a BI. An example of threat-to would be in the considered case a derailment of a train with chemical freight. During HI it is also important to check for interdependencies between hazards and elements.

The hazards capable of ultimately generating BIs due to traffic disruptions were understood to be:

- Earthquakes, i.e. large magnitude events considered in this study, potentially resulting in:

  - widely spread liquefaction;
  - seismically-induced slope landslides;
  - flooding of large portions of terrain by as much as 1.5m of water if the quake happens simultaneously with high tide.

- Rivers and creeks flooding and dam failures,potentially resulting in:

  - scouring of foundations of bridges and walls along the watercourse;
  - washouts (slips, erosion) of watercourse banks during flooding, as determined by experience (both historically: past, and subject expert opinions: future).

- Static gravitational instability of man-made and natural slopes, potentially resulting in:

  - landslides, avalanches;
  - rock-falls (of large size);
  - possible submarine landslides (those occurring underwater).

- Intrinsic railroad traffic and trackage anomalies, possibly resulting in:

  - derailments of trains with major chemical freight and large environmental impacts;
  - fires;
  - fires of railroad assets (trestles).

- Natural and man-made (not traffic and trackage related) events, possibly resulting in:

  - forest fires;
  - fires of railroad assets;
  - rock-falls and landslides
  - scouring and erosion (seen under rivers and creeks above);
  - derailments;
  - barges and boats hitting against bridge pillars and other assets.

Climate change divergence could obviously affect several of the hazards and their outcomes above. However, it became evident that only a few of the hazards mentioned above would be capable of generating significant BIs. Sensitive areas along the railroad lines were understood to be:

1. Bridges and old trestles;
2. Retaining walls of large size in difficult topography;
3. Tunnels and tunnel portals;
4. Shipping docks and their access structures (causeways);

5.   Soft soil areas subject to liquefaction (where bridges are present). Track is
     sensitive to liquefaction, but repairs would likely be quickly carried out unless
     a large magnitude quake has occurred).

Why we do it.
Without performing a proper hazard identification (HI) the risk assessment is
almost meaningless. Jumping straight into risk identification, a very unfortunate
habit, is prone to lead to neglect of important details. Brainstorming, energy-based
analyses, and resource-flow analyses are all valid ways to complete the hazard identification. This is not the time to censor the list, any idea is good, as apparently "crazy"
scenarios will disappear by virtue of the analysis later on.

14.7   Risk Model Design

In compliance with the failure criteria (Sect. 14.2) we are looking at the evaluation of
the risk of BIs due to rail transportation along the last stretch of the chemical plants
to Terminal. Thus, as stated earlier, a hazard to the railroad does not necessarily enter
in this analysis. Thus, scenarios have to be filtered according to the failure criteria.

   Table 14.1 displays the scenarios which lead to cases of capacity reduction and
service suppression and therefore BI. For the BIs, the table gives the maximum
and minimum estimated disruption duration for all hazards specifically involving
Kryptonite Corp.

   The estimated return periods displayed in Table 14.1 were derived subjectively,
based on RR and other experts' opinions, while including the scarce factual data
available. In some case experience from other countries was used. Because weather
patterns are changing due to climate change, and railroad maintenance and risk
abatement programs follow cyclical patterns, it is advisable to periodically review
the pertinent information, possibly leading to a shortening of the estimated return
periods (increased estimated frequencies). Because of the broad range of subjective
assessment inherent in the generation of the data in the table, they must be taken as
order-of-magnitude prior estimates only, to be updated later on using new information
and Bayes theorem (see Sect. 8.1.4).

   Some hazards exhibit a significant correlation between consequence and probability, that is, some scenarios with high probabilities have lower consequences,
whereas others with low probabilities have higher consequences. However, that is
not a general rule and many display a constant probability estimate throught the
consequence range. Figure 14.2 provides a graphic illustration of the risk scenarios
of Table 14.2.

   Why we do it. Being able to compare and plot all the pertinent identified events
together in terms of probabilities and consequences allows a first understanding of
the relative risks.

   BI costs. Table 14.2 shows a worksheet made available by the client stating the
total cumulative costs (M USD) as a function of BI duration expressed in months.

Table 14.1 Scenarios leading to capacity reduction and service suppression (BI) cases

Hazard scenario	BI (Months duration)		Probability per year	
	Min	Max	Max	Min
Natural Earthquake	2	18	$2.5*10^{-3}$	$2.0*10^{-3}$
Natural High wind (Windstorm & Hurricane)	1	6	$5.0*10^{-2}$	$3.3*10^{-2}$
Natural Lightning	0.25	1	$1.0*10^{-4}$	$1.0*10^{-4}$
Natural Snowstorm	0.25	0.25	$1.0*10^{-2}$	$1.0*10^{-3}$
Natural Volcanic ash	2	12	$1.0*10^{-5}$	$1.0*10^{-6}$
Natural Flooding	0.25	0.5	$5.0*10^{-1}$	$1.0*10^{-1}$
Natural Extreme cold, freezing rain	3	4	$2.0*10^{-1}$	$1.0*10^{-2}$
System Outage System of communication	0.25	0.25	$5.0*10^{-2}$	$5.0*10^{-2}$
System Outage Power electric	0.25	0.25	$1.0*10^{-4}$	$1.0*10^{-4}$
System Outage Power hydrocarbons	0.25	0.25	$1.0*10^{-4}$	$1.0*10^{-4}$
Equipment Failure from Natural cause such as "wear"	2	4	$5.0*10^{-2}$	$5.0*10^{-2}$
Fire, explosion due to any hazard EXCEPT terrorism, pandemic, outages and natural disasters	3	9	$5.0*10^{-3}$	$5.0*10^{-3}$
Spill external Hydrocarbons	3	6	$1.0*10^{-4}$	$1.0*10^{-4}$
Spill external Chemical	4	9	$1.0*10^{-4}$	$1.0*10^{-4}$
Personnel Mishandling	0.25	1	$1.0*10^{-3}$	$1.0*10^{-3}$
Personnel Pandemic	0.25	1	$1.0*10^{-3}$	$1.0*10^{-3}$
Personnel Employee Dishonesty	0.25	1	$1.0*10^{-2}$	$1.0*10^{-2}$
Terrorism Riots	0.25	0.25	$1.0*10^{-3}$	$1.0*10^{-3}$
Terrorism Arson	2	4	$5.0*10^{-3}$	$5.0*10^{-3}$
Terrorism Cyber-attacks	0.25	1	$1.0*10^{-2}$	$1.0*10^{-2}$
External Wildfire, Airplane crash, Nuclear accident	6	12	$1.0*10^{-6}$	$1.0*10^{-6}$
Maritime/Fluvial Traffic Bridge collapse	4	12	$5.0*10^{-2}$	$3.3*10^{-2}$
Hydro Dam generated flooding	12	18	$3.3*10^{-2}$	$2.0*10^{-2}$

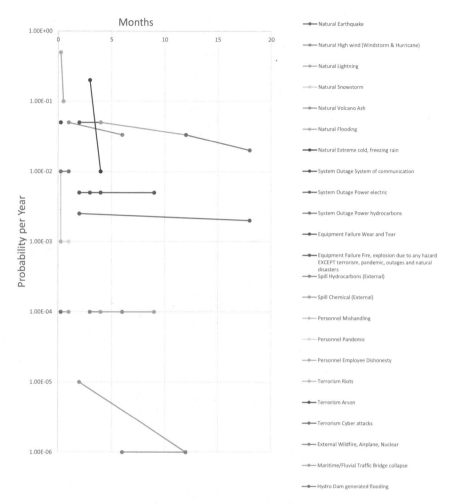

Fig. 14.2 Graphic representation of the risk scenarios of Table 14.2. BI expressed in months, horizontal axis; annual probability on the vertical axis. Min-max are shown and linked with lines to highlight uncertianties

It was understood that the cumulative cost of BI would also be affected by uncertainties, including, of course, the price of Kryptonite products, currency rates, etc. Then, using the values delivered by the client, it was possible to perform the following analysis summarized in Table 14.3.

Figure 14.3 displays the cumulative cost of BI (vertical axis) vs BI duration in months (horizontal axis). Note that in this specific case the relationship is approximated by a linear equation, but in other case non-linear effects could take place.

Table 14.2 Cumulative costs as a function of BI duration

Cost (M USD)	Exposure	Approx. BI duration (Months)
$0	during first 1 week after event	0
$37M	after 2nd week after event	0.5
$73M	after 3rd week after event	0.75
$98M	after 4th week after event	1.0
$194M	after 8th week after event	2.0
$290M	after 12th week after event	3.0
$387M	after 16th week after event	4.0 = minimum BI for major earthquake
$628M	after 26th week after event	6.5
Linear increase	for periods above 26 weeks	Up to 18 months

Table 14.3 Cumulative costs as a function of the BI duration parametrization

Months BI	Cost of BI (M USD)
0.25	0.00
0.5	37.60
0.75	75.30
1	98.20
2	190.00
3	281.80
4	373.50
6.5	603.00
18	1,658.40

Below we explore two possible developments of the study. Case A (Sect. 14.7.1) shows how to combine hazards which are not well-known. Here we work distribution-free. Indeed, as we discussed in Sect. 8.1.4 if the distribution type (shape) is not well known, working distribution-free allows us to reduce errors. Let's note that gaining enough knowledge to be able to select a distribution type is generally not possible (lack of data) or unsustainable (requires too much testing). Case B shows how to proceed if a single hazard is well-known. Here we approximate reality using an empirical distribution. We also use it to show how the selection of a distribution

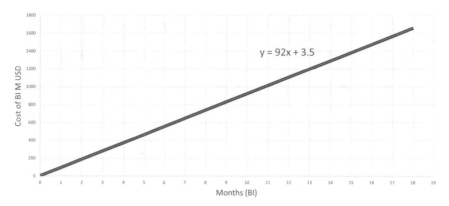

Fig. 14.3 Cumulative costs as a function of the BI duration in months

can influence the results and sometimes hide the risk landscape under heavy use of mathematical tools.

14.7.1 Case A: Distribution-free Computation of BI

As stated earlier, we selected the month as a metric for BI and define M= months duration of BI

For each impinging hazard, considering the uncertainties underlying any evaluation, we define a pair of duration M (min, max) and their related probability p and (i.e., of failure as defined in the Success metric section above) namely min M_1, p_1,;max M_2,p_2,.

Therefore the BI risk formula for a given hazard is:

$$R_M = Cost_M * p_{\text{(event causing the BI duration M)}} \tag{14.1}$$

Where $Cost_M$ is defined in Table 14.3, respectively Fig. 14.3.

The hazards have to be combined in order to obtain an evaluation of the overall risk.

Probability to see an event per year.

If we look at the probability to see a BI (of any duration) in any given year, we can compute the serie of the optimistic probabilities and the pressimistic probabilities, respectively as the series of all the probabilities related to to (min) M_1 and to (max) M_2. These yield respectively $2.90*10^{-1}$ (rounded to 0.3) and $6.99 * 10^{-1}$ (rounded to 0.7), which are somewhat easy to agree with, based on historical values perceived by personnel. However these probabilities tell us there is a relatively high chance of a BI, but do not say anything about the duration. Thus they are of limited use.

Caveat: when dealing with probabilities and relatively high frequencies (above 1/10) the frequency number cannot be equated to the probability value. For example, 1/100 frequency is almost identical to 0.01 probability, but 1/5 does not equate to 0.2 probability (see Sect. 8.1.3 and Appendix A).

Suppose that the min/max results of the estimates for the volcanic ash scenario are respectively:

$$M_1 = 2 \text{ months, } p_1 1 * 10^{-5},$$

and

$$M_2 = 12 \text{ months, } p_1 = 1 * 10^{-6}.$$

We are looking at an event with a certain magnitude, so the statement above implicitly states that a BI from 0 to 2 months will be certain, obviously in case the hazard occurs. Therefore we assume the max value of p_1, p_2 which is $1.*10^{-5}$ for a BI of 0 months to $M_1-1=2$ months. If $p_1=p_2$, there is no correlation between p and the magnitude of the consequences, then the above indicates a uniform allocation (NB: this is not a probability distribution) of the probability of BI per month, with probability $p_1=p_2$ (see Fig. 14.4).

Then in between M_1 and M_2 (both included) we step linearly toward $p_{min}=p_1$. As we do not know the true distribution other options could be assumed.

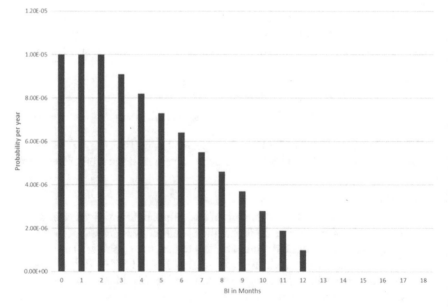

Fig. 14.4 Initial probability estimates and their discretization enabling the BI computation

We then evaluate the probability p of BI at every month, considering the series system (see Sect. 8.2.1) of all impinging hazards (see Table 14.1). The graph in Fig. 14.5 shows 0.7 as the probability of having a BI of less than a month in any given year, in compliance with the values given at the beginning of this section.

We can then plot a graph (Fig. 14.6) representing the annualized BI risk (due to the full set of impinging hazards and their consequences) per month of BI duration, to truly understand the exposure. This graph is built considering there is no buffer stock at Terminal, thus leaving Kryptonite exposed to the full impact of any BI.

We can see in Fig. 14.6 that there are two peak risks, one at 3 months and one at 12 months. We can infer from this graph that the longest BIs (past 12 months) do not correspond to the highest risks. This is a common finding: as risk is the combination of probability and consequences, oftentimes the largest consequences do not correspond to the largest risks due to the interplay between their respective probabilities. If the client does not want to evaluate any a priori mitigation, a rational

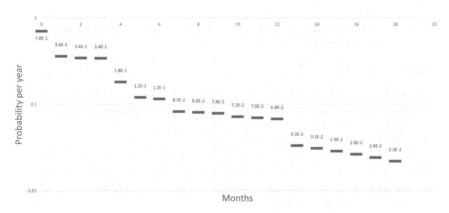

Fig. 14.5 Annualized probability of having a BI of a certain duration (months)

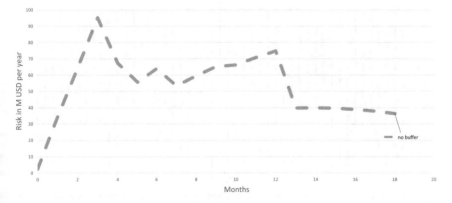

Fig. 14.6 BI annualized risk due to the full set of impinging hazards and their consequences, per month of BI. Assumes no buffer stock at the terminal

choice would be, for example, to transfer risks for 60M USD/yr for BI up to 18 months, and in addition to insure a layer up to 100M USD (40M USD additional) for BIs between 1 and 13 months.

Although it is possible to buy separate layers for BI with different indemnity periods, this appears not to be done very often. Interviewed insurers confirmed this and wondered what the insurance pricing impact would ultimately be. Furthermore, let's note that insurers generally don't offer separate BI coverage without insuring the property damage, hence there is very little market available to insure against only the BI or an additional BI layer, which in theory would likely be the most efficient way to structure the coverage. Thus, the insured would likely need to buy primary coverage for property damage and BI and then, for the additional layer(s) change the BI conditions as an alternative.

In this day and age of climate change, divergent risks and insurance denial (Technical note 5.1) the layering system may offer the insured an additional element of negotiation and, in the end, a win-win solution. As we will see in the next section, considering tolerance to risk and implementing buffers will add a layer of refinement and finally lead to remarkably interesting and efficient solutions, even when dealing with divergent and interdependent risks.

Tolerance and buffer stock

If the buffer stock at Terminal is worth one month of shipping, then the probability of the first month is negated and the first month of BI becomes the second month.

Using this principle, we are able to model the effect of 1, 2, 3, and 4 months' worth of buffer stock at Terminal (Fig. 14.7).

This will enable the company to design a volume of buffer appropriate to the risk they deem tolerable. It is interesting to note that a one-month buffer already significantly reduces the peak at three months, making it possible to completely change the possible risk transfer program.

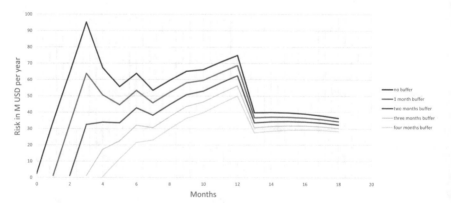

Fig. 14.7 Annualized risk due to the full set of impinging hazards and their consequences, per month of BI, based on amounts of buffer stock from none to 1, 2, 3, and 4 months' worth at Terminal

If, for example, the corporate risk tolerance is constant at 50M USD/yr then we can see that a 4-month buffer allows Kryptonite to stay below that threshold at the 12-month peak. It is obviously necessary to find a compromise between buffer stock costs, tolerable risks and risk transfer, and these graphs make it possible to discuss and communicate this in a transparent way.

We will see in Case Study 2 (Terminal, Chap. 15) how to deal with sophisticated and more realistic tolerance thresholds that are not constant.

Divergent risks

What about divergent risks? Let's suppose a scenario of climate change. If flooding becomes five times more likely to occur due to climate change, the results change as discussed below.

First of all, what does "five times more likely" mean? It means that the frequency is five times greater. This has a direct impact on the probability of occurrence and an indirect impact on BI. This is due in part to the repairs, which will take longer as the system becomes more and more wounded (see Sect. 4.2, 4.4) and therefore fragile (see Sect. 1.1, 11.1.1).

Initial estimates in the natural flooding scenario in Table 14.1 were respectively:

$$M_1 = 0.25 \, \text{months}, \, p_1 = 5 * 10^{-1}$$

and

$$M_2 = 0.5 \, \text{months}, \, p_2 = 1 * 10^{-1}.$$

The fivefold divergence then alters those numbers to (abbreviated as w-div flood):

$$M_{1 \, w-\text{div flood}} = 1 \, \text{months}, \, p_{1 \, w-\text{div flood}} = 9.2 * 10^{-1},$$

and

$$M_{2w-\text{div flood}} = 2 \, \text{months}, \, p_{2w-\text{div flood}} = 4 * 10^{-1}.$$

NB: An event with a probability of 0.5 per year that becomes five times more likely will have a frequency of 2.5, which is linked to a probability of 0.92 per year (see Poisson, Sect. 8.1.3).

Figure 14.8 represents the annualized BI risk (due to the full set of impinging hazards and the consequences) per month of BI duration. The orange dotted line shows the risk profile with no buffer and no divergent flooding; the gold dashed line shows the new scenario: no buffer with divergent flooding due to climate change (no buffer w-diver flood).

As can be seen by comparing this graph with that in Fig. 14.7, the peak of risk around 2 months is now up to 114M USD, whereas the peak at 12 months remains identical.

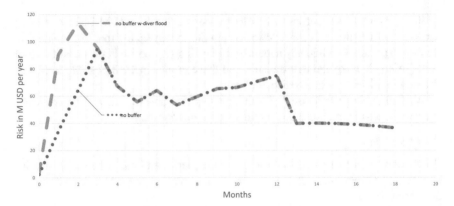

Fig. 14.8 Annualized BI risk (due to the full set of impinging hazards and the consequences) per month of BI duration. Orange dotted line: risk profile with no buffer and no divergent flooding; gold dashed line: no buffer and divergent flooding scenario due to climate change (no buffer w-diver flood)

Fig. 14.9 displays the analysis of the impact of buffer stock. Notice that while the effect of flooding due to climate change (w-diver flood) intervenes for BIs shorter than two months, there is no difference between graphs beyond that duration.

If the client had an insurance coverage for the full BI before divergence, we can assume that the premium would be the object of renegotiation after the first couple of hits due to the flooding, and that insurance denial (Technical note 5.1) could quickly become a critical issue. The company might want to explore the cost of the buffer expansion as we see that alternative would reduce the risk dramatically.

But this is not all! Due to interdependency, the divergence may alter the number of the Hydro Dam generated flooding from the original values of:

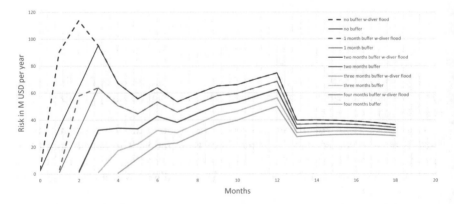

Fig. 14.9 Analysis of the impact of buffer stock. While the effect of flooding due to climate change intervenes for BIs shorter than two months, there is no difference between graphs beyond that duration

$$M_1 = 12 \text{ months}, \ p_1 = 3.33 * 10^{-2},$$

and
$$M_2 = 18 \text{ months}, \ p_2 = 2 * 10^{-2}$$
to the new values (abbreviated as w-div flood +dam) estimated at:

$$M_{1w-\text{div flood }+\text{dam}} = 12 \text{ months}, \ p_{1w-\text{div flood}+\text{dam}} = 8.33 * 10^{-2}$$

and

$$M_{2w-\text{div flood }+\text{dam}} = 18 \text{ months}, \ p_{2w-\text{div flood }+\text{dam}} = 5 * 10^{-2}.$$

This is because in this particular case it is estimated that only half of the five times more likely natural flooding is assumed to impact the possible failures of Hydro dam.

In Fig. 14.10 we see the corresponding BI risk curves. Again, the base case (no buffer, blue dotted line) is displayed for comparison. As can be seen, the case with divergent flooding and its effect on dam flooding (orange dotted line: (w-diver flood+dam) diverges from the original values at two months and significantly alters the risk profile towards the longest BI durations. The: dashed brown covers no buffer with divergent flooding (no buffer w-diver flood) case already shown in Fig. 14.8.

Figure 14.11 compares the impact of 1, 2, 3 and 4 months' worth of buffer stock for the original case and for the divergent flooding and dam scenarios.

We can see a significant increase of the risks at both peaks, thus a risk divergence. As a result, the cost of the buffer implementation may become competitive against more traditional risk transfer techniques such as insurance.

In summary, armed with this type of analyses, Kryptonite can find a path to sustainable operations even in presence of divergent and interdependent exposures. The roadmap may include a thoughtful blend of:

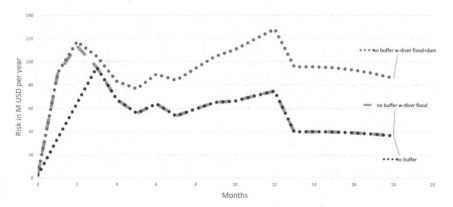

Fig. 14.10 Comparison between BI risk curves: blue: no buffer (base case); brown dashed line: no buffer with divergent flooding on railroad (no buffer w-diver flood); orange dotted: no buffer with divergent flood and dams effects (no buffer w-diver flood+dam)

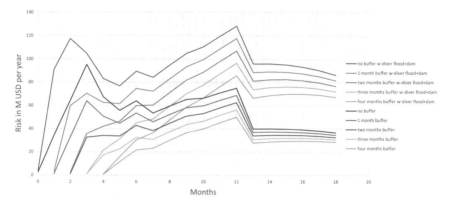

Fig. 14.11 Comparison of various cases of buffer stock: 1, 2, 3, 4 months' worth and business-as-usual and divergent scenarios defined in Fig. 14.10

- risk acceptance (setting the tolerance at a certain level and thus accepting some losses);
- upkeep of buffer stock (if feasible at Terminal, of course);
- layered BI insurance.

The approach explained above makes it possible to make perform RIDM, and ultimately adds value to the company while providing a firm basis for negotiation with insurers, if denials are lurking.

14.7.2 Case B: When the Distribution is Known

As stated earlier, Case B assumes that only one hazard has to be analyzed and therefore constitutes a very small subset of the Case A. Case B uses BI in months and delivers the final results in terms of M USD based on the cumulative values defined in Fig. 14.3. Here one might be tempted to approximate the hazard probability with an empirical distribution, especially if there is some comfort with the significance and completeness of data but a true distribution is not known. Methods such as Monte Carlo and many commercial software programs force users to assume distributions of all kinds, neglecting to make evident the severe limitations and possible errors arising from this practice (Sect. 8.1.4, 8.1.5).

This section shows how to proceed with the evaluation of BI with a beta (empirical) distribution. The beta distribution can be used in project planning to model probable completion times given an expected completion time and variability, so its use in this study is perfectly justified.

In the event of a large earthquake hitting the system, expert consensus was reached that the estimates of BI would be the following:

- minimum duration: 2 months (less than what the client stated in Table 14.1, min BI for major earthquake 4 months);
- maximum duration: 18 months;
- average: 8 months;
- standard deviation: 2 months.

We should note that no attempt was made to match the results of case A (Sect. 14.7.1). The two procedures were carried out independently as one objective was to show how similar or different the results would be under seemingly reasonable assumptions for each.

The BI values from 2 to 18 months are plotted as horizontal axis in Fig. 14.12. BI values in months are depicted on the horizontal axis, the cumulative probability (0-1) on the left vertical axis; the probability function values on the right vertical axis. The cumulative beta distribution of BI is shown in blue and the BI mass distribution in orange.

The cumulative beta distribution describes the BI in the event of the hazard hitting the system (i.e., the cumulative probability distribution reaches 1). Thus, to understand the annualized risk we have to combine the likelihood of the given hazard per year ($p_{(event)}$).

The full BI risk formula is then:

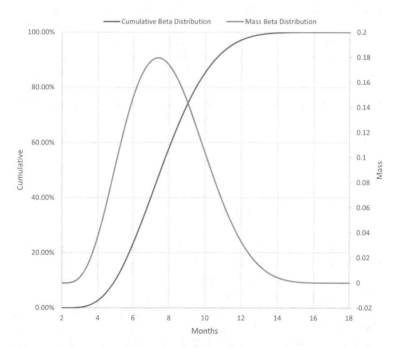

Fig. 14.12 BI distribution (orange) and cumulative values (blue). Horizontal axis, BI; left vertical axis, cumulative distribution; right vertical axis, mass distribution values

Table 14.4 Selected values of the cumulative probability distribution of the beta function in Fig. 14.12

Month	Cumulative beta distribution (%)
6.75	36
7.5	50
10	85
10.50	90
11.75	96

$$R_M = Cost_M * p_{(event\ causing\ the\ BI\ duration\ M)} * p_{(BI=duration)} \qquad (14.2)$$

Why we do it? The Beta distribution is a probability distribution on probabilities. It is the conjugate prior for the Bernoulli, binomial, negative binomial and geometric distributions which are the distributions that involve success and failure in Bayesian inference. The difference between the binomial and the beta is that the former models the number of successes (x), while the latter models the probability (p) of success.

Computing a posterior using a conjugate prior is very convenient, because it makes it possible to *avoid expensive numerical computation*(https://towardsdatascience.com/beta-distribution-intuition-examples-and-derivation-cf00f4db57af). The beta distribution belongs to a family of continuous probability distributions that have been applied to model the behavior of random variables limited to intervals of finite length in a wide variety of disciplines. Beta distributions have been used for time allocation in project management / control systems, variability of soil properties, proportions of the minerals in rocks in stratigraphy etc. Thus, they are widely accepted empirical distributions and their mathematics is well known (even through Excel worksheets formulae).

Table 14.4 shows a tabulation of selected values of the cumulative distribution (blue line in Fig. 14.12) of BI. Excel worksheet formulas allow calculation of these values. The values were selected to allow for specific examples we will show in Sect. 14.8.

14.8   Results and Communications

As shown in Table 14.4 we selected specific values of BI in order to present how to use the cumulative distribution of BI. The corresponding durations were Case 1: 6.75 months, Case 2: the interval between 7.5 months and 10 months, and Case 3: longer than 11.75 months, then 7.5 and 10.5 months.

The results of the three cases are shown graphically in Table 14.5 under the assumption that BI can extend to a maximum of 18 months (first row).

Case 1: there is a probability of 36% that the BI will be shorter than 6.75 months, reciprocally 100-36=64% that the BI will be longer than 6.75 months.

Case 2: there is a probability of 85-50=35% that the BI will last between 7.5 months and 10.0 months.

Table 14.5 Comparing values of the probability for two different upper bound

Max BI	Case 1 shorter than 6.75 month	Case 1 longer than 6.75 months	Case 2	Case 3
18.00	36	64	35	3
24.00	24	76	35	2
Difference %	32%	−18%	5%	29%

Case 3: There is 100-96=3% chances that a BI will last longer than 11.75 months. Furthermore note that there is a probability of 50% that the BI will be 7.5 months or less and a probability of 90% that the BI will be 10.5 months or less.

As all the parameters of the proposed distribution, but in particular the values of the standard deviation and the upper bound (maximum BI duration) are estimates based on third party reports, interviews, etc., a sensitivity analysis has to be performed.

First the maximum BI duration was raised to 24 months. The results of the three cases described above changed as follows:

Case 1: there is a probability of 24% that the BI will be shorter than 6 months, respectively 100−24=76% that the BI will be longer than 6 months.

Case 2: there is a probability of 35% that the BI will last between 7.5 and 10.0 months.

Case 3: There is 2% chances that a BI will last longer than 12 months.

Second, the standard deviation (st_{dev}), a measure of uncertainty of the estimate, is studied.

An increase of the standard deviation (for example, an increase by one month from the original value of two months to three months) requires the modification of the minimum duration (lower bound) to three months in order to allow the mathematical definition of the distribution. These changes will provoke a complete change in the shape of the distribution, which is then less likely to reflect the possible BI duration as it shifts the likelihood of the durations too far towards the lower values (see Fig. 14.12). Figure 14.13 shows how the graph of the beta distribution of BI is impacted if standard deviation is increased from two months to three months. It uses the same conventions as Fig. 14.12.

This is an excellent example to show that good sense and rational approaches are valid also when using this type of mathematics to derive numbers from poorly defined/scarce information.

In these cases the mathematics are certainly not a magic pill, but rather are to be considered as a support to bring clarity to the assumptions and help form a more informed understanding of a future situation.

Now returning to our three examples, we now make an assumption about a smaller standard deviation (reflecting possibly lower uncertainties), evaluated by setting the standard deviation to one month. The results are:

Case 1: there is a probability of 2% that the BI will be shorter than 6 months, respectively 98% that the BI will be longer than 6 months.

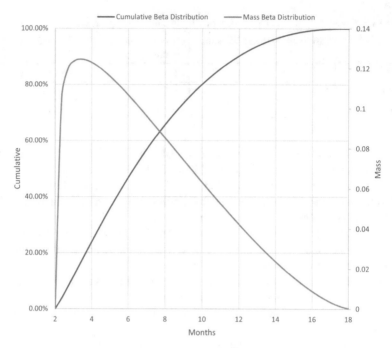

Fig. 14.13 Effect of the alteration of the standard deviation on the probability distribution when $st_{dev} = 3$

Case2: there is a probability of 65% that the BI will last between 7.5 and 10 months.

Case 3: There is 0.01% chances that a BI will last longer than 11.75 months.

These results call for the following comments: it is deemed very unlikely that a coefficient of variation (i.e., standard deviation divided by the expected duration, a measure of volatility or the uncertainty) of less than 2/8=0.25 would be a reasonable assumption, and it is on the safe side to overestimate it.

With the same BI assumptions, we can now draw a curve showing cost of BI vs probability (Fig. 14.14). The bounds of the distribution are respectively 200M USD and 1,700M USD. The curve allows us to see that the highest likelihood of approximately 0.18 corresponds to a loss of approximately 750M USD and a BI duration of 8 months and 1 week.

We can now extract interesting values for management. For example in the event of an earthquake, there is a 40% probability that the BI will remain below 722M USD. We want to reiterate that this is still in the event of the hazard hitting the system and is not yet the annualized probability of the BI.

By modifying the BI-cost function it is possible, in a RIDM approach, to study how proactive mitigation and/or step-by-step resumption of operations would influence the risk function, thus the probability-loss function. If climate change has the potential to alter the parameters of the distribution (e.g., increase the average BI because larger,

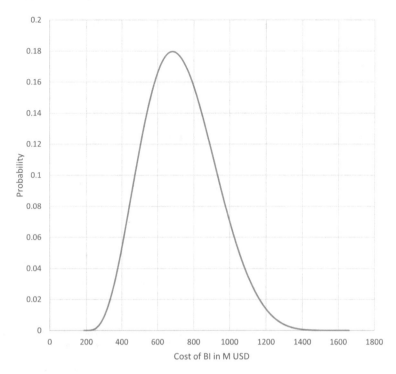

Fig. 14.14 Probability versus cost of BI

more extreme events may be expected) the same methodology can be applied to evaluate these scenarios.

Why we do it. We want to evaluate and draw a beta distribution that depicts our BI and, based on our assumptions, discuss the chance of different durations. This is where a lot of good judgment must be applied, as the blind application of formulas may lead to absurd outcomes. We do not operate in a perfect world, with perfect knowledge of the variables. There are plenty of uncertainties and we must continually exert common sense when generating support for RIDM.

So far we calculated the BI values in the event of an earthquake. In other words we answered the question: if an earthquake occurs, what will be the BI distribution? Of course management is interested to find the annualized risk of that event, for example of an earthquake with a likelihood of 1/475 per year (Fig. 14.15).

The cumulative probability (Fig. 14.16), which obviously sums up to 1/475, and not to 1, allows to find the value of the BI at different probabilities thresholds; for example if the BI risk tolerance threshold corresponds to a probability of 1/1000 then the associated value is 780M USD.

If we now consider the beta approach (Sect. 14.7.2) (recall that in Case B the beta was evaluated subjectively for earthquake only), we have a quake risk of 0.16M USD/yr at 6 months BI, and 0.25M USD/yr at ten months BI.

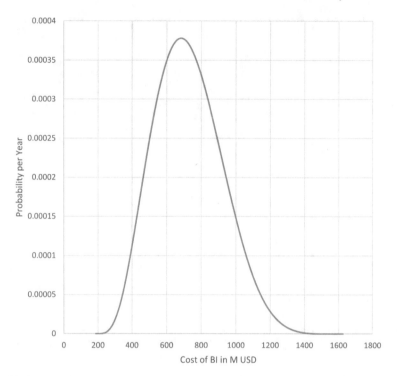

Fig. 14.15 Annualized risk with an event (earthquake) that has the likelihood of 1/475 per year probability curve

The direct calculation (Sect. 14.7.1) for earthquake gives respectively 1.31M USD/yr at 6 months and 2.08M USD/yr at 10 months. It is not at all shocking to see a difference of approximately one order of magnitude as there was no attempt to calibrate one option with the other. The calibration of the beta parameters would likely allow to obtain even more similar results between the two approaches.

14.9   Recommendations and Conditions of Validity

In conclusion, the following RIDM general recommendations are made:

1. Maintain sizeable amounts of buffer stock both at Terminal and another selected location, upkeeping a volume to be optimized by analysis.
2. Maintain open lines of communication with RR, including making sure that the line receives due care and maintenance.
3. Purchase insurance limits covering the extra costs possibly incurred in case of long-term severe traffic capacity reduction (depends on how RR contracts are

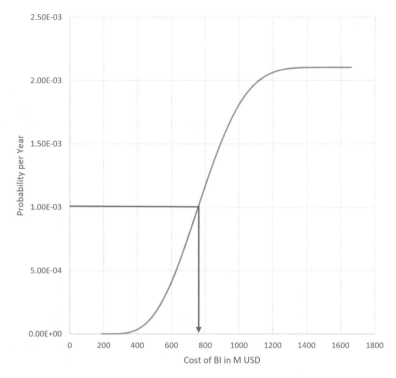

Fig. 14.16 Cumulative distribution of the cost of BI for an event that has 1/475 probability of occurrence per year

written), depending on the risk appetite/risk tolerance of management and the insurance premium vs cost of implementing a larger amount of buffer stock.

Condition of validity

The studies leading to the quantification of the BI and associated frequencies are as unbiased as possible given the available data.

If climate change were going to impact significantly the frequencies established for winter hazards (freezing rain, snowstorm etc.) and flooding in the selected time frame, we would have to alter those values and update the study. Of course, the procedures just outlined make it possible to play with what-if scenarios.

The BI cost function might not be linear in other cases, due to the loss of contract/client and reputational damages. Due attention has to be devoted to this.

Why we do it. Adding Conditions of validity is a step unfortunately oftentimes forgotten in risk assessment. Conditions of validity help the reader to understand the assumptions that shore up the study. They also work as a reminder that the risk assessment is simply a model that by design cannot replicate reality.

Appendix

Links to more information about the Key terms from the Authors	
A,B	*Act of God* (https://www.riskope.com/2020/12/09/act-of-god-in-probabilistic-risk-assessment/) *Black swan* (https://www.riskope.com/2011/06/14/black-swan-mania-using-buzzwords-can-be-a-dangerous-habit/) *Business-as-usual* (https://www.riskope.com/2021/01/13/business-as-usual-definition-in-risk-assessment/)
C,D	*Convergent* (https://www.riskope.com/2021/01/20/convergent-risk-assessments/) *Divergent* (https://www.riskope.com/2020/11/18/tactical-and-strategic-planning-to-mitigate-divergent-events/) *Drillable* (https://www.riskope.com/2020/01/15/probability-impact-graphs-do-not-fly/)
F	*Foreseeability/foreseeable* (https://www.riskope.com/2021/01/06/foreseeability-and-predictability-in-risk-assessments/) *Fragile/fragility* (https://www.riskope.com/2020/04/01/antifragile-resilient-solutions-for-tactical-and-strategic-planning/)
P,R	*Predictability/predictable* (https://www.riskope.com/2021/01/06/foreseeability-and-predictability-in-risk-assessments/) *Resilient, Resilience* (https://www.riskope.com/2016/11/23/resilience-cannot-based-instinctual-decision-making/)
S	*Scalable* (https://www.riskope.com/2015/04/16/how-system-definition-and-interdependencies-allow-transparent-and-scalable-risk-assessments/) *Societal risk acceptability* (https://www.riskope.com/2014/01/09/aspects-of-risk-tolerance-manageable-vs-unmanageable-risks-in-relation-to-critical-decisions-perpetuity-projects-public-opposition/) *Sustainability/sustainable* (https://www.riskope.com/2019/01/16/improving-sustainability-through-reasonable-risk-and-crisis-management/) *Survivability* (https://www.riskope.com/2011/03/17/ale-fmea-fmeca-qualitative-methods-is-it-really-what-we-need/) *System* (https://www.riskope.com/2017/07/26/three-ways-to-enhancing-your-risk-registers/)

(continued)

(continued)

Links to more information about the Key terms from the Authors	
T,U	*Tolerance* (https://www.riskope.com/2020/04/29/risk-tolerance-thresholds/) *Uncertainty/uncertainties* (https://www.riskope.com/2015/12/10/3-decision-making-truths-derived-from-uncertainty-taxonomy-scheme-of-classification-and-a-road-sign/) *Updatable* (https://www.riskope.com/2020/01/07/climate-adaptation-and-risk-assessment/)

Other linked information (https://www.riskope.com/blog-news/) search Riskope blog and use the search box

Chapter 15
Case Study 2: Terminal

This chapter is devoted to the risk assessment of Kryptonite Terminal. A convergent quantitative risk assessment is developed allowing for risk informed decision making and giving quantitative support for the definition of insurance limits.

15.1 The Client's Request

Kryptonite's Terminal operates 24 h a day, seven days a week, 362 days a year. The "product" of Terminal is the safe unloading of arriving loaded trains, temporary storage of bulk, ship mooring at wharves, ship loading, and safe departure of empty trains. To deliver its products Terminal uses automated and semi-automated train unloading/dumping equipment, conveyor belts, and various safety systems. Terminal also uses various computerized systems to ensure its operational efficiency. The Data Center is located on top of the Terminal's power-substation. It is linked to all the site (IoT, SCADA) with a redundant loop of fiber optics. The terminal has multiple berths.

The cargo is shipped to markets around the world, including Asia, South America and Europe. Terminal is said to be risk averse, maintains high standards of safety and reliability as to minimize BI as far as possible. The client requires the preparation of a study including:

- A holistic Risk Assessment bearing on a wide spectrum of natural, man-made, and technological hazards organized in a coherent and logical hazard and risk register to be prioritized allowing an efficient approach to the steps that follow. The risk assessment to be used to point out which are the risks that require detailed business interruption (BI) evaluation and subsequent planning. The risk assessment should also, of course, yield data to be used for the definition of insurance limits.
- An analysis of BI under a variety of scenarios including earthquake and other natural and man-made events scenarios. BI evaluations support decision making for insurance limits, but also allow proactive planning for Business Continuity

© The Author(s), under exclusive license to Springer Nature Switzerland AG 2021   299
F. Oboni and C. H. Oboni, *Convergent Leadership—Divergent Exposures*,
https://doi.org/10.1007/978-3-030-74930-9_15

Plans (BCP), Disaster Recovery and Business Resumption Plans (DRP) (see Sect. 15.8.2).

- Quantitative support for the definition of insurance limits based on assets and operations present and conducted at Terminal. A parallel study by a separate outside engineering firm has delivered Maximum Probable Losses (MPL) under the 1/475 earthquake at various degrees of confidence.

Like in all our studies we ensured the client request was correctly understood and presented our *glossary* (https://www.riskope.com/knowledge-centre/tool-box/glossary/) and specified key terms (see links in the references).

15.2   Success Metric (Failure) and Consequences Dimensions

The success metric (failure criteria) can be defined as follows:

- System must be working 24/7/362, reliably loading ships without delays.
- Buffer stocks have to be sufficient to cover potential BI due to transportation (to be agreed upon with Kryptonite, see Chap. 14).

In contrast to Case Study 1 (Chap. 14) any failure at the Terminal would directly impact the client. Just as a reminder, in Case Study 1 we were not looking at RR intrinsic risks but just at the BI it would generate to Kryptonite. Thus, for example, a H&S issue would not be considered. Now, as on Case Study 2 we are studying the convergent risks at terminal, the failure criteria and consequences dimensions are completely different and need to be clearly stated as shown in Table 15.1.

The scope of the risk assessment we were asked to perform does not cover specifically legal and financial risks, as these were discussed by other parties engaged in Terminal's management process, but we recognized that legal and financial risks are intimately linked to operational hazards consequences. For example, being fully responsible for intermediate storage of bulk materials and related ship loading, with

Table 15.1 Consequences metric (see Chap. 9)

Types of consequences	
Code	Name
PL	Physical losses
HS	Health and safety, Including potential casualties by accident or disease
BI	Business interruption
ED	Environmental damages
RD	Reputational damages, Including public outcry (Community outrage) damages
CR	Crisis potential

contracts possibly poorly defining force majeure clauses (see Sect. 4.4), may lead to undue or unreasonable losses in case of an exceptional or divergent natural event. Legal risks also intersect with environmental damages, hence environmental risks, in case of non-compliance and damages of various kinds.

Why We Do It The definition of the success/failure criteria (Sect. 1.3, Anecdote 1.2) is a necessary preliminary step to any risk assessment. Without clearly defining success/failure, the risk assessment is likely to slide off theme and lose focus. Numerous risk assessments initiatives fail because failure is not properly defined.

15.3   System Definition

Based on Terminal's asset management plan (ISO 55000) the system was split into four internal elements (unloader, conveyor belts, buffer storage (including stacker), loader (including berth)) and two external ones (wharves and navigation, and neighboring facilities threatening Terminal's functionalities), as shown in Fig. 15.1. While the buffer is an element of the system in regard to the risks, its initial volume, without expansion (as per Chap. 14) gives insignificant mitigation for business interruption purposes.

Table 15.2 describes the battery lines (see Sect. 6.2) of the system considered in the analyses.

Why We Do It Without clear definition of the system (see Chap. 6) the risk assessment will fail. By defining the system, we understand where our job starts, where it ends, and what the limits of the field are. It is necessary that system limits be developed together with the client.

Fig. 15.1 Terminal system split in four internal elements. External elements not shown

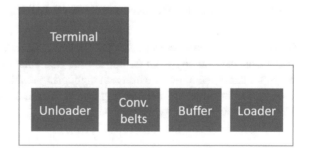

Table 15.2 System elements definitions

Battery lines	Comments
Terminal "perimeter"	Inside the perimeter, where operations and personnel are under Terminal direct supervision and responsibility
Environment	Outside the perimeter, where external environmental damages can occur, public is present
Personnel	Any person working directly or indirectly (subcontractors) for Terminal within the perimeter
Cyber space	The realm of information, the web, data
Transportation	The transportation system(s) outside the perimeter, on which Terminal depends for its operations

15.4   Gathering Existing Information

We carried out interviews with management and personnel to gather information regarding:

1. site buffer capacity, alternative ship loading, production rates, etc.
2. general transportation network information, including critical nodes, and related true and false redundancies;
3. operational details about wharves, machinery portfolios, etc.
4. history of incidents and/or accidents.

Why We Do It Terminal management and personnel have significant knowledge of most aspects of the operation, both in documentation and personal knowledge. We almost always interview employees individually to get the most honest answers possible at all different levels to be sure to grasp the truest picture.

15.5   Requesting Further Necessary Information

The information firstly delivered by the client was operation centric (see Sect. 6.2.2). It did not mention any element outside the immediate perimeter of the Terminal. Thus we had to conduct research on neighboring residential areas, industrial facilities, infrastructures and neighboring hydro dam and then extend the analyses to meteorological conditions and navigation.

15.6   Hazard Identification (HI)

The external hazard identification (see Chap. 7) includes neighboring hazardous industries, possible hydrocarbon (oil) spills in the harbor, RR and highways major

Table 15.3 Classification of hazards type

Hazards type
Natural
System outage
Equipment failure
Spill internal (caused by Terminal)
Spill external (caused by others)
Personnel
Terrorism

bridges (including of course their related natural geohazards and man-made hazards), other critical infrastructures that could possibly impact operations based on the information described in Sect. 15.4, 15.5.

A section of the study was devoted to climatological events such as winds and typhoons, heavy rains, snow, lightning, then epidemics, terrorism and riots. Of course power and telecom outages have also been identified as possible hazards. In order to be systematic a list of "hazard type" (Table 15.3) was prepared.

Many risks exist and in order to avoid double counting we have to systematically approach the hazard identification. In order to be complete, a systematic and rigorous approach to identify the threats-from/threats-to must be implemented.

Thus a "threat from" (Table 15.4) and "threats-to" list were established.

Table 15.4 shows the list of threat-from identified for this study classed by the selection of families adopted for this study. Each threat has to be related to the system's element it may hit. For each element-threat couple the hazard and risk register must record the estimated probability (see Chap. 8) and cost consequences (see Chap. 9).

The following threat-to list was developed: elements in the Terminal (see Fig. 15.1) and also residential neighborhood, marine environment, neighboring bulk food storages, etc.

Not all threats-from affect every element of the system or constitute a threat to some targets, but deciding to omit them from the analysis has to be a conscious, documented decision.

Why We Do It Without doing a proper hazard identification (HI) (see Chap. 7) the risk assessment is almost meaningless. Brainstorming, energy-based analyses, resources flow-analyses are all valid ways to complete the hazard list. This is not the time to censor the list, any idea is good, as "crazy" scenarios will disappear by virtue of the analysis later on.

15.7   Risk Model Design

A risk model should enable:

Table 15.4 List of threats-from for the Terminal convergent risk assessment

Family	Threat from
Natural	Earthquake
Natural	Tsunami
Natural	High wind (Windstorm and Hurricane)
Natural	Lightning
Natural	Snowstorm
Natural	Volcano ash
Natural	Flooding
Natural	Extreme cold, freezing rain
System outage	System of communication
System outage	Power electric
System outage	Power hydrocarbons
Equipment failure	Wear and tear
Equipment failure	Fire, explosion due to any hazard EXCEPT terrorism, pandemic, outages and natural disasters
Spill	Hydrocarbons (Internal)
Spill	Chemical (Internal)
Spill	Hydrocarbons (External source)
Spill	Chemical (External source)
Personnel	Health and safety
Personnel	Mishandling
Personnel	Succession planning
Personnel	Pandemic
Personnel	Employee dishonesty
Terrorism	Riots
Terrorism	Arson
Terrorism	Cyber attacks
External hazards	Wildfire, Airplane, Nuclear
External hazards	Maritime/Fluvial traffic bridge collapse
External Hazards	Hydro Dam generated flooding
Availability	Key material

- a semantically robust hazard and risk register;
- a quantitative convergent risk assessment;
- a quantitative evaluation of insurance volume.

In case of a catastrophic event we understood there were force majeure clauses (see Sect. 4.4) in Terminal-client contracts as well as in client-buyer contracts that would probably limit Terminal's own business damage. Furthermore, Terminal has great flexibility in deploying personnel and its fixed costs were relatively low.

However, the damage for Kryptonite (Terminal's owners) would be loss of sales, overcosts for shipping (delays, rerouting if possible), and ultimately, by domino effect (failure interdependencies) stoppage of chemical plant production.

Figure 15.2 shows the maximum BI per element per area.

The relatively short BI predicted for most elements is due to the exceptional state of preparedness and historical record at Terminal. Another reason for such a short estimate was the impressive spare parts stock and the overall organization geared toward minimizing BI.

As we will see in Sect. 15.8.2 the risk model also allows to evaluate, for example, general liability coverage by looking at each family of pertinent threats excluding of course those which are not under Kryptonite's responsibility, as, for example, a spill from a third-party ship in the harbor.

The estimation of the probability of extreme events is difficult because these are events that have not yet happened or have happened only very rarely, so relevant data are scarce. The theory helps us to evaluate, from a given sample of a given random variable, the probability of events that are more extreme than any previously observed (exceedance probabilities, see Sect. 8.1.2, Technical note 8.2).

For Terminal we considered for each threat-from the average BI for every scenario, and probabilities were assigned based on what is outlined in Chap. 8 and Appendix A.

Consequence Function

As we saw earlier in this section, the consequences need to include physical loss, environmental damages and BI but also health and safety.

Therefore, our study includes estimates of the various dimensions of consequences for each scenario. The value of human life is a contentious subject (see Chap. 9, Sect. 10.2.1) and in our practice, we advise our clients not to use such a value on ethical grounds (see Chap. 5). Thus, here we do not attribute a cost to human life but instead look at accepted capital expenditure in various countries (Marin 1992; Mooney 1977; Jones-Lee 1989; Lee and Jones 2004; Pearce et al. 1995) to save the life of a citizen potentially exposed to natural hazards to find a "bounding value". In other words, our report uses the mitigative investment a society is ready to make to save a life, that is, its Willingness To Pay (WTP) attitude (see Sect. 10.2.1).

Reputational damages were dealt with, in compliance with our long-established practice, with a multiplier linked to public outcry, outrage and qualitatively/subjectively estimated duration/depth of the resulting crisis.

Risk Tolerance Thresholds

Tolerable risk thresholds (see Chap. 10) are always project- and owner-specific and indicate the level of risk which has been deemed acceptable by an owner for a specific project or operation (possibly taking into account public opinion). This means, as an example, that within large companies' corporate risk tolerability may differ quite substantially from that of a branch operation.

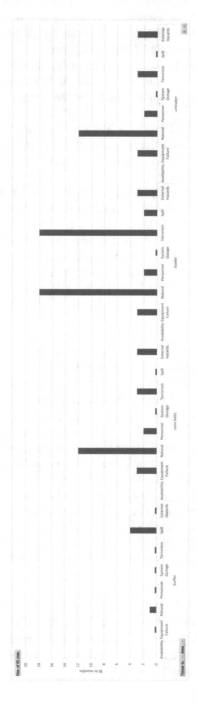

Fig. 15.2 Maximum BI (in months) per element per internal area of terminal

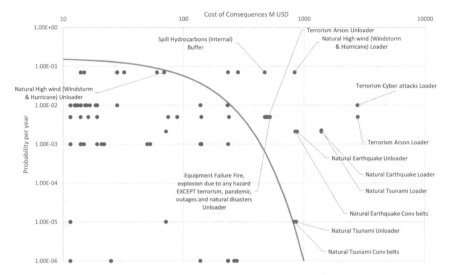

Fig. 15.3 p-C graph with selected scenarios and corporate risk tolerance threshold (orange curve). Vertical axis: annual probability of events; horizontal axis: cost of Consequences in M USD. NB: the families of scenarios correspond to those in Table 15.4

When data are available company/project specific curves can be developed and then discussed with key personnel. The development of empirically estimated tolerability curves requires caution and continuous calibration; they should always be defined by a group, and not by an individual. In the example below we have a tolerance threshold that was developed by the operation manager.

The p-C Graph

Figure 15.3 shows only selected scenarios identified on the p-C graph, because if all had been displayed the graph would look too crowded. Risks are compared to the corporate tolerance threshold (orange curve).

All the risk scenarios that are below the tolerance threshold are, by definition, tolerable and therefore not a priority. The risks to the right and above the threshold can be split into families (Sects. 10.2.1, 11.3, Chap. 12) as follows:

- Manageable, tactical risks can be brought below tolerance (e.g., high wind on the loader, or earthquake on the conveyor belts) by reducing the probability (not of their occurrence, but of the damage they create);
- Unmanageable, strategic risks (e.g., cyber terrorism on the loader, or arson on the unloader) can only be brought under tolerance by lowering the identified consequence (shifting toward the left) through a change in the system, as reducing the probability is inefficient.

Making these distinctions allows for RIDM.

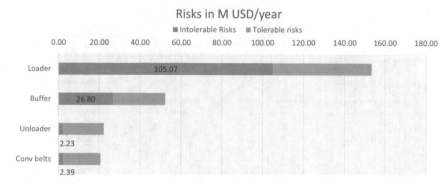

Fig. 15.4 Terminal's risk landscape with the distinction between tolerable risk (orange + grey bar) and intolerable part of the risks (orange bar), per element. Horizontal axis: consequences expressed in M USD/year. The sum of the orange bars: 105.07, 26.8, 2.23 and 2.39 is the total intolerable risk evaluated at 136.49M USD/year

15.8   Results and Communications

15.8.1   Risk and Intolerable Risks

At this point it is possible to understand the threat-to risk per element.

We can see in Fig. 15.4 that even though the annual convergent risk to the buffer and unloader are roughly equal, the buffer has a greater portion of intolerable risk leading to a higher urgency of mitigation.

As we discussed in Chap. 12, the correct risk prioritization should be performed as a function of what is deemed intolerable. This prioritization allows Terminal's manager to understand which equipment needs replacement and or should be stocked with more spare parts.

We can also look at the threat-from risks (Fig. 15.5). The highest overall risk, hydrocarbons, is not the highest intolerable one, which is instead high wind. In prioritizing hazard mitigation using RIDM, the correct roadmap is the one that addresses intolerable risks part (the orange bars in Fig. 15.4 and 15.5); in this case it would place high wind as the top priority.

15.8.2   Roadmap

The RIDM risk mitigation roadmap should be based on aggregated intolerable risks generated by the various convergent hazards. Indeed, as we saw in the first case study, each element is threatened by multiple hazards. Thus, if we sort by hazard and look at the aggregated intolerable risks (an aggregate of all the exposed system's elements),

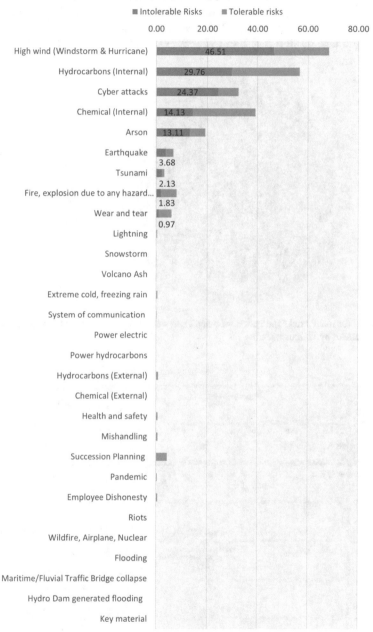

Fig. 15.5 Terminal's risk landscape with the distinction between tolerable risk (orange + grey bars) and intolerable part of the risks (orange bar), per threat-from. Horizontal axis: consequences expressed in M USD/year. The sum, of all the orange bars is the same as in Fig. 15.4 (136.49M USD)

we can prioritize a roadmap for the mitigation actions (prioritization based on the red arrow in Fig. 15.6)) from the most intolerable to the least intolerable part.

Caveat: As shown in Fig. 15.7, an individual risk scenario might be strategic, requiring a change in the system to become tolerable (e.g., a cyber-attack on the

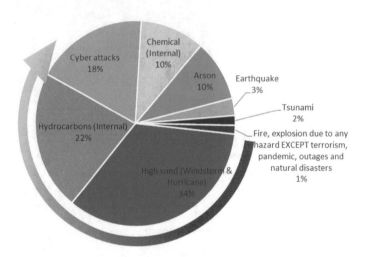

Fig. 15.6 Terminal's risks mitigation roadmap expressed as relative weight of the intolerable risks parts generated by various hazards

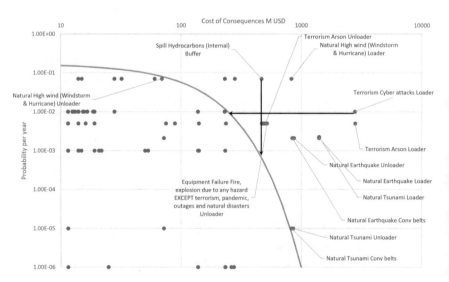

Fig. 15.7 Operation's risk landscape with intolerable risks defined as those right and above the corporate risk tolerance threshold (orange curve). Horizontal axis: consequences expressed in $M; vertical axis annual probability. Notice the vertical (down arrow) indicating a possible mitigation of tactical risk, the horizontal arrow (toward the left) indicating a possible mitigation of a strategic risk (see Sect. 10.2.3)

loader). Indeed, no mitigation (reduction of the probability, i.e., a vertical downward shift) could bring the scenario below credibility. Other risk scenarios may be intolerable and manageable (e.g., high wind on the loader), meaning that capital expenditure can reduce the probability and bring the risk down below the tolerance threshold.

We can now evaluate, beside the roadmap for mitigation, some interesting insurance related values. For example, P_{75} and/or P_{90} values of the probable maximum loss (PML) and/or maximum possible loss (MPL). If the corresponding premium is too high, convergent RIDM will make it possible to make valid decisions based on mitigative actions.

Why We Do It Plotting all the identified events together is necessary to enable their comparison and to define a roadmap to sustainable mitigation including, for example rational risk transfers. At this point if the success/failure metric is incoherent, some events cannot be plotted. Thus if the plot can be drafted, then the selected metric was/is correct.

Application of the Risk Model to Business Continuity Planning (BCP)
Business Continuity Planning (BCP) is the process of developing prior arrangements and procedures that enable an organization to respond to an event in such a manner that critical business functions can continue within planned levels of disruption. BCP identifies an organization's exposure to internal and external threats, synthesizes hard and soft assets to provide effective prevention and recovery for the organization, while maintaining competitive advantage and value system integrity. BCP is also called Business Continuity and Resiliency Planning (BCRP). BCP is a roadmap for continuing operations under adverse conditions such as extreme storms or a cyber-attack. BCP is often used to refer to those activities associated with preparing documentation to assist in the continuing availability of property, people and information and processes (AS/NZS 2020). In the US, governmental entities refer to the process as Continuity Of Operations Planning (COOP).

The end result of BCP can help coping with divergent exposures if they become part of the overall mitigative tactical and strategic planning.

BCP starts precisely with a convergent quantitative risk assessment as shown above. However, the example above did not explicitly include lost sales and income, delayed sales or income, increased expenses (e.g., overtime labor, outsourcing, expediting costs, etc.), regulatory fines, contractual penalties or loss of contractual bonuses, customer dissatisfaction or defection, delay of new business plans which one may include in a full pledged convergent risk assessment.

Related terms are disaster recovery planning (DRP), i.e., the process of restoring the ability to operate, and business resumption planning (BRP), i.e., the process of re-opening each of the facility components. Here we will not delve into these, as they would also use exactly the same scenarios drawn from the convergent risk assessment.

15.8.3   Possible Mitigation Tactics

This study made evident four major intolerable threats to Terminal. Here we list them, along with possible mitigation tactics.

- High wind: can be easily mitigated by means of capital expenditure.
- Hydrocarbon and chemical spills: a spill would generate intense scrutiny and therefore a BI. It seems advisable to foresee some kind of spill containment system along the wharves such as a special drainage system, or a special sump in order to prevent direct spill into inlet waters.
- Cyber-attack: this is deemed unmanageable (no tactical mitigation is possible) at Terminal and thus requires strategic mitigation by altering the system (for instance, removing the Supervisory Control and Data Acquisition (SCADA) system from the loader, etc.).
- Arson and Fire: given historic information on the effects of extinction water, possibly loaded with contaminants of various natures, it seems advisable to foresee some kind of extinction water barrier along the wharves, or a special drainage, or a special sump, in order to prevent direct spill into the inlet waters. Special training of the personnel in case of a fire would also be advisable. General site security could be increased.
- In Chap. 14 we analyzed the buffer to mitigate the RR BI. Here we are not considering the buffer to also be a mitigation of the Terminal to avoid double counting and because in the event of a Terminal failure (as per definition) the Terminal would not be able to load the ship and therefore fail anyways the success criteria.

15.9   Recommendations and Conditions of Validity

The study allowed the determination, identification and characterization of various natural and man-made hazards/hazardous situations taking into account presently known and reported conditions and possible dominos effects (chains of failure).

1. It then lead to the evaluation of the potential types of consequences generated in each scenario considering extant mitigative measures and their assumed potential efficiency, transportation mechanisms, ripple effects on the consequences), and vulnerabilities (health, socio-economical, physical, production, etc.).
2. The cost of consequences arising from crises (i.e. potential turmoil and increased scrutiny, etc.) and reputation damages was included in the cost of consequences by using a specific multiplier defined on the basis of experience and, when possible, available data about similar crises. The multiplier allows the quantification of the impact of soft consequences such a public-opinion, public outcry, and potential media impact.
3. Finally, the convergent risk assessment yielded the evaluation of the risk associated with each accident scenario as the product between the probability of

occurrence of the mishap and the value of the evaluated damages (the object of this Sect.). The probability-cost for each scenario was drafted in a p-C diagram and superimposed to a carefully selected published risk tolerance criteria or with the client's own tolerance curve.

Validity Conditions
The quantification of the hazards and associated frequencies are as unbiased as possible. For example, the maximum BI was identified for the 1/475 earthquake, at a value estimated in the earthquake vulnerability study carried out by a separate engineering firm. There it was set equal to 9 months and was not developed with a hidden agenda such as selling more insurance coverage.

The external hazard identification was as exhaustive as possible, based on the data and interviews conducted with Kryptonite's (Terminal's owner) personnel.

The risk tolerance thresholds were developed with Terminal's key personnel.

Why We Do It Conditions of validity, unfortunately oftentimes forgotten in risk assessments, help the client understand the assumptions that shore up the study. They also work as a reminder that the risk assessment is just a model that by design cannot replicate reality.

Appendix

Links to more information about the Key terms from the Authors	
A, B	*Act of God* (https://www.riskope.com/2020/12/09/act-of-god-in-probabilistic-risk-assessment/) *Black swan* (https://www.riskope.com/2011/06/14/black-swan-mania-using-buzzwords-can-be-a-dangerous-habit/) *Business-as-usual* (https://www.riskope.com/2021/01/13/business-as-usual-definition-in-risk-assessment/)
C, D	*Convergent* (https://www.riskope.com/2021/01/20/convergent-risk-assessments/) *Divergent* (https://www.riskope.com/2020/11/18/tactical-and-strategic-planning-to-mitigate-divergent-events/) *Drillable* (https://www.riskope.com/2020/01/15/probability-impact-graphs-do-not-fly/)
F	*Foreseeability/foreseeable* (https://www.riskope.com/2021/01/06/foreseeability-and-predictability-in-risk-assessments/) *Fragile/fragility* (https://www.riskope.com/2020/04/01/antifragile-resilient-solutions-for-tactical-and-strategic-planning/)

(continued)

(continued)

Links to more information about the Key terms from the Authors	
P, R	*Predictability/predictable* (https://www.ris kope.com/2021/01/06/foreseeability-and-pre dictability-in-risk-assessments/) *Resilient, Resilience* (https://www.riskope.com/2016/11/ 23/resilience-cannot-based-instinctual-dec ision-making/)
S	*Scalable* (https://www.riskope.com/2015/04/ 16/how-system-definition-and-interdepende ncies-allow-transparent-and-scalable-risk-ass essments/) *Societal risk acceptability* (https:// www.riskope.com/2014/01/09/aspects-of-risk-tolerance-manageable-vs-unmanageable-risks-in-relation-to-critical-decisions-perpetuity-pro jects-public-opposition/) *Sustainability/sustainable* (https://www.ris kope.com/2019/01/16/improving-sustainab ility-through-reasonable-risk-and-crisis-man agement/) *Survivability* (https://www.riskope. com/2011/03/17/ale-fmea-fmeca-qualitative-methods-is-it-really-what-we-need/) *System* (https://www.riskope.com/2017/07/26/three-ways-to-enhancing-your-risk-registers/)
T, U	*Tolerance* (https://www.riskope.com/2020/04/ 29/risk-tolerance-thresholds/) *Uncertainty/uncertainties* (https://www.ris kope.com/2015/12/10/3-decision-making-tru ths-derived-from-uncertainty-taxonomy-sch eme-of-classification-and-a-road-sign/) *Updatable* (https://www.riskope.com/2020/01/ 07/climate-adaptation-and-risk-assessment/)

Other linked information (https://www.riskope.com/blog-news/) search Riskope blog and use the search box

References

[AS/NZS] (2020) AS/NZS 5050.1: Managing disruption-related risk. Australian and New Zealand Standards for business continuity management. Standards Australia. https://www.techstreet.com/ sa/standards/as-nzs-5050-int-2020?product_id=2194877
Jones-Lee MW (1989) The economics of safety and physical Risk. Blackwell
Lee EM, Jones DKC (2004) Landslide risk assessment. Thomas Telford
Marin A (1992) Costs and benefits of risk reduction. Appendix in risk: analysis, perception and management. Report of a Royal Society Study Group, London
Mooney GM (1977) The valuation of human life. Macmillan
Pearce DW, Cline WR, Achanta AN, Fankhauser S, Pachauri RK, Tol RSJ, Vellinga P (1996) The social costs of climate change: greenhouse damage and the benefits of control. In: Climate change

1995: economic and social dimensions of climate change. contribution of working group III to the second assessment report of the IPCC, Cambridge, Cambridge University Press

Chapter 16
Case Study 3: Convergent Enterprise Risk Management (ERM) on Divergent Risks

This chapter closes the development of the Case Studies showing how the railroad (Chap. 14), the Terminal (Chap. 15) and three Kryptonite chemical plants can be evaluated convergently for business-as-usual risks as well as divergent ones. The result is a convergent-divergent ERM supporting risk informed decision making.

16.1 The Client's Request

As stated earlier Kryptonite has three chemical plants, whose products are then sent via RR to Terminal to be loaded on ocean shipping vessels taking products to the market. The request was simple: where are our more significant exposures and risks or, in other words, can you build an ERM for our system including the plants, the logistic and the Terminal?

Like in all our studies we ensured the client request was correctly understood and presented our *glossary* (https://www.riskope.com/knowledge-centre/tool-box/glossary/) and specified key terms (see links in the references).

16.2 Success Metric (Failure) and Consequences Dimensions

The success metric is exactly the same as for Case Study 2 (see Chap. 15) but extended to cover the chemical plants. Please refer to Sect. 15.2 for details.

F. Oboni and C. H. Oboni, *Convergent Leadership—Divergent Exposures*,
https://doi.org/10.1007/978-3-030-74930-9_16

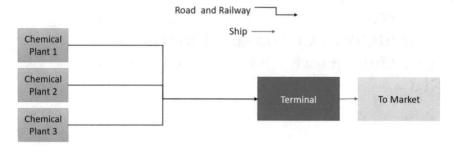

Fig. 16.1 ERM System definition with its elements, namely, the three chemical plants, the logistic network (Road and Railway), Terminal

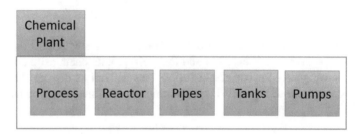

Fig. 16.2 Chemical plants' elements

16.3 System Definition

.

The system definition is shown in Fig. 16.1.

The elements and sub-elements of RR and Terminal are identical to the those in Chaps. 14 and 15, while the chemical plants each include the sub-elements shown in Fig. 16.2.

N.B.: In Chap. 14 we analyzed the buffer to mitigate the RR BI. We are not considering the buffer to also be a mitigation of the Terminal to avoid double-counting and because in the event of a Terminal failure (as per definition) the Terminal would not be able to load the ships and therefore fail anyways the success criteria.

16.4 Gathering Existing Information

We carried out interviews with the Chemical Plants key personnel for both the operations and other potential ancillary elements. The system was defined as explained in Chap. 6 and interviews were conducted as discussed in Sect. 7.1.2.

16.5   Requesting further necessary information

Similarly to the findings of Sect. 15.5 the first draft of the system definition was process centric (see Sect. 6.2.2). That made it necessary to dig deeper in regard to the plants supply chain potential disruption, possible threats-to around the plants and potential threats-from neighboring facilities.

16.6   Hazard Identification (HI)

The list of threats-from is identical to the Case Study 2 (Chap. 15, see Table 15.4).

16.7   Risk Model Design

Please refer to Sect. 15.7 for risk model design details. A significant difference of this chapter risk model lies in the tolerance thresholds. Indeed, as expected the tolerance defined by Terminal's operational manager differs substantially from the one selected by Kryptonite top management and each chemical plant manager. It is indeed completely normal to have different tolerances at different levels of the organization: the tolerance at operation's level is necessarily more stringent than at corporate level. This explains why tolerance has to be evaluated in a reproducible and swift manner at all levels pertinent with the enterprise risk management.

In Fig. 16.3 we chose to display two risk tolerance thresholds, i.e. the corporate (orange) and the Terminal (purple) as well as the risks present in the ERM risk register. Not all risk tolerance thresholds (we omitted the chemical plants) and risk scenarios have been labeled to avoid overcrowding. The graph of Fig. 16.3 bears on the horizontal axis the cost of consequences C (see Eq. 2.1) and on the vertical axis the annual probability of occurrence of the scenarios. The figure displays risk scenarios in green if they are corporately tolerable, respectively in blue if they are corporately intolerable. A number of risks are corporately tolerable but intolerable at Terminal operation's level. We can see a selection of risks that are above Terminal operation tolerance but below corporate tolerance.

16.8   Results and Communications

In Fig. 16.4 we display selected corporately intolerable scenarios on the p-C graph. Again only selected labels are shown to avoid overcrowding.

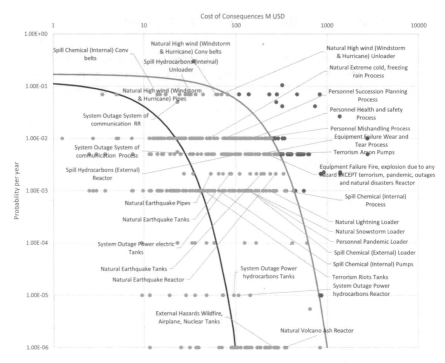

Fig. 16.3 ERM probability-cost of consequences graph with Corporate tolerances (curve in orange) and Terminal's operational (curve in purple) tolerances. Horizontal axis: consequences expressed in M USD; vertical axis: annual probability. It can be immediately noticed that intolerable risks for an operational manager may differ from Corporate, thus leading to possibly contradictory RIDM objectives. Thanks to the graph, which makes these differences clear, discussions can take place in a more serene communication environment.

We can look at intolerable risks per element (each chemical plant, RR and Terminal) convergently, starting with the assumption that no buffer stock is present at Terminal (Fig. 16.5).

From this graph we can see that the RR and Terminal generate a greater portion of intolerable risk than the chemical plants do. Using the built-in granularity (see Sect. 6.2.2), we can zoom into the risk register to show the intolerable risk per sub-elements within the operations (Fig. 16.6). In addition, if Kryptonite were to implement a four-months' worth of buffer stock, the system's risks would change as shown in Fig. 16.7.

The buffer risks shown here stem out of Chap. 15 and more particularly Figs. 15.2, 15.4.

The four-month buffer results in a three-fold reduction of RR intolerable risks part and overall risks (from 89.33M USD to 27.59M USD) and 150MUSD to 50M USD and changes Kryptonite's risk prioritization of the entire system. This shows that top management should consider maintaining buffer stock in Terminal, since it mitigates the RR risks, as demonstrated in Chap. 14 and above.

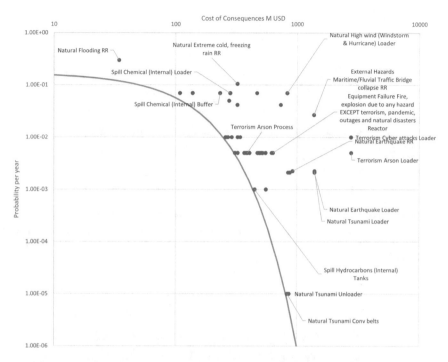

Fig. 16.4 Probability-cost of consequences graph with selected corporately intolerable scenarios. Horizontal axis: consequences expressed in M USD; vertical axis annual probability. The blue dots are the ones that will be the object of RIDM as they are intolerable

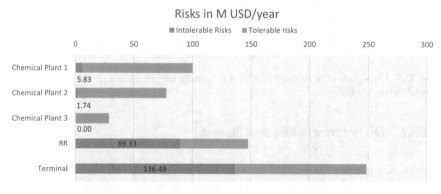

Fig. 16.5 Intolerable risks per element (operation) (horizontal axis M USD) assuming no buffer is present at the Terminal. The values of Terminal for intolerable and tolerable are consistent with what we have shown in Chap. 15 (Figs. 15.4, 15.5). However an attentive reader will see that the values of RR risks cannot be found in Chap. 14. This is because Chap. 14 client's request context was specific to BI and not geared toward ERM.

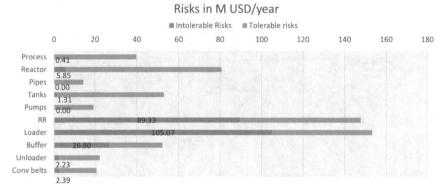

Fig. 16.6 Zoom of the intolerable risks (horizontal axis M USD) per sub-elements (3 chemical plants, RR and Terminal). Process, reactor, pipes, tanks and pumps are the aggregated risk of those elements in the three chemical plants; unloader, conveyor belts, buffer stock, and loader are present in Terminal

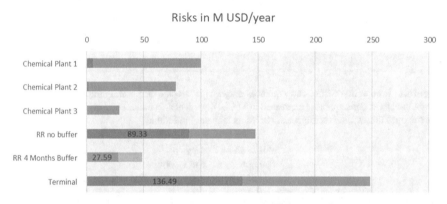

Fig. 16.7 Simulation of 4-month buffer at the Terminal and its effects on the system's risks (horizontal axis M USD)

16.8.1   Divergence: Climate Change

Even if Kryptonite were to implement the four-month buffer, climate change divergence (effects of flooding and dams: see discussion of divergence in Sect. 14.7.1), would still remain above tolerance within one order of magnitude of the no buffer scenario (Fig. 16.8). Indeed, the intolerable part of risk with divergence and four months buffer is evaluated at 63.27M USD, respectively 27.59M USD without divergence and four months buffer. As we saw in Chap. 12, the intolerable risks parts are those on which we need to focus our attention in a RIDM approach.

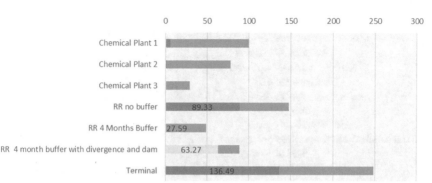

Fig. 16.8 Results of the study in terms of system's intolerable risks only (horizontal axis, M USD)

16.8.2 Divergence: Cyber-Attack, Communication And Consequences

We have purposefully not chosen BI due to quarantine as a divergent risk as we do not want to overshadow the concept illustrated in this section with emotionally heavy events linked to the 2019 pandemic.

A cyber-attack can hit all levels of operation, from SCADA systems and IOT, to IT and telecommunications. In this section we simulate a case where the likelihood of a successful attack is assessed at 1/10 or (one every 10 years). The number might seem high, but for high profile corporation it might actually be underestimated.

The result on the system is a significant increase of risks. The annualized risks for each plant more than double; intolerable risks for each chemical plant multiply by almost one order of magnitude. The risk related to RR would still be mitigated by the four-month buffer and therefore appear unaffected by the simulated cyber-attack.

Terminal's risk jumps by a factor of almost 4, with intolerable risks that reach the combined corporation pre-divergence risks. This is mostly due to Terminal's loader, which in case of a mishap would disrupt the supply chain.

Figure 16.9 displays a full comparison of various scenarios studied in this chapter. For each of the three chemical plants we see the intolerable risks with/without cyber-attack. It also displays the intolerable risks for three scenarios of the RR (no buffer, 4-month buffer, 4-months buffer with divergent flood and hydro dams' effect (see Sect. 14.7.1), and finally the intolerable risks part for Terminal with/without cyber-attack.

Because of the nature of divergent risks due to cyber-attack, which are distributed across the system, they are bound to hit the weakest point, in this case the Terminal, highlighting its vulnerability by pinpointing the disproportionate effects on the overall risks.

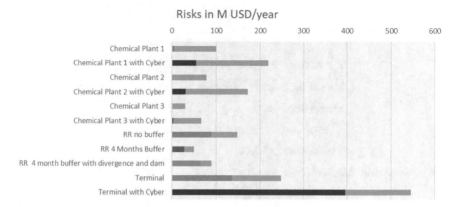

Fig. 16.9 Full comparison of the system's intolerable risks scenarios studied in this chapter (horizontal axis risks in M USD)

If we dig deeper into Terminal's sub-elements (unloader, conveyor belts, buffer, loader), we can see that its risks are primarily linked to the loader. Figure 16.10 displays the tolerable and intolerable portions of each element's risks.

Faced with such an increase of consequences and based on experience, a communication plan should be implemented to inform all employees, contractors and subcontractors regarding cyber-hazardous behaviors and common rules to avoid their being taken unaware, even before IT solutions and other countermeasures are considered and implemented.

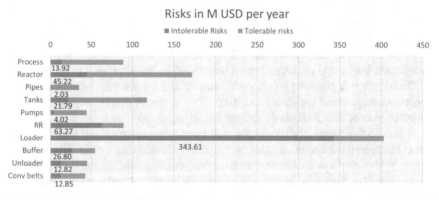

Fig. 16.10 Zooming in on the risks for each sub-element (divergent cyber-attack scenario) (horizontal axis risks in M USD)

16.9 Recommendations and Conditions of Validity

Based on the findings the following sample recommendations can be formulated:

1. Remove external access to Terminal's loader. This has to be done both physically and cyber-security-wise.
2. Use the adjacent berths and makeshift loading cranes. The causeway or jetty may also be used to load ships, provided they offer adequate temporary berthing, are still in service, and the necessary (emergency) permits are available. If it is anticipated that receiving permits could take longer than, say, thirty days, proactive measures could include asking for a conditional permit ahead of time. There are, however, hazards that could damage and destroy the causeway or jetty, such as tsunami, earthquake, sinkholes, and hurricanes. Other hazards that could provoke shorter BI to the causeway or jetty include snow and ice, and these should therefore also be evaluated.
3. Acquire a contingency permit for using another causeway at an alternative location in a different harbor. This could also be a mitigation of RR risk, as Kryptonite could bypass the critical last stretch that lacks redundancy as we saw in Chap. 15. However attention should paid to the possible side effects such as delays, lack of sufficient rolling stock, etc.

Appendix

Links to more information about the Key terms from the Authors	
A, B	*Act of God* (https://www.riskope.com/2020/12/09/act-of-god-in-probabilistic-risk-assessment/) *Black swan* (https://www.riskope.com/2011/06/14/black-swan-mania-using-buzzwords-can-be-a-dangerous-habit/) *Business-as-usual* (https://www.riskope.com/2021/01/13/business-as-usual-definition-in-risk-assessment/)
C, D	*Convergent* (https://www.riskope.com/2021/01/20/convergent-risk-assessments/) *Divergent* (https://www.riskope.com/2020/11/18/tactical-and-strategic-planning-to-mitigate-divergent-events/) *Drillable* (https://www.riskope.com/2020/01/15/probability-impact-graphs-do-not-fly/)

(continued)

(continued)

Links to more information about the Key terms from the Authors	
F	*Foreseeability/foreseeable* (https://www.ris kope.com/2021/01/06/foreseeability-and-pre dictability-in-risk-assessments/) *Fragile/fragility* (https://www.riskope.com/ 2020/04/01/antifragile-resilient-solutions-for-tactical-and-strategic-planning/)
P, R	*Predictability/predictable* (https://www.ris kope.com/2021/01/06/foreseeability-and-pre dictability-in-risk-assessments/) *Resilient, Resilience* (https://www.riskope.com/2016/11/ 23/resilience-cannot-based-instinctual-dec ision-making/)
S	*Scalable* (https://www.riskope.com/2015/04/ 16/how-system-definition-and-interdepende ncies-allow-transparent-and-scalable-risk-ass essments/) *Societal risk acceptability* (https:// www.riskope.com/2014/01/09/aspects-of-risk-tolerance-manageable-vs-unmanageable-risks-in-relation-to-critical-decisions-perpetuity-pro jects-public-opposition/) *Sustainability/sustainable* (https://www.ris kope.com/2019/01/16/improving-sustainab ility-through-reasonable-risk-and-crisis-man agement/) *Survivability* (https://www.riskope. com/2011/03/17/ale-fmea-fmeca-qualitative-methods-is-it-really-what-we-need/) *System* (https://www.riskope.com/2017/07/26/three-ways-to-enhancing-your-risk-registers/)
T, U	*Tolerance* (https://www.riskope.com/2020/04/ 29/risk-tolerance-thresholds/) *Uncertainty/uncertainties* (https://www.ris kope.com/2015/12/10/3-decision-making-tru ths-derived-from-uncertainty-taxonomy-sch eme-of-classification-and-a-road-sign/) *Updatable* (https://www.riskope.com/2020/01/ 07/climate-adaptation-and-risk-assessment/)

Other linked information (https://www.riskope.com/blog-news/) search Riskope blog and use the search box

Chapter 17
Conclusions and Path Forward

To date, industries and societies have focused on increasing efficiency and creating interconnected systems in order to foster productivity. In the process, they have forgotten the side effect which is deeply rooted in the interconnectivity and the interdependency (see Chaps. 7, 8): a reduction of resilience and a proneness to systemic risks such as climate change and cyber related ones. Systemic shocks therefore become not only more common and intense but also propagate further due to cascading probabilities (dominos effects, see Sect. 9.1) and amplification of consequences (ripple effects, see Sect. 9.1).

To prevent this, corporate and governmental leadership as well as competitive advantage should be rooted in resilient behavior (see Sect. 11.1.1) and driven by sensible convergent quantitative risk assessments amid mass shocks such as climate change, pandemics, cyber-attacks, financial crises, and disinformation campaigns.

Throughout this book we have seen case histories and examples (some of which include two almost simultaneous occurrences, or two sites at different dates) where deficiencies leading to widespread consequences were technical, behavioral/cognitive and informational and related to voluntary or involuntary ignorance of risk exposures. The consequences ranged from severe BI to private and public losses, infrastructure losses, human losses (fatalities, evacuations) and environmental damages and were thus clearly multidimensional. Below is a summary of the conclusions:

- Cognitive biases of any kind may be very costly and are paramount in explaining many failures of human endeavors and projects. In this book we have explored how to foster sustainable thinking by reducing biases and better evaluate risks.
- Clients can receive support from hazard specialists, including for example climate change experts, who take a siloed approach. However, experience shows that specific risk management support is necessary to avoid wasted money and corporate blunders.
- Recent harmful events have generated more casualties than expected, showing that past trends and mitigation programs can be easily superseded. Past does not equal future in this type of analyses: divergence is lurking.

© The Author(s), under exclusive license to Springer Nature Switzerland AG 2021   327
F. Oboni and C. H. Oboni, *Convergent Leadership—Divergent Exposures*,
https://doi.org/10.1007/978-3-030-74930-9_17

- If a system is designed to withstand all the credible worst-case accidents, the system is NOT "by definition" safe against any credible accident. By believing it to be so, we might be exposing the system, the environment and the public to unwanted dangers.
- There is no way to escape reality: even the world of crypto mining gets hit by real-life natural disasters. We are sure very few thought that flooding could cripple the synthetic world of cryptocurrencies!
- There are no risk-free alternatives in any industry or endeavor. The risks linked to any new solution or mitigation alternative need to be transparently evaluated including possible change in the systems. That is especially true in the current conditions of rapid climate change. We have seen some projects inadvertently taking risks because of the implementation of new, "safer" solutions that seemed to constitute a universal panacea. Caution has to be exerted and the risk profiles of all alternatives have to be presented so that the decisions are as risk-informed as possible.

Above are the reasons why, in every risk assessment performed, it is necessary to look convergently at natural and man-made hazards. The common siloed approach, which keeps risks divided according to source and treats them separately, will inevitably lead to costly misunderstandings and an unclear roadmap to sustainable mitigation especially when facing divergent climate change and cyber threats.

Here are guidelines to enhance survivability and sustainability within the context of divergence:

- fight biases and censoring as your worst enemies;
- consider consequences as they indeed are, multi-dimensional, thereby reducing un-foreseeability;
- maintain corporate responsibility and business ethics at the highest standards;
- build scenarios by thinking of the unthinkable and thereby reduce unpredictability.

In our present world, clogged with codes and rules, it is not advisable to design any system with the aim of merely passing compliance tests. Our advice is to always think in term of durability and sustainability, thus maximizing value creation and profits. By doing that you will not be among the authors of that future "Ten Plagues" book (see Sect. 3.1.1).

If unmanaged, climate-related and other divergent hazard risks will inevitably weaken company's finances and damage their CSR (Corporate Social Responsibility) and SLO (Social License to Operate) (see Chap. 5) in addition to their ESG (Environmental, Social, Governance) and overall legal stance (see Sect. 6.1.2). We noted that risk assessments must deliver enough granularity to allow meaningful prioritization when facing these scenarios. In addition, they need to support decision-making on multiple simultaneous fronts. As a result, they will make it possible to decide where and when to allot resilience enhancement funds. This will in turn make it possible in some cases to pursue new opportunities. Ill conditioned resilience enhancements, such as those resulting from siloed approaches, will leave their legacy of unpleasant side effects to the future.

A case history from Lao PDR (Anecdote 4.1) illustrated how adverse conditions, but not necessarily divergent hazards, may lead to failure. The example also showed why many countries have adopted the concepts of ALARA, ALARP, and BACT (Sect. 4.2), in particular for hydroelectric dams and other infrastructural projects. We discussed an interesting metaphorical taxonomy of risk by German researchers based on uncertainties of likelihood and consequences, and why using return periods may be a disservice, and concluded with a discussion on force majeure contractual clauses and insurance denial issues (Anecdote 5.1). Again, quantitative risk assessments can help solving some conundrums and finding alternative ways to deal with corporate and private risks.

Chapter 6 entered into the details of building a sound convergent quantitative risk assessment pointing out that the system definition in a convergent platform needs to cover all the dimensions in which the system exists. A first macro taxonomy can be based on the social dimensions of the system, the legal environment, and the physical system (Sect. 6.2), including the various possible interdependencies. Chaps. 7, 8 and 9 followed the same train of thought and delivered the information supporting the creation of better hazards and risk registers. We pointed out that hazards are sometimes blatant, sometimes scary, or difficult to identify. Hazard-based prioritizations generally lead to poor decisions because what is scary, or big, does not necessarily generate large consequences. Furthermore hazards do not act alone, one reason why convergent risk assessments are needed.

Probabilities are perishable goods. They change as the considered system evolves, hence they require updates. Frequencies, which are linked to long-term averages of occurrences, also vary but are slower to change. In heavy industrial applications we may well consider an horizon of two to five years and assume initially probabilities will remain fresh that long, but we need to keep an eye on this and perform updates as soon as needed, especially if divergence is likely.

In (Oboni and Oboni 2013) we stated, "Especially for very large projects, risk assessments generally consider too simplistic consequences and ignore "indirect/life-changing" effects on population and other social aspects" (see Technical note 9.1). We noted that simplistic consequences are misleading, unrealistic and can significantly affect SLO, CSR and ESG.

Papers discussing this ubiquitous phenomenon of oversimplification are rare. However, some have recently discussed ESG risks for deep sea mining (Kung et al. 2020) pointing out that those risks are poorly defined, and we would have liked to see a clearer distinction between risks and their consequences in their discussion.

Tolerance/acceptability thresholds must be developed independently from risks to ensure unbiased results and transparent RIDM. Everyone has a different pain threshold and likewise, everyone has a different risk tolerance threshold. We use the plural because each one of us has various thresholds, for example, a perceived one and a financial one. Each of us decides every day to undertake some activities and consciously or unconsciously assumes risks. It is paramount that industries and corporations adopt clear risk tolerance thresholds in order to support their decision-making.

Appendix B plays the counterpoint to the chapters on building risk assessment and many points delivered throughout the book, as it concentrates on the DON'Ts related to risk assessments.

A modern, convergent quantitative risk assessment should empower decision makers with answers to questions about different aspects of the risks, such as which risks tolerable, intolerable but manageable, and intolerable and unmanageable. Defining tolerance thresholds makes it possible to provide a transparent definition of what constitutes a manageable risk: if a risk above tolerance can be brought below the selected tolerance threshold before hitting the credibility limit of, say, 10^{-6}, by means of mitigative investments and risk transfer that still preserve the economic livelihood of a company, then that risk is manageable.

The key element here is a corporate/government choice of what level of effective mitigative investment preserves the economic livelihood of an entity. That is why improving project cost evaluation, including risk adjustments, is paramount. If the risk cannot be brought below the tolerance threshold as described, then it must be considered unmanageable. Unmanageable risks cannot be simply mitigated; they require strategic shifts in the corporation/government. Insurers and lenders may use this notion to select projects and clients, or create bundles of risks that solve the unmanageability of one or more specific projects in a portfolio.

A risk analysis can be either precise or accurate, but not both. Oftentimes in the desire to be numerically as precise as possible, we forget that the world is mostly random and thus probabilistic by nature. This means that if we want to remain descriptively accurate the model needs to include uncertainties and ranges should be used to describe the risk landscape and reach rational prioritization (see Chap. 10). Then the risk prioritization, the classification of risks as operational, tactical, strategic and families of metaphors and other Olympian menageries (Chap. 1) can be rationally performed.

The roadmap for mitigating an operation's risks, expressed as relative value of the intolerable risks, helps maximize the time and resources of managers by focusing on the most important issues and not necessarily on immediate issues. That is the ultimate goal of RIDM.

A change in the system oftentimes solves a family of risks looming over an operation, thus constitutes the mitigation of highest value. Again that has to be tested with a quantitative convergent risk assessment, playing divergence scenarios.

Divergent risks, which are distributed across siloes, are bound to hit the system at its weakest point, and generally affect the risk landscape of an organization disproportionately.

References

Kung A, Svobodova K, Lebre E, Valenta R, Kemp D, Owen JR (2020) Governing deep sea mining in the face of uncertainty. J Environ Manag 279: https://doi.org/10.1016/j.jenvman.2020.111593

Oboni F, Oboni C (2013) Factual and foreseeable reliability of Tailings Dams and nuclear reactors: a societal acceptability perspective. https://www.riskope.com/wp-content/uploads/2013/06/Factual-and-Foreseeable-Reliability-of-Tailings-Dams-and-Nuclear-Reactors-a-Societal-Acceptability-Perspective.pdf

Appendix A
Making Sense of Probabilities and Frequencies

A.1 Defining Probabilities and Frequencies

Frequency is a measure of how often an event occurs on average during a unit of time (for example, how many times an engine supposed to start every morning fails to start per year). It ranges from 0 to infinite. For people in charge of performing risk assessments a common unit of time is per year, but other time units can be used, such as day, quarter, or even a generation. One can measure frequency by long-term observations (building a statistic).

Probability is, by definition, a number between 0 and 1, measuring the chances some event may or may not happen. 0 means there is no possibility of occurrence, 1 means that occurrence is certain.

A.2 Making Sense of Probabilities and Frequencies

Table A.1 can be used for estimating a large-probability, high-frequency event x. A non-expert should stop at the level of probabilities 0–0.01 (the bottom row of Table A.1), leaving estimates for lower probabilities to trained analysts.

Table A.2, suggested for use only by trained analysts, starts where Table A.1 finishes (0.1) and goes down to 10^{-6}, as this value is commonly considered to be the threshold value of human credibility. Going below credibility would require solid data that are normally not readily available, thus it is highly recommended to stay away from that range.

The last two columns to the right in both tables display the frequency equivalent and the corresponding probability of event x occurring (at least once) "next year". For small frequencies [up to 1/10 years (or another unit as explained in Sect. A.1)], one can assume annual probability is approximately equal to frequency, as shown in Fig. A.1. However, at frequency = 1/5 years (f = 0.2) the error of the approximation

© The Editor(s) (if applicable) and The Author(s), under exclusive license to Springer Nature Switzerland AG 2021
F. Oboni and C. H. Oboni, *Convergent Leadership—Divergent Exposures*,
https://doi.org/10.1007/978-3-030-74930-9

Table A.1 Estimation of large probabilities

Colloquial vocabulary used to describe the event x occurrence.	Event x	Frequency equivalent of x N.B. If these events occur with a known average rate and independently of the time since the last event.	P_x of the event occurring next year p_x $_{min}$ — p_x $_{max}$
Usually, Almost always	Finding at least one container of ice cream in a family freezer	≥ 1	0.63 to ~ 1.0
	At least one sunny weekend in the next year		
Common, Must be considered, Not always	A member of the family gets a cold next year	0.7–1	0.5–0.63
	Getting stuck in a traffic jam for at least 20 min next year (exclude commuting)		
Not uncommon	A person between the age of 18 and 29 does NOT read a newspaper regularly	0.36–0.7	0.3–0.5
May be, Possibly	Getting stuck for more than one hour in traffic (exclude commuting)	0.23–0.36	0.2–0.3
	A celebrity marriage will last a lifetime		
Not usually, Occasionally	Odds of dying from heart disease or cancer in the US (1/7)	0.11–0.23	0.1–0.2
	Chance of rolling 1 when throwing a fair die (1/6 = 0.16)		
Rarely Almost never Never	NB: A nonexpert should stop at this level of scrutiny Experts can develop more in depth estimates for lower probabilities levels using the next Table A.2		0–0.1

Table A.2 Lower-range probabilities

Likelihood of rare Phenomena	Event x	Return period (years) probabilities ≈ frequencies	P_x of the event occurring next year $P_{x\ min} - P_{x\ max}$
High	Having an income of more than $700 k USD in the US in 2017 (1 in 100)	100–10	0.01–0.1 $(10^{-2}-10^{-1})$
	Higher bound of likelihood of having an event of magnitude 7.0 or even higher along the San Andreas Fault		
Moderate	Assault by a firearm in the US (237 in 100,000 inhabitants)	1'000–100	0.001–0.01 $(10^{-3}-10^{-2})$
Low	Death from influenza (1 in 5000 to 1 in 1000) per person	10'000–1'000	0.0001–0.001 $(10^{-4}-10^{-3})$
	An major earth tailings dam breach somewhere on earth per year		
Very Low	Fatal accident at work (1 in 43,500 to 1 in 23,000) per worker	100'000–10'000	0.00001–0.0001 $(10^{-5}-10^{-4})$
	Class 5 + nuclear accident somewhere on earth		
Extremely Low	Person stricken by lightning (1 in 161,856)	1'000'000–100'000	0.000001–0.00001 $(10^{-6}-10^{-5})$
Credibility threshold Lower likelihoods exist	Fatality in railway accident (travelling in Europe) (0.15 per billion km)	N/A	Unless data abound, lower values should not be used.
	Meteor landing precisely on your house; a major Swiss hydroelectric dam breaching		

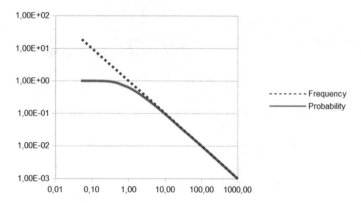

Fig. A.1 Frequency versus probability. Vertical axis: probabilities (limited, by definition, to 1); horizontal axis: corresponding value of 1/frequency. Frequency expressed in events per year

raises to 20%. After that, mathematics is needed to evaluate the probability if the frequency is known or vice versa.

Appendix B
Risk Assessments Don'ts

B.1 Don't Declare a System "Safe"

As clearly shown by the case histories in Chapter 2, for companies, engineers and regulators to portray, describe, or assess a system of any kind as being safe is misleading and incomplete. Furthermore, this can be unethical and reduce SLO, CSR (Chap. 5) and ESG (Sect. 6.1.2).

One of the main problems is the complacency engendered by the use of the terms "safe" or "safety" when referring to system designs and operating procedures. It may be an excellent aspirational goal but, as discussed in Chap. 2, long-term history (Chap. 3) shows that human systems are not necessarily as safe as some parties may claim. All are exposed to natural and man-made hazards and all are exposed to design and operational anomalies that may be created by shortcomings, singly or in combination, with regard to the development and application of knowledge, the level of corporate commitment and the strength of regulatory oversight.

A paradigm shift in the approach is necessary to alter the course of the events. An urban legend attributes to Albert Einstein the following saying: "We cannot solve our problems with the same thinking we used when we created them". Even if the quote was not from Einstein, we think it is wise enough to remember.

While running a safe system may be a noble objective, having a design or operating management plan declared safe is misleading to the public and to those who have to make decisions related to the integrity of individual systems' management facilities. We will now discuss several points related to the "safety" discussion.

B.2 Don't Accept Incremental Answers

The design, construction, operation and closure of any human endeavor is carried out within a complex system requiring high levels of expertise, commitment and diligence. The owner, which must accept ultimate responsibility, sometimes shared with the operator, will generally retain professional consultants and experts to assist them in meeting their responsibilities and may rely on the regulatory system to add rigor through government permitting and compliance responsibilities.

Whatever the situation, an intricate relationship exists between owners, operators, regulators and consultants with regard to ensuring that the highest standards of design and management are identified and implemented. Many individual elements have to come together within a system that examines the interrelated nature of the activities that each must perform in meeting their respective responsibilities. For the purposes of this book we call this integrated relationship the System's Responsibility Framework (SRF).

Recent major accidents discussed in Chaps. 2, 3 and 5 have led to the identification of incremental improvements in the design, management and regulation of human systems that should be adopted globally to reduce the likelihood and consequences of future catastrophic failures. However, we must move beyond the incremental answers provided by the narrow accident investigation approach and use advanced risk assessment methodologies to identify the gaps, hidden flaws or hazards that have been created within existing SRFs that could contribute to future catastrophic failures.

B.3   Don't Call Unpredictable What Indeed Is Predictable

The immediate response to a failure of any kind is generally to call for an investigation to determine the cause. This is established by expert consensus and generally described as a mix of causalities hitting some parts of the system and possibly propagating to involve the whole. It may then be traced back to a hazard that generated a fatal blow to the design, an operating practice that caused the integrity of the design to be compromised and/or the failure of government to enforce the conditions of its operating permit.

A failure may be triggered by a particular natural or man-made event or action but the root cause of failure is not what happened at the moment. The root cause is what decisions were made that allowed that hazardous situation to exist in the first place. In the case of a particular hazard or a combination of hazards, the fact that they were allowed to exist within the SRF needs to be examined from a number of perspectives as described below. Oftentimes the first reactions are to call any failure a "unforeseeable event", "unheard of", "impossible to predict because of the uncertainties, complexities", etc.

At this point it becomes necessary to go back to *basic definitions* (https://www.riskope.com/knowledge-centre/tool-box/glossary/) and to state what a non-credible failure may be (Oboni and Oboni 2018). This is an important addition to the discussion when the concepts of negligence, which are often used in the aftermath of a disaster, come into play as discussed in Sect. 4.1. As stated earlier (see Discussion of Key Terms and Sect. 2.2.2), using a general consensus from various horizons of the hazmat industry this book considers a threshold of credibility in the range of around 10^{-5} to 10^{-6}. Any event with lower probabilities is considered non-credible, or an act of God.

B.4 Don't Jump to Risks: Hazards Come First!

Most tables of *risk comparisons in the literature* (https://books.google.ca/books? id=hrOvBQAAQBAJ&lpg=PA80&ots=cnAv40fwQE&dq=risk%20comparisons% 20in%20the%20literature&pg=PA82#v=onepage&q&f=false) contain a mix of hazardous events, not risks, characterized by different levels of uncertainty and expressed in terms of annualized probability of occurrence. In addition, most comparisons in those tables offer only single number estimates, with no range or error term. These tables are toxic!

For risks such as driving, where fatalities can be counted on large samples, the number is likely to be reliable, at least in some countries. However, even if the risk comparison data are carefully and accurately reported, they can be misleading. Indeed, the risk calculation for driving does not differentiate different driving situations and hazards: for example, speeding home after a long party in the wee hours of the morning is two orders of magnitude more dangerous than driving during the day. This type of reasoning may explain why in our experience people have trouble relating such general risk comparisons tables to the specific risks posed by a project.

Useful risk comparison should be accurate and pertinent. For example, comparing the health risks of chemical processing plants to voluntary actions such as smoking or driving without a seatbelt is neither accurate nor pertinent. Analysts attempting comparisons must stress that values are relative and not absolute. This is because we cannot know the future but can evaluate relative values with the information we have at hand.

However, relative risk evaluations and benchmarking are paramount, as are complete analyses of consequences. Furthermore, the comparison of results can be completely biased if one poorly states the failure criteria or omits the causality analysis. Well-conditioned benchmarking becomes a powerful tool of comparison. It also is a powerful tool for road mapping, enabling managers to know where to act. Whereas risk comparison amounts to showing activities of comparable risks, benchmarking corresponds to rate of a given activity related to other similar activities.

A hazard is defined (see Chap. 2) as a condition outside the business-as-usual conditions with the potential to cause undesirable consequences. Thus, a hazard can be an event, a person or a group of persons, a behavior, etc., with a certain likelihood of occurrence and potential consequences for the system. Natural hazards such as earthquakes and meteorological events are important, but hazards may also be created due to a wide range of actions and decisions throughout the mining cycle related to technical, management and regulatory factors.

Technical hazards may be related to inadequacies in the technical body of knowledge, unresolved uncertainties in the design process, lack of an appropriate level of engineering experience and judgment and unprofessional conduct. Management-related hazards can stem from weak corporate commitment to the oversight of material issues, inadequate management systems, cost and production pressures, lack of adherence to key operating parameters, and poor management. Regulatory hazards can be introduced as a result of design approvals and amendments being given without an adequate focus on risk tolerance, the lack of strict permit operating and construction conditions, and the absence of an enforcement culture.

Human factors can also result in hazards not being recognized or addressed adequately (see Anecdote 1.1 and Chap. 7). The underlying drivers of failure that may lead to inadequate risk management are primarily due to any or a combination of:

(1) Ignorance—not being sufficiently aware of risks (Sects. 2.3 and 4.2);

(2) Complacency—being sufficiently aware of risks but being overly risk tolerant (Sects. 7.1.2, 10.2.3);

(3) Overconfidence—being sufficiently aware of risks, overestimating the ability to deal with them (Sect. 2.3).

If the focus is not, at all times, on the identification and reduction of the root causes of the hazardous situations, they may go unrecognized, uncertainties may not be understood, and the resulting risks may not be properly managed. Here is an example: if, when asked to approve the development of a new project, a company's board of directors is told that the design has been declared safe by their professional consultants and it has been approved by the appropriate regulator, the corporate board members have little reason to probe further. However, a significant set of dynamics are put into play if the board members are told that:

(1)   the system's management has been identified as a material/corporate risk, and the process system represents an intolerable but manageable risk exposure, with the intolerable risk being, for example, the fourth largest corporate risk;

(2)   the consequence classification of the proposed new project evaluated in terms of lives, direct costs, environmental damages, share capital loss, reputation is the third within the corporate portfolio; and

(3)   a risk-based approach to mitigation has identified a roadmap to reducing risks to a tolerable corporate and societal levels.

These dynamics will result in a more disciplined and rigorous approach that will, if taken seriously, work towards significant improvements in the rational management of hazardous situations, hence a reduction of risks related to the design and operation of the system. Uncertainties will be identified and reduced, remaining hazards will be addressed, and critical control procedures will be developed with the objective of mitigations of risks that may have been created by bad management practices.

Maintaining the status quo while merely changing the words to claim a goal of zero failures, and merely talking about robust, resilient and responsible practices without taking appropriate action will not do.

B.5 Do Not Consider Consequences of Failures as One-Dimensional

No accident has one-dimensional consequences (consequences that can be described with one category only, for example: Business Interruption (BI)) of failure which can be described with a single magic number (see Chap. 9). The consequences of a system's failure are always multidimensional, including environmental damage AND harm to people AND BI AND legal costs, etc. Any attempt to rate the consequences based on the worse dimension (an unfortunate common practice) must be rejected as simplistic and prone to biasing the perception of safety.

Let's start with an example drawn from recent history, the cumulative damages effects of the *Hurricane Katrina* (https://www.sciencedaily.com/terms/hurricane_katrina.htm#:~:text=Hurricane%20Katrina%20was%20the%20costliest,landfall%20in%20the%20United%20States) (2005) which reportedly was the largest, third strongest hurricane ever recorded to make landfall in US. The levees in the region were designed for Category 3 hurricanes, but Katrina was Category 5. The consequences were:

- 1836 fatalities with a number of fatal starvation cases, AND
- 20-ft (six meters) high storm surge and its related impacts AND
- 15 million people affected AND
- $81B USD in damages, but the total economic impact may exceed $150B.

Clearly any attempt to classify consequences based on a single dimension would have been biased and misleading.

Oftentimes we see failure mode and effects analyses (*FMEAs*) (http://www.riskope.com/2015/01/15/failure-modes-and-effects-analysis-fmea-risk-method ology/#_blank) performed where the costs of consequences associated to each risk scenario are described with a single value deriving from a mono-dimensional worst clause. It is indeed common practice methodology (though certainly not a recommended practice) to divide the consequences into categories—for example reputational, environmental, cost, and number of lives lost—then ask the risk assessor to select the worst and use that single value for the determination of the risk. Over the years we have shown many of our clients why this approach is bogus.

Two important facts must be kept in mind (Oboni and Oboni 2016):

- Cost of consequences associated to each risk scenario are multi-dimensional;
- Consequences are an aggregate of costs including but not limited to:

    - direct & indirect;
    - health and safety;
    - environmental, image and reputation;
    - legal;
    - reparations and moral damages.
    - To cite an example in the oil and gas arena, BP learned the lesson in the hard way with their $62B USD additive estimate for the 2010 Deepwater Horizon

oil spill in the Gulf of Mexico. This is the reason why, when we perform any risk assessment, even a very simple one, we use the consequences category listed above in an additive way!

Once it is understood that consequences are multidimensional it becomes self-evident that qualifying safety with a single number—be it a Factor of Safety (FoS) or a probability of failure p_f—is severely insufficient.

So, let's now look at consequences in relation to so-called materiality:

• From an internal corporate perspective, materiality is primarily defined in financial terms.
• For a government, materiality is largely defined in non-economic terms considering possible loss of life, environmental damage and economic loss (see Sect. 5.3).
• For the public, materiality is primarily defined in terms of personal impact with their personal safety being paramount.

The potential economic losses to a company for even a partial failure of a system include BI, loss of profits and costs related to reconstruction, environmental rehabilitation, lawsuits and government fines. Economic factors alone will usually dictate that plants and processes be considered a material risk for a corporation. In addition, a company must also consider potential impacts such as the loss of human life, environmental damage and public economic loss in its materiality ranking. When all potential impacts are considered, including a company's loss of public credibility, it is hard to visualize a company not considering the design and operation of any one of facilities a material risk issue.

B.6 Don't Forget to Define Performance, Success and Failure Criteria

In our experience, industrialists very often skip a first, very important step: defining the failure criteria (see Anecdote 1.2) to be considered in the risk assessment. Indeed, without failure (or malfunction) there is no possible risk definition. Thus, this is paramount from the very beginning in order to avoid confusion, conflict of interest, biases and censoring from this very point on. Censoring and biases occur when, for example, experts convene to look only at credible failures, without defining what "credible" means; they neglect to properly define the system (see Chap. 6) and use ill-conditioned categorizing exercises (FMEA, PIGs) with arbitrary indexes and related values. Any statement of safety based on a blatantly censored and biased scenario is obviously devoid of any meaning and unethical.

Part of this effort also goes into using a well-defined and clear *glossary* (https://www.riskope.com/knowledge-centre/tool-box/glossary/).

The failure criteria and performance criteria are linked. Failure stems from performances not being attained. It is a combination of them made with "AND" and "OR" statements. The probability of the failure criteria to occur is the probability of failure p_f. Failure will generate consequences C evaluated by combining the various dimensions of C. The risk assessment's dimensions are, by definition, the dimensions of C.

The idea that "processes are in place to recognize and respond to impending failure of systems and mitigate the potential impacts arising from a potentially catastrophic failure" (ICMM 2016) obviously requires clear identification and communication of success criteria and their opposite: the failure criteria.

Because consequences are multidimensional, the system's performance should encompass all of them, hence the failure criteria are expected to be multidimensional as well. Thus, the successful performance of a systems could be declared, for example, if:

- Service parameters will be constantly maintained within $+$-5% of target value x;
- Spills of certain fluids does not exceed a Q m^3/year;
- Discharge of wastewater with certain characteristics does not exceed q m^3/event and v m^3/yr.

The points above constitute an example of success criteria for the system once all the missing parameters x, Q, q, v, etc. are defined.

Also, health, culture, environment, customs as well as ways of life of affected social groups are all considered to be valid success metrics that should be taken into account in the evaluation of multidimensional failure criteria/consequences of potential accidents. Multicriteria decision models are present in the literature, incorporating various techniques, such as Multi-Attribute Utility Theory (MAUT), which considers points such as decision makers' preferences. Various dimensions of consequences (for example, human, financial and environmental) can be expressed in terms of probabilities (Marsaro et al. 2014).

Disaster risk reduction (DRR) and community resilience are emerging as key priorities in the agendas of governments and businesses. The trigger may be the apparent acceleration in the frequency and magnitude of weather-related hazards, such as hurricanes, fires and floods, compounded by the increasing exposure of population and infrastructure (Peel and Fisher 2016). This is, of course, the realm of divergent exposures.

A successful system is one that can be considered "unfailed" based on all the assumptions and conditions used in its evaluation.

The success/failure criteria are also necessary to ensure that "internal and external review and assurance processes are in place so that controls for system's risks can be *comprehensively assessed and continually improved*" (https://www.riskope.com/2013/11/07/social-acceptability-criteria-winning-back-public-trust-require-drastic-overhaul-of-risk-assessments-common-practice/).

It is important to distinguish between the performance criteria of a project, operation, or company and the success/failure criteria to be used in the risk assessment. The performance criteria are the set of criteria for which the system is designed/created. The success/failure criteria are observer-dependent, that is, they depend on the viewing angle of the various entities involved (corporate, investor, regulators, public).

Consequence						
		Minor	Moderate	Significant	Major	Catastrophic
		1	2	3	4	5
Likelihood	Rare 1	3	1	2	6	15
	Unlikely 2	4	2	15	22	8
	Moderate 3	2	10	24	8	5
	Likely 4	3	9	9	6	0
	Almost Certain 5	1	2	2	1	0

Fig. B.1   A typical qualitative/indexed PIG used by many common practice approaches

In a specific risk assessment addressing the concerns of a particular stakeholder, the success/failure criteria may differ from the corporate performance criteria. This is why it is important that the hazard and risk register must be drillable from different angles to evaluate risks.

B.7   Don't Use Common Practice Matrix Approaches (PIGs, FMEAs)

A paper entitled "The Risk of Using Risk Matrices" (Thomas et al. 2014) showed that oil and gas are also victims of risk matrix. The authors come to very similar conclusions to those reached by other academics and practitioners, and ourselves.

However, we are very aware that most companies, owners and operators may already have Probability-Impact Graphs (PIGs), risk matrices (indexed or qualitative), heat maps and the like at hand (Fig. B.1).

Heat maps and PIGs are ubiquitous. This is a fact, and the reasons are:

- the apparent, but actually *misleading* (https://www.riskope.com/2018/01/31/oil-and-gas-are-also-victims-of-risk-matrix/), simplicity of the approaches;
- the abundance of oversimplified software dealing with them;
- the perception that they provide a correct screening level prioritization;
- the sometimes appealing characteristic of being amenable to saying what the boss wants them to be saying!

Perhaps mostly because of biases (see the last bullet above), and the misunderstanding of the underlying limitations of PIGs, accidents have continued to occur. As a result, many have lost confidence in risk assessments. We have written many times that no one should attempt to perform a risk assessment without first carefully defining the system (Chap. 6) and identifying hazards (Chap. 7). Yet, numerous PIG-based attempts are performed through un-prepared workshops. Oftentimes participants do not even share a correct common glossary.

These kinds of matrices are oftentimes the result of boiler-plate searches in the web. They sport arbitrary colors and arbitrary cell limits, that is, there appears to be no understanding that each one of these elements has an impact on the results.

Finally, the use of indexes (scores) in the definition of probabilities and consequences prohibits the use of any sensible mathematical approach. That is true even with simplified but rational attempts to make sense of prioritization.

B.7.1 Probability Impact Graphs Deceitful

"PIGs do not fly" was part of the title of a risk management course we gave at *TMW 2012* (https://www.riskope.com/wp-content/uploads/Is-it-true-that-PIGs-fly-when-evaluating-risks-of-tailings-management-systems.pdf), during which we explained why PIGs and heat maps are obsolete and misleading and should be abandoned. We are not alone, but part of a small group of researchers and authors who share our point of view, and cites some of the same authors we have also cited above.

Indeed, we started over ten years ago to point out the numerous *misleading* (https://www.riskope.com/2013/11/28/car-accidents-cold-more-examples-of-biased-and-misleading-fmea-pigs-results/) aspects of PIGs and to show that there are numerous issues PIGs cannot solve and make evident their liabilities. In particular, PIGs fail to support unbiased decision making.

Technical note B.1 compares FMEA and ORE results for Joey's corporation, already discussed in Anecdotes 7.1 and Technical note 11.2.

Technical Note B.1: Comparing FMEA and ORE for Joey's Corporate Level (ORE)

At Joeys corporate level, following FMEA ranking, the Action Required category included 113 records out of 406 records, i.e. 27.8%. If on one hand, at present level of knowledge we would consider that the difference between ORE intolerable and FMEA Action required (23.4% vs 27.8%) results is insignificant, some readers may balk at the idea of mitigating 20% more records if they follow FMEA results. However, what really makes the difference between ORE and FMEA is that the FMEA Action Required records are undifferentiated and therefore FMEA does not help prioritizing mitigative investments among them.

Furthermore, FMEA evaluated a further Action Recommended category for 78 records on 406 records, i.e., an additional 19.2% of the total number of records.

If we now add FMEA's Action Required and Action Recommended percentages, we get to roughly 50% of records requiring attention.

ORE's approach was more focused. By looking at the intolerable tactical or strategic risks ORE helped steering Joey's corporation into safe areas, allot mitigative funds in a rational and transparent way, protect its user from internal and external challenges because all evaluations and assumptions are open and transparent and can be discussed before, during and after implementation, and, finally, be adapted along with the business evolution.

Closing point. Obviously, as risks are mitigated, they will drop below the tolerability threshold, and ORE will focus the attention to other areas, until all the risks are hopefully finally below the tolerability curve (they will of course never be reduced to zero, as zero risk does not exist).

We use a classic example in our courses (in-house courses at corporations, MBA courses, etc.) to point out the inability of FMEA, PIGs to differentiate between risks. Think of a high probability/low consequence event (say, the seasonal cold of your boss) and a low probability/high consequence event (say, a severe explosion at your plant). Most of the time the architecture of the matrix you use will lead to the same coloring of these very different events. You may even neglect a risk that proves to be intolerable when properly studied.

Let's look at some details.

Risk-Acceptance Inconsistency. The regions depicted in common practice PIGs feature arbitrary stepped borders which generally bear no relation with actual corporate or societal risk tolerance thresholds. Thus those risk matrices lead to misleading decision-making support.

Range Compression. Risk matrices and their probabilities and consequences (losses) indexes using scores to mimic expected-loss calculations do not reflect actual "distances between risks", i.e., specifically, the difference in their expected loss. In Figs. B.1 and B.2 you can see consequences (losses) divided into five classes, with M USD used as a metric. In the usual log-scale (Fig. B.2) the intervals seem somewhat equally spaced. However, when displaying the same values in decimal scale (Fig. B.3) we can see a significant range compression.

Users generally avoid extreme values or statements if they can. For example, if a score range goes from 1 to 5, many will select values in the 2 to 4 range as a result of the centering bias.

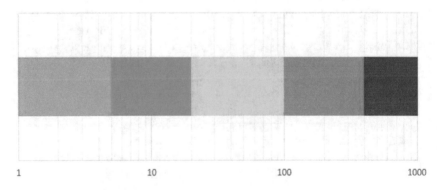

Fig. B.2 Loss classes in M USD display a somewhat uniform width in log scale

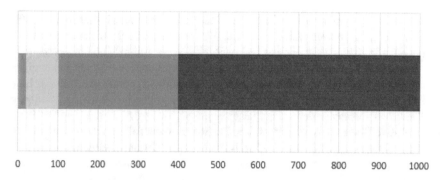

0   100   200   300   400   500   600   700   800   900   1000

Fig. B.3 Once the same classes are displayed in decimal scale the range compression phenomenon (for the lower classes) becomes evident

Category-Definition Bias. Users oftentimes confuse frequency and probability (see Appendix A). The confusion is not solved by the risk matrix compilation guidelines generally offered to users. The results are confused communications, as pointed out by *Canadian CEOs* (https://www.riskope.com/2016/03/23/canadian-ceos-are-looking-to-improve-measurement-risks-communication/#_blank). Researchers even talk about the illusion of communication generated by the category-definition bias.

What Risk is and is not. Risk are not colored cells. Risks are events with a probability (range) leading to complex ranges of consequences. Risks are not a dot in a colored box. They are complex entities that may hurt your operation and the public around it in various ways (the "dimensions" of the consequences), such as BI, health and safety issues, environmental damage, *reputational damage* (https://www.riskope.com/2016/10/05/samarco-dam-disaster-cumulative-damages-effects/), etc. Risk may be tolerable or intolerable. The classification is not arbitrary but rather the result of the analysis performed once we define the tolerance threshold. Intolerable risks may be tactical or strategic. The classification depends on whether or not mitigation is capable of bringing a given risk to below the tolerance threshold (Chap. 10). Tactical and strategic risks, as well as tolerable ones, allow companies to develop *operational, tactical and strategic planning* (https://www.riskope.com/2019/02/06/perform-operational-tactical-and-strategic-planning-avoiding-limbic-brain-traps/). The key is to allow rational, sensible and sustainable decision-making. This, of course works for traditional risks as well as for new risks such as cyber-attacks and climate change-induced risks.

B.7.2 Newly Recognized Risk Matrices Deficiencies

Ranking is Arbitrary. If the risk matrix uses an index approach (rankings from 1 to *n*, see Fig. B.1), then indexes can be in ascending or descending order. Both approaches are used in various industries. This adds to risk-matrix-generated confusion. Ascending or, even worse, descending indexes increase the distance between a matrix-based risk assessment and reality. Rankings are arbitrary and misleading.

Instability because of Categorization. Categories are the cardinal index given to ranges of probabilities and consequences. Categories generate instability as they influence the categorization exercise necessary to compile a risk matrix. Arbitrary range limits and categories are not necessary.

Relative Distance is Distorted. The fact that risk-matrix axes display, explicitly or not, values in logarithmic scale (Fig. B.2) distorts the distance between points. This is a phenomenon that occurs in many instances in science and economics. Unless properly explained and conveyed to users, this distance distortion leads to significant misunderstandings.

B.7.3 Can We Solve the Deficiencies of Risk Matrices?

There are two ways in which the deficiencies of risk matrices can be addressed.

Correcting a Series of Common Mistakes. Oboni et al. (2016) explores the idea that correcting a series of common mistakes, which can be done quite easily and inexpensively, would lead to an honest and more representative way to assess risks.

Housekeeping in the Risk Register. In many instances clients have called us for help with meaningless third-party risk matrices. We define the cause of those calls as the "two hundred yellow syndrome" as clients find themselves in front of a unprioritized plethora of risks. Indeed, the risk matrix's numerous deficiencies lead to many risks being categorized in the central, generally yellow, area of the matrix.

Furthermore, confusion in the hazards, consequences and risk scenario descriptions leads to added fuzziness. The housekeeping of the risk register consist in eliminating confusion by first strictly abiding to a *well-defined glossary* (https://www.riskope.com/knowledge-centre/tool-box/glossary/#_blank). Once the register is consistent, we vet the risks one by one, look at possible common cause failures, interdependencies and rationalize probabilities and consequences. The final result is a p-C graph unencumbered by arbitrary cell limits.

B.7.4 The Final Word

The final word for clients is, even after a good housekeeping, if they want to be sustainable, maintain their CSR, SLO and ESG while remaining profitable, they will have to allow their operations to align with twenty-first-century standards.

Third Parties links in this section	
Riskope's glossary	https://www.riskope.com/knowledge-centre/tool-box/glossary/

(continued)

(continued)

Third Parties links in this section	
Risk comparisons	https://books.google.ca/books?id=hrOvBQAAQBAJ& lpg=PA80&ots=cnAv40fwQE&dq=risk%20compari sons%20in%20the%20literature&pg=PA82#v=onepage& q&f=false
Hurricane Kathrina	https://www.sciencedaily.com/terms/hurricane_katrina. htm#:~:text=Hurricane%20Katrina%20was%20the%20c ostliest,landfall%20in%20the%20United%20States
FMEA	http://www.riskope.com/2015/01/15/failure-modes-and-effects-analysis-fmea-risk-methodology/#_blank
Gaining back public trust	https://www.riskope.com/2013/11/07/social-acceptabi lity-criteria-winning-back-public-trust-require-drastic-overhaul-of-risk-assessments-common-practice/
Misleading aspects of PIGs	https://www.riskope.com/2018/01/31/oil-and-gas-are-also-victims-of-risk-matrix/
PIGs can't fly	https://www.riskope.com/wp-content/uploads/Is-it-true-that-PIGs-fly-when-evaluating-risks-of-tailings-manage ment-systems.pdf
Canadian CEOs want improvements	https://www.riskope.com/2016/03/23/canadian-ceos-are-looking-to-improve-measurement-risks-communica tion/#_blank
Reputational damages	https://www.riskope.com/2016/10/05/samarco-dam-dis aster-cumulative-damages-effects/
Tactical and strategic planning	https://www.riskope.com/2019/02/06/perform-operat ional-tactical-and-strategic-planning-avoiding-limbic-brain-traps/

References

[ICMM] International Council on Mining and Metals (2016) Position statement on preventing catastrophic failure of tailings storage facilities. https://www.resolutionmineeis.us/sites/default/files/references/icmm-position-statement-2016.pdf

Marsaro MF, Alencar MH, de Almeida AT, Cavalcante CAV (2014) Multidimensional risk evaluation: assigning priorities for actions on a natural gas pipeline. In: Proceedings of Probabilistic Safety Assessment and Management PSAM 12

Oboni C, Oboni F (2018) Geoethical consensus building through independent risk assessments. Resources for Future Generations 2018 (RFG2018), Vancouver BC, June 16–21, 2018, https://www.riskope.com/wp-content/uploads/2018/06/rfg-2018-06-18_rev2.pdf

Oboni F, Oboni C (2016) The long shadow of human-generated geohazards: risks and crises. In: ed Arvin F (ed) Geohazards caused by human activity. InTechOpen. ISBN 978-953-51-2802-1, Print ISBN 978-953-51-2801-4

Oboni F, Caldwell J, Oboni C (2016) Ten rules for preparing sensible risk assessments, risk and resilience 2016. Vancouver Canada, 13–16 Nov 2016. https://www.riskope.com/wp-content/uploads/2016/11/Ten-Rules-For-Sensible-Risk-Assessments.pdf

Peel J, Fisher D (2016) International law at the intersection of environmental protection and disaster risk reduction. In: The role of international environmental law in disaster risk reduction. Brill Nijhoff, pp 1–25

Thomas P, Bratvold RB, Bickel JE (2014) The risk of using risk matrices. SPE Econ Manag 6(2):56–66

Printed in the United States
by Baker & Taylor Publisher Services